Intrinsically Disordered Proteins and Disease

Intrinsically Disordered Proteins and Disease

Edited by Rudyard Ramsey

hayle
medical

New York

Hayle Medical,
750 Third Avenue, 9th Floor,
New York, NY 10017, USA

Visit us on the World Wide Web at:
www.haylemedical.com

ISBN: 978-1-64647-540-7

Cataloging-in-Publication Data

Intrinsically disordered proteins and disease / edited by Rudyard Ramsey.
    p. cm.
Includes bibliographical references and index.
ISBN 978-1-64647-540-7
1. Proteins--Structure. 2. Proteins--Separation. 3. Proteins--Conformation. 4. Proteins--Pathophysiology.
5. Proteins--Metabolism--Disorders. I. Ramsey, Rudyard.
QP551 .I58 2023
572.6--dc23

Table of Contents

Permissions

List of Contributors

Index

Preface

An intrinsically disordered protein (IDP) is a protein in which the fixed or ordered three-dimensional structure is absent, generally when its macromolecular interaction partners, such as other proteins or RNA are absent. The flexible structure of these proteins allows them to accomplish their biological functions. IDPs play an important role in DNA regulation and cell signaling owing to their ability to participate in weak multivalent interactions that are highly cooperative and dynamic. There are various IDPs that can adopt a fixed three-dimensional structure after they bind themselves to other macromolecules. IDPs are associated with numerous diseases such as cancer, cardiovascular disease, amyloidoses, neurodegenerative diseases and diabetes. This book includes some of the vital pieces of works being conducted across the world, on various topics related to the relationship between intrinsically disordered proteins and diseases. Its extensive content provides the readers with a thorough understanding of the subject.

This book is a result of research of several months to collate the most relevant data in the field.

When I was approached with the idea of this book and the proposal to edit it, I was overwhelmed. It gave me an opportunity to reach out to all those who share a common interest with me in this field. I had 3 main parameters for editing this text:

1. Accuracy – The data and information provided in this book should be up-to-date and valuable to the readers.

2. Structure – The data must be presented in a structured format for easy understanding and better grasping of the readers.

3. Universal Approach – This book not only targets students but also experts and innovators in the field, thus my aim was to present topics which are of use to all.

Thus, it took me a couple of months to finish the editing of this book.

I would like to make a special mention of my publisher who considered me worthy of this opportunity and also supported me throughout the editing process. I would also like to thank the editing team at the back-end who extended their help whenever required.

Editor

Intrinsically Disordered Linkers Impart Processivity on Enzymes by Spatial Confinement of Binding Domains

Beata Szabo [1], Tamas Horvath [1], Eva Schad [1], Nikoletta Murvai [1], Agnes Tantos [1], Lajos Kalmar [1], Lucía Beatriz Chemes [2], Kyou-Hoon Han [3,4] and Peter Tompa [1,5,*]

[1] Institute of Enzymology, Center of Natural Sciences, Hungarian Academy of Sciences, 1117 Budapest, Hungary; szabo.beata@ttk.mta.hu (B.S.); hotafin@gmail.com (T.H.); schad.eva@ttk.mta.hu (E.S.); murvai.nikoletta@ttk.mta.hu (N.M.); tantos.agnes@ttk.mta.hu (A.T.); lk397@cam.ac.uk (L.K.)

[2] Instituto de Investigaciones Biotecnológicas IIB-INTECH, Consejo Nacional de Investigaciones Científicas y Técnicas (CONICET), Universidad Nacional de San Martín, Buenos Aires 1650, Argentina; lchemes@iibintech.com.ar

[3] Genome Editing Research Center, Division of Biomedical Science, Korea Research Institute of Bioscience and Biotechnology (KRIBB), Daejeon 34113, Korea; khhan600@kribb.re.kr

[4] Department of Nano and Bioinformatics, University of Science and Technology (UST), Daejeon 34113, Korea

[5] VIB Center for Structural Biology, Vrije Univresiteit Brussel, 1050 Belgium, Brussel

[*] Correspondence: peter.tompa@vub.be

Abstract: (1) Background: Processivity is common among enzymes and mechanochemical motors that synthesize, degrade, modify or move along polymeric substrates, such as DNA, RNA, polysaccharides or proteins. Processive enzymes can make multiple rounds of modification without releasing the substrate/partner, making their operation extremely effective and economical. The molecular mechanism of processivity is rather well understood in cases when the enzyme structurally confines the substrate, such as the DNA replication factor PCNA, and also when ATP energy is used to confine the succession of molecular events, such as with mechanochemical motors. Processivity may also result from the kinetic bias of binding imposed by spatial confinement of two binding elements connected by an intrinsically disordered (ID) linker. (2) Method: By statistical physical modeling, we show that this arrangement results in processive systems, in which the linker ensures an optimized effective concentration around novel binding site(s), favoring rebinding over full release of the polymeric partner. (3) Results: By analyzing 12 such proteins, such as cellulase, and RNAse-H, we illustrate that in these proteins linker length and flexibility, and the kinetic parameters of binding elements, are fine-tuned for optimizing processivity. We also report a conservation of structural disorder, special amino acid composition of linkers, and the correlation of their length with step size. (4) Conclusion: These observations suggest a unique type of entropic chain function of ID proteins, that may impart functional advantages on diverse enzymes in a variety of biological contexts.

Keywords: enzyme efficiency; polymeric substrate; processive enzyme; disordered linker; binding motif; binding domain; spatial search; local effective concentration

1. Introduction

Processivity is a kinetic phenomenon widespread among enzymes that act on polymeric substrates, such as DNA, RNA, polysaccharides, and proteins [1]. Once committed, processive enzymes engage in multiple rounds of modification instead of releasing their substrate after modifying it once. Served by different sliding mechanism(s), very effective enzymatic modifiers arose in evolution that can carry out hundreds or thousands of elementary steps upon a single engagement with the substrate [1].

Processivity occurs in: (i) synthesis (e.g., DNA by DNA polymerase [2], RNA by RNA polymerase, and protein by the ribosome [3]); (ii) degradation (e.g., DNA by DNAse [4], RNA by RNAse [5], polysaccharides by glycohydrolases [6] or proteins by the proteasome [7,8]); (iii) structural modification (e.g., DNA by helicase [9]); (iv) chemical modification (e.g., ubiquitination of proteins by ubiquitin ligases [10,11]); or (v) cargo transport (e.g., movement by mechanochemical motors kinesin, dynein and myosin [12–15] along actin and tubulin tracks).

A compilation of domain-linker-domain (DLD)-type monomeric processive enzymes is taken from the comprehensive list given in Supplementary Table S1. Important parameters including the length of predicted disordered linker, mean linker length of orthologous proteins (see Table S2 for species), κ value describing charge distribution, and the level of processivity (such as the length of processive move, the number of steps taken or the number of elementary substrate units covered, if determined at all), are given.

Given the extreme diversity of substrates upon which these processive enzymes act and also the variability of the chemical/mechanochemical changes they make, it is of little surprise that the molecular details of processivity are rather diverse, yet they are based on combinations of two basic designs principles. The classic and amply studied mechanism relies on structural confinement by circular/cylindrical or asymmetric binding domains or subunits of the enzymes. The former occurs, for example, when the PCNA subunit of DNA polymerase encircles the template DNA (Figure 1A) to ensure that the enzyme adds a practically unlimited number of nucleotides [16,17] to the growing DNA polymer. A closely related solution is used by HIV reverse transcriptase [18], which has an asymmetric binding domain that strongly favors sliding along the RNA substrate over dissociating from it (Figure 1B). A completely different mechanism has evolved in mechanochemical motors, such as kinesin and dynein, which move along polymeric protein tracks of tubulin [15]. These dimeric proteins have long coiled-coil stalks and ATPase binding domains, which undergo conformational changes that result in a strong preference for rebinding following dissociation due to a proximity effect, i.e., spatial confinement (Figure 1C). The region connecting the dimerization domain with the binding domain may even undergo transitions between ordered and disordered states [19]. The latter class of processive motors suggests that the presence of two binding elements (motifs or domains) connected by long, conformationally adaptable/flexible linker region(s) appears to be a key element of processivity, which combines deterministic and probabilistic elements of binding [20].

Here we generalize this concept by observing and analyzing that proteins in which binding domains are connected by a disordered linker may show probabilistic bias for re-binding over dissociation from their substrate, due to which they possess processive capacity. As structural disorder of proteins (intrinsically disordered protein/region, IDP/IDR) is widespread in eukaryotic proteomes [21,22], this may be a frequently applied mechanism. IDPs/IDRs often engage in protein-protein interactions [23,24] but their function may also directly stem from the disordered state, termed entropic-chain functions [25]. Binding and entropic-chain functions can actually be combined because often part of the IDP remains disordered even in the bound state, a phenomenon termed fuzziness [26]. Of particular relevance to the observed processivity is that binding motifs embedded in disordered regions, due to the arising "proximity effect" or "optimal effective concentration" around binding sites, may feature facilitated binding, which is central to the concepts of: (i) acceleration of binding by "fly casting" [27], (ii) reduction of binding dimensionality by the "monkey-bar" mechanism [28], and (iii) "ultrasensitive" binding by repetitive binding motifs in signaling proteins [29,30].

By statistical-physical modeling and bioinformatics analysis we show that this kinetic proximity effect is also a widespread inherent property of many monomeric processive enzymes that are capable of multiple rounds of modification of their polymeric substrate. These enzymes, such as a variety of glycohydrolases (e.g., cellulases) [6,31,32], Ribonuclease H1 (RNAse-H1) [5] and matrix metalloproteinase-9 (MMP-9) [33], need no ATP energy for processivity, which makes it a robust and widespread mechanism in the proteome. Here we have selected 12 such monomeric (ATP-independent) processive enzymes from the literature and provide a comprehensive analysis of their physical and structural properties. We show that once engaged with their substrate, their structural organization kinetically biases binding of their free binding domain over dissociation of both its domains, resulting in multiple successive binding events without ever fully releasing the polymeric partner (Figure 1D). We suggest that this type of processivity represents a unique type of "entropic chain" function enabled by the structural disorder of their linker region [25,34], which may be a general mechanism that arises in a broad range of biological contexts.

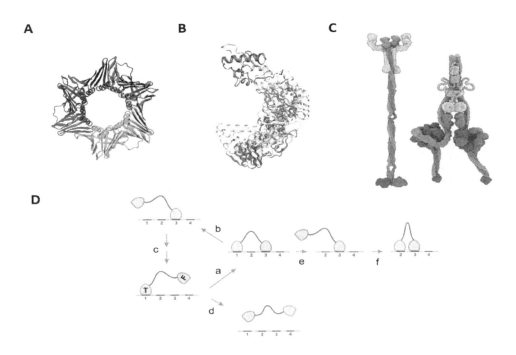

Figure 1. Basic mechanisms of processivity. The figure illustrates the two basic types (and four subtypes) of the mechanism of processivity. The classical mechanism based on structural confinement are represented by folded proteins that either (**A**) completely surround their partner by an oligomeric structure of toroidal shape, such as PCNA (PDB: 1AXC) [16,17], or (**B**) use an asymmetric binding domain to restrict its dissociation, such as in HIV reverse transcriptase (PDB: 1REV) [18]. Basically, different mechanisms are based on spatial confinement allowed by two binding motifs connected by a long, adaptable or flexible linker, as appears in (**C**) the ATP-dependent dimeric mechanochemical motors kinesin-1 and dynein (adapted from [20]), or (**D**) monomeric processive enzymes of domain-disordered linker-domain arrangement. These types of enzymes analyzed here in detail (for cases, see Table 1) bind their polymeric substrate via two binding domains, termed "bound" or "tethered" (T) for the one that anchors the enzyme to the substrate and "unbound" or "free" (F) for the one that is in search for substrate "target" binding sites), connected by a structurally disordered linker. We show by statistical-kinetic modeling that binding via the tethering domain kinetically favors binding via the free domain (a) over full dissociation of the protein (d), which may then result in processive diffusional moves (c) or directed movements driven by energy-dependent binding and/or modification of the substrate (e,f).

Table 1. ATP-independent monomeric domain-linker-domain (DLD)-type processive enzymes.

	Protein Name	UniProt ID	ATP	Partner	Linker Length	Kappa Value (Plot Region)	Processivity
1	*H. sapiens* **RNAse H1**	O60930	-	RNA	50 aa (78–127)	0.254 (2)	
2	*H. sapiens* **XPF**	Q92889	-	DNA	22 aa (821–842)	0.187 (1)	60 nucleotides
3	*T. reesei* **Cel7A**	P62694	-	cellulose	33 aa (445–477)	0.503 (1)	21 catalytic steps
4	*H. insolens* **Cel6A**	Q9C1S9	-	cellulose	46 aa (68–113)	0.288 (1)	
5	*C. cellulolyticum* **Cel48F** *	P37698	-	cellulose	28 aa (106–133)	0.069 (2)	
6	*C. thermocellum* **1,4-beta-glucanase** *	Q5TIQ4	-	cellulose	103 aa (688–790)	0.238 (1)	
7	*H. sapiens* **Telomerase**	O14746	-	DNA	94 aa (231–324)	0.252 (1)	
8	*X. laevis* **XMAP215**	Q9PT63	-	tubulin	121 aa (1079–1199)	0.189 (1)	25 tubulin dimers
9	*H. sapiens* **Chitotriosidase-1**	Q13231	-	chitooligosaccharides	31 aa (387–417)	0.263 (1)	8.6 cleavage steps
10	*B. circulans* **Chitinase A1**	P20533	-	crystalline-chitin	23 aa (444–466)	0.353 (1)	
11	*O. sativa subsp. Japonica* **Chitinase 2**	Q7DNA1	-	chitin	17 aa (74–90)	0.848 (1)	
12	*H. sapiens* **MMP-9**	P14780	-	gelatine	76 aa (434–509)	0.112 (1)	

* no sufficient number of orthologous proteins.

2. Results

2.1. The Classical Mechanisms of Processivity

For rationalizing the diverse mechanisms of processivity, we suggest that they fall into two broad mechanistic categories (cf. Table S1). The structural underpinning of the mechanism is straightforward when the enzyme uses structural confinement to make dissociation from the substrate highly unfavorable [1]. Complete confinement may result from ring-shaped oligomeric structures (e.g., PCNA [16,17] (Figure 1A)), whereas asymmetric structures of a single polypeptide chain can also either fully (e.g., exonuclease I [1]) or partially (e.g., HIV reverse transcriptase [18] (Figure 1B)) enclose the substrate. These mechanisms can be interpreted in terms of a preferred 1D sliding of the substrate (template) within the well-defined structural element of the enzyme.

Processivity of a completely different structural rationale can be observed in motor enzymes that use chemical energy for unidirectional movement along cytoskeletal tracks [12,13]. These motors usually have a dimeric structure, with their dimerization region and ATPase domains connected to their substrate-binding domains by long and extended structures (stalk) (Figure 1C). Large-scale conformational changes elicited by ATP hydrolysis in the ATPase domain(s) propagate to these binding domains, which result in a preference for the re-binding to the substrate track vs. full dissociation [14,15]. In these mechanisms, passive diffusional moves and energy-driven directional steps are combined, i.e., they represent a combination of confining the sequence of events by structural and spatial means. As outlined in the next paragraph, confinement by the limitation of search space by a disordered linker connecting binding domains (Figure 1D) can also account for processivity of enzymes, which appears to be widely applied in biology.

2.2. Statistical Physical Modelling of Domain-Linker-Domain Enzymes

In order to determine how the disordered linker influences (re)binding kinetics of binding domains within a DLD-type enzyme, we used a statistical-kinetic approximation of their binding/unbinding behavior. As the effect of linker length will depend on distances between binding sites and on/off rates of binding domains, we used as a representative example the cellulose/cellulase (Cel7A in Table 1) system. To describe the kinetic behavior of the system, we used a Gaussian approximation of the exact Freely Jointed Chain (FJC) model (see Supplementary Methods and Figure S1). Figure 2 shows the results of varying parameters of a sample case where the tethering domain (cf. Figure 1D) is bound at a substrate site, and we calculate the average binding time (the time it takes for half the free domains to bind a target binding site on the substrate; cf. Supplementary Methods, Equations (S9) and (S10)). By considering the distribution of concentration of the free domain around the bound tethering domain (Figure S1) and integrating binding events (kinetics) based on the binding rate of cellulases (Table S3) over all binding sites within the reach of the free domain, it appears (Figure 2A) that the average time required for re-binding (Supplementary Equation (S10)) increases with increasing linker length. By assuming a threshold set by the kinetics of the dissociation of the tethering domain (for illustration, dissociation half-time (i.e., the time taken for half the bound domains to dissociate) taken as 3×10^{-3} s), the system is processive below a certain linker length (re-binding will be preferred over dissociation), and becomes non-processive for longer linkers (e.g., the threshold linker length is 50 residues in Figure 2A). It should not be forgotten here that the domains in this modelling are dimensionless, due to which there is no minimum on the curve (although there appears to be a minimum imposed by the separation between binding sites, setting a minimum to Kuhn segments).

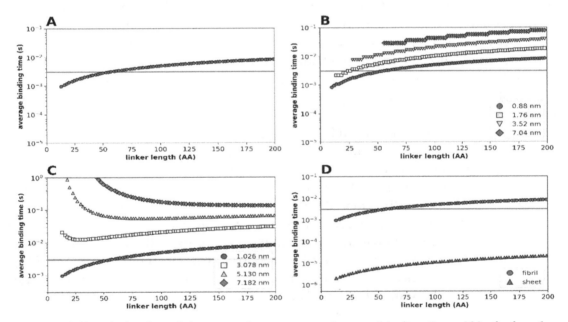

Figure 2. Modelling linker length in processive enzymes. Average binding times (tb) of a free domain linked to the tethering domain already bound to the substrate by a disordered linker of the given length (cf. Figure 1D, and Supplementary Equations (S9) and (S10)). The substrate is modelled based on cellulose geometry: it is assumed to contain binding sites spaced equidistantly every 1.026 nm (1 cellobiose unit) in the X dimension for a thread, and every 2 nm in the Y dimension in case of a sheet. (**A**) Average binding time of the free domain with a random-coil linker (length of Kuhn segment (lk) = 0.88 nm) and binding domains with no physical dimensions. (**B**) Lengthening the Kuhn segment length from 0.88 nm (random-coil) to 7.04 nm (PPII helix) significantly slows binding and reduces processivity. (**C**) "Diluting" binding sites on the substrate (by lengthening the distance between binding sites from 1 cellobiose unit to 7) has a dramatic effect on binding time. (**D**) Binding to a 2D substrate (sheet) is much faster than binding to a 1D substrate (fibril), making the enzyme more processive. On all the panels, if we assume a dissociation half-time of 3×10^{-3} s (limited by catalysis), the enzyme is typically processive at shorter, but not at longer, linker lengths (see text for details).

Therefore, spatially confined diffusional search by the free domain can result in processivity under certain circumstances, when (re)binding by the free domain is kinetically favored over dissociation of the tethering domain. Next, we asked how the flexibility of the linker affects binding time by the free domain. To this end, we ran the statistical kinetic model by varying the length of Kuhn segments (and therefore the persistence length of the chain, see Supplementary Methods) from 0.88 nm (characteristic of random coil chains) to 7.04 nm (characteristic of a polyproline II (PPII) helix), and found a marked effect (Figure 2B), with a more rigid linker providing longer binding times, making the enzyme less processive (e.g., at a length of 30 residues, the enzyme is processive with a linker of 0.88 nm, but not of 3.52 nm, Kuhn-segment length), which may be a prime factor in determining the amino acid composition and sequence conservation of processive linkers, as shown later.

As the calculated binding time is an aggregate value (integrating binding events over all substrate binding sites that can be reached by the free domain, see Supplementary Equation (S10)), we intuitively expect that processivity is increased when possible binding sites are closer to each other, i.e., there are more sites within the reach of the free domain. This is formally demonstrated by varying the spacing of sites (Figure 2C), showing that a processive enzyme can be made non-processive by moving the target sites farther away (this will depend on linker length and could actually be a tuned feature of each system). Along a similar logic, one might expect that the level of processivity is higher when target sites are spread on a two-dimensional surface, by making more sites available for binding. This is formally shown in Figure 2D, where clearly the enzyme is much more processive with a two-dimensional substrate.

Another caveat to the model calculations is if, besides qualitatively assessing whether an enzyme is processive or not, we can draw quantitative conclusions on the level of processivity (average number of steps taken before releasing the substrate). For this, one has to note that the extent of processivity (average number of elementary steps upon engagement with the substrate) is straightforward to define, but not trivial—and is probably not unequivocal—to measure. Furthermore, being a kinetic phenomenon, it may show high stochastic fluctuations and may be very sensitive to experimental conditions.

Nevertheless, one can infer the typical linker-length range where a particular enzyme may behave processive (say, 10–100 residues, cf. intersection of red and blue traces in Figure 2A). This inference may also suggest that linker length and the distance between substrate binding sites must have co-evolved. As an additional note, whereas preferential binding (over dissociation) follows from the kinetic setup of the system, its capacity for unidirectionality does not. As a diffusive move can equally well occur in the backward direction (Figure 1D), directionality may stem from additional mechanistic elements, such as the use of energy and/or post-translational modifications of the substrate. This may even include its degradation, such as that of extracellular matrix proteins in the case of MMP-9 [33,35] or cellulose in the case of cellulases [31,32,36]. This may hinder backward movement and result in rapid unidirectional, forward translocation (Figure 1D).

2.3. Multiple Examples of DLD-Type Processive Enzymes

The foregoing modelling studies show the potential for processivity encoded in the DLD arrangement of enzymes. Next, we demonstrate that there are many such enzymes in biology. Out of 47 processive enzymes of various mechanisms (Table S1), a simple literature search identified 12 processive systems that appear to rely on the DLD domain arrangement, such as MMP-9 [33,37], RNAse H1 [5], or a variety of glycohydrolases [6,31,32]. These ATP-independent enzymes enlisted in Table 1, are analyzed further.

2.3.1. Structural Disorder of Linkers in Monomeric Processive Enzymes

A critical element of processivity in these DLD-type of processive enzymes is the structural disorder of the linker region connecting the binding domains, which has been experimentally demonstrated in only a few cases. For example, the cellulose-binding domain can be effectively separated from the catalytic domain of cellobiohydrolase I by limited proteolysis [38], in agreement with the extreme proteolytic sensitivity of IDPs [34]. Structural disorder was directly observed in cellulase Cel6A and

Cel6B by small-angle X-ray scattering (SAXS) [39], in xylanase 10C by X-ray crystallography [40], and in MMP-9 by atomic-force microscopy (AFM) [33]. Besides these few examples, however, structural disorder has not yet been systematically analyzed in monomeric processive enzymes.

To this end, we applied bioinformatic predictions for the local structural disorder of the linker regions of DLD enzymes in Table 1 (Figure 3). Prediction of structural disorder of three processive enzymes MMP-9, Cel6A and RNAse H1 by IUPred [41] shows a distinctive pattern of a very sharp transition from local order in the binding domains to structural disorder within the linker region. Given the reliability of disorder prediction [42], we may conclude that the linker region in processive enzymes is always disordered, as confirmed for all the cases collected from literature (cf. Table 1, predicted disorder values). Interestingly, the length of the linkers in these processive enzymes always falls within the critical range suggested by model calculations above (cf. Figure 2).

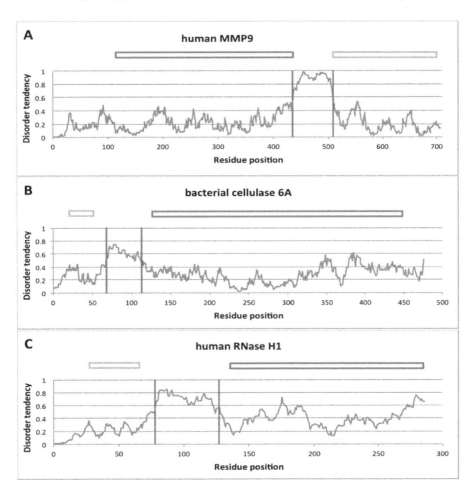

Figure 3. Structural disorder of linker regions in processive enzymes. The linker region in monomeric processive enzymes tends to be highly disordered, as shown here for three illustrative examples by the IUPred algorithm [41]. Traces of disorder score are given for the human and matrix metalloproteinase-9 (MMP-9) sequence (**A**), bacterial cellulase 6A (**B**) and Ribonuclease H1 (RNAseH1) (**C**). In each case, the sharp transition from order to disorder (IUPred score > 0.5) and again to order clearly delimits the linker as a disordered element connecting two globular domains. Globular domains are visualized on top of the diagrams, with blue rectangles representing binding domains and red ones representing catalytic domains.

2.3.2. Conservation of Sequence, Length and Dynamics of Linkers

Modelling (Figure 2) suggests that the length, structural disorder and rigidity of the linker are key elements of processive behavior, which may be in (co)evolutionary link with the typical distance between binding sites (step size) of the given system. This inference also suggests evolutionary constraints on the length and physical properties of the linker regions in these enzymes. We address this issue next.

Regarding evolutionary conservation, IDPs/IDRs have been roughly classified into three classes [43], constrained (where both sequence and structural disorder are conserved), flexible, where sequence varies but structural disorder is conserved, and non-conserved where both lack evolutionary conservation. The underlying assumption in this classification is that disordered regions that function by molecular recognition tend to have conserved sequences, whereas those having linker function are free to evolve, as long as they preserve their structural disorder. As shown in our modelling studies (Figure 2), however, spatial confinement does limit the acceptable length and flexibility of the linker. We assessed these features of the linkers for the 12 DLD-type processive enzymes in Table 1.

In agreement with this expectation, their length shows notably narrower distribution than that of all disordered regions and all disordered linker regions in the DisProt database [44]. Processive enzymes have no short (<30 residues) or long (>150 residues) linkers, although there are many such examples of IDRs in general (Figure 4A). Furthermore, there are characteristic differences between the different DLD enzyme families (Figure S2), which also suggests a co-evolutionary relationship with the typical step size the enzyme takes. When the mean of the linker length of different families is plotted as a function of unit size of different substrates (Table S2), we can see an increase in linker length with the lengthening of processive steps (Figure 5).

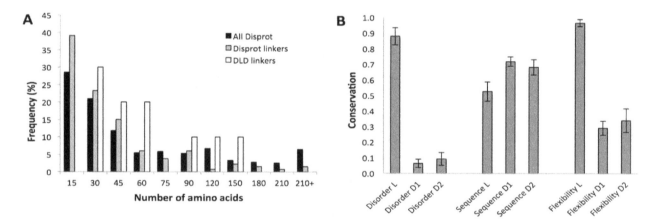

Figure 4. Length distribution and conservation of linker regions in DLD type processive enzymes. (**A**) Length distribution of linkers in DLD enzymes (Table 1), in comparison with that of all disordered regions and disordered linkers in the DisProt database [44]. (**B**) Comparison of the variance (mean values of the data ± SD) of structural disorder (predicted by IUPred [41]) flexibility (as approximated by the ratio of flexible residues predicted by DynaMine [45]) and sequence (assessed by DisCons [22]) of the linkers (L) and their flanking domains (D1 and D2) of the processive DLD type of enzymes (from Table 1) calculated for sequences in species given in (Table S2). Sequence conservation is defined in Section 4 Data and Methods.

This suggests an adaptation of linker length to the geometry of the actual substrate, which also explains: (i) very similar linker length of different processive enzymes functioning on the same substrate, and (ii) the lack of very short and very long linkers in this functional class (Figures 4A and 5).

Their particular function also suggests that selection pressure may also act on their flexibility. As suggested by the above classification [43], classical entropic-chain linker functions are manifested in flexible disorder, where the sequence of the disordered region is rather free to vary, but structural disorder itself is conserved; this is what is expected for the linkers of DLD-type processive enzymes. Therefore, we analyzed the evolution of these features next (Figure 4B). First, we have shown that structural disorder of DLD linkers is highly conserved (as defined in Section 4 Data and Methods), i.e., it shows very little variation. This does not necessarily entail conservation of the sequence (as suggested by flexible disorder [43]), in fact we observe that linker sequences are rather free to vary. Even though structural disorder of the linkers is conserved, it may not necessarily mean that their level of flexibility is maintained at the same level, although this is a critical feature of linkers for the level of processivity (cf. Figure 2). Actually, it was experimentally shown for a similar linker by NMR that despite extreme sequence variation, the flexibility of a linker is maintained [46]. To formally address this issue in DLD linkers, we applied the DynaMine tool developed for assessing local dynamics of IDP backbones [45]. As expected, the overall flexibility of the linker is very high and hardly varies in any of the processive enzymes (Figure 4B).

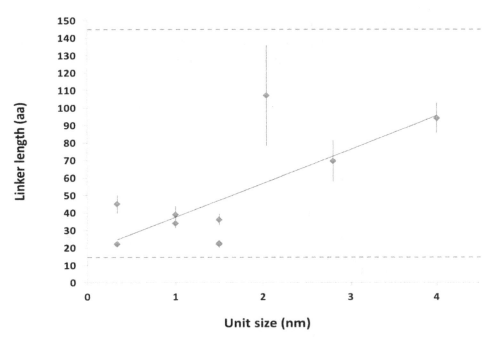

Figure 5. Linker length in DLD enzymes correlates with step size. Linker length in amino acids of the DLD-type processive enzymes (Table 1) is plotted as a function of the unit (step) size in the given substrate. The unit size is the size of the elementary unit (e.g., cellobiose in cellulose, nucleotides in RNA and DNA cf. Table S4) derived from the geometry of the substrate, which is the first approximation of the size of elementary steps the enzyme may take along the given substrate. The linear fit shows the correlation between the two ($R^2 = 0.4998$), whereas horizontal dashed lines show the shortest and longest linker that occurs in DLD processive enzymes (Figure 4A).

Another characteristic closely linked with flexibility of linkers is their charge state, i.e., net charge and charge distribution, because they are among the primary determinants of the chain dimensions and conformational classes of IDPs [47], and even in the lack of hydrophobic groups, polar IDPs/IDRs may favor collapsed ensembles in water. To evaluate sequence polarity, usually the net charge per residue (NCPR), total fraction of charged residues (FCR) and the linear distribution of opposite charges (characterized by κ value) [48] are considered. Interestingly, for all the DLD linkers, their NCPR is low and their FCR is below the threshold of 0.2 (Figure S3), suggesting that they tend to have very similar behavior (they are weak polyampholytes), preferentially populate collapsed states [48]. Their low κ value (Table 1), however, suggests that they tend to have coil-like conformations. It is of note that high proline content may make the structure more extended than simply suggested by charge distribution suggests. In our case, eight out of 12 proteins have high proline content, with the exception of the two proteins in the boundary region (1: Human RNAse H1 and 5: *Clostridium cellulolyticum* Cel48F, cf. Table 1), which do not have high proline content.

2.3.3. Specific Sequence Features of Processive Linkers

Disordered linkers can also be classified by their amino acid composition [49]. Processive linkers in DLD enzymes may also be under special pressure in this regard, because their potential to interact with the flanking domains and/or with other protein partners, or to undergo regulatory post-translational modifications (PTMs), may be of paramount importance. To assess these features, we analyzed the amino acid composition of disordered linkers in DLD enzymes and compared them to that of DisProt linkers and all disordered regions and annotated disordered linkers in the DisProt database [44] (Figure 6). Our results show that processive linkers have significantly less hydrophobic residues than other linkers and disordered proteins in general, which suggests they have to avoid hydrophobic collapse (cf. restraints on κ value stated above) and/or interactions with partners, which most often is mediated by motifs of hydrophobic character [50]. On the other hand, they are enriched in Pro and Gly (denoted as special residues, Figure 5A only shows P under 'special'), which entails that they have to remain extended and flexible and have a balance in oppositely-charged residues (D + E vs. R + K). Probably also for the same reason, they are, on average, more polar.

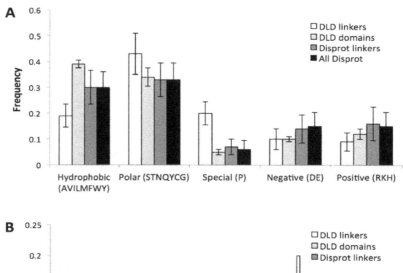

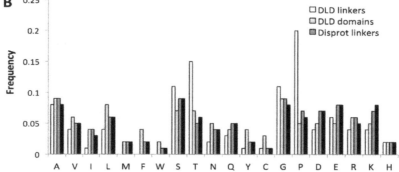

Figure 6. Special features of amino acid composition of linkers. Amino acid composition of linkers in DLD processive enzymes was analyzed and depicted with reference to similar measures of other data. (**A**) Amino acids of linkers were grouped into five categories and compared to the composition of non-linker (binding domain) regions of DLD enzymes (in Table 1) and also of all disordered linkers and assigned disordered linkers in the DisProt database [44]. (**B**) The abundance of amino acids in linkers and non-linker regions in DLD processive enzymes and in all disordered regions and assigned linker regions in the DisProt database.

A further notable feature of DLD linkers is their enrichment in Ser and Thr, which may be indicative of frequent O-linked glycosylation and/or regulatory phosphorylation. A search in UniProt [51] for post-translational modifications (PTMs) of the DLD linkers shows several such modifications in these enzymes (Table 2).

These modifications may impact their kinetic and structural parameters and may tune their interaction with one of the domains of the flanking domains or with external partners. For example, the linker of cellulase emerges from a point not proximal to the cellulose substrate, rather from a point behind, i.e., the kinetic behavior of the enzyme is fine-tuned by the binding of the linker to the surface of the catalytic domain (see next section). Regulated linker-domain interactions are also instrumental in MMP-9, in which the linker has two short binding motifs, that bind the catalytic domain of the enzyme [35].

Table 2. Additional functions of linkers in DLD processive enzymes. Cases where the linker was shown to bind to its adjacent domain are marked with "+".

Enzyme	UniProt ID	PTMs	Domain Binding	Ref.
H. sapiens RNASEH1	O60930	Phosphorylation: S74, S76		[52]
T. reesei Cel7A	P62694	Glycosylation: T461, T462, T463, T462, T469, T470, T471, S473, S474	+	[53]
H. sapiens Telomerase	O14746	Phosphorylation: S227		[54–56]
H. sapiens Nedd4-1	P46934	Phosphorylation: S670, S742, S743, S747, Y785, S884, S888. Ubiqutination: K882		
H. sapiens MMP-9	P14780		+	

The primary function of linkers in DLD processive enzymes is to ensure relatively unrestricted spatial search of domains for binding sites along a multivalent (polymeric) substrate partner. They, however, are also often involved in the regulation of the functioning of the enzyme, as witnessed by additional binding functions and/or PTM events within the linkers themselves (for PTMs, data are either taken from UniProt or from the reference given).

2.3.4. Modelling Cellulase, a Processive Enzyme

Based on all the foregoing analyses, it appears compelling that the DLD arrangement makes enzymes processive. This seems a general phenomenon, which can be demonstrated by low-resolution statistical-kinetic modelling (Figure 2). Here we proceed to show that by incorporating structural details, i.e., atomistic structural models of the domains, into the model and considering domain-linker interactions (Figure 7), we can quantitatively describe the mechanistic and kinetic behavior of one of the most-studied DLD processive enzymes, that of bacterial cellulase (*Trichoderma. reesei* Cel7A, cf. Table 1). Cel7A has two domains of different size, a larger catalytic domain (CD) that confines the linear cellulose substrate, i.e., in itself tends to be processive, and a smaller cellulose binding domain (also termed motif, CBM) attached with a disordered linker of 33 amino acids in length (Figure 7A). The enzyme is processive, typically carrying out about 20–100 cleavage events before dissociating form its substrate. By modeling all parameters of: (i) linker length and flexibility, (ii) catalytic parameters of the enzymatic domain (for the range of kinetic parameters within the Cel7A family, cf. Table S3) and binding parameters of the free (binding) domain, (iii) structural hindrance arising from the actual structures of the domains and domain-linker interaction, and (iv) distance of cellulose binding sites, we show that average binding time of the CBM domain (Figure 7B) undergoes a minimum at a linker length range that is very close to the observed linker lengths in cellulases (Table 1). Furthermore, binding of the linker to the CD has an effect on the behavior of the system (Figure 7B, cf. blue region in color scheme) as it restricts the freedom of movement of the domains, making it less processive. Since all the known cellulase linkers are highly flexible and contain little or no secondary structural elements, changing the Kuhn-segment length is not applicable in this system. The level of processivity that can be approximated as the ratio of the time of binding of CBM to the time of the catalytic reaction (for the CD of cellulase, Table S3, measured with rather artificial substrates) is on the order of 10–100, which agrees with the values reported (Table 1).

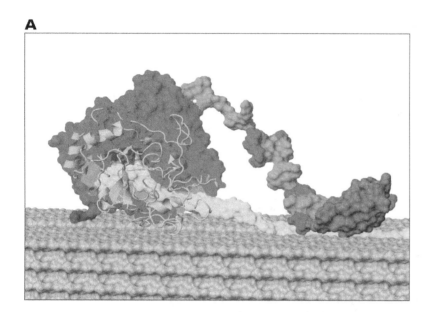
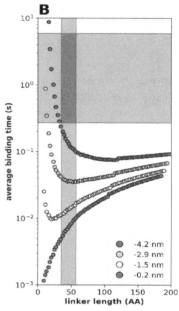

Figure 7. Cellulase: a model processive enzyme. (**A**) Model of the Cel7A cellulase based on the structure PDB 8cel for the catalytic domain (CD) and PDB 2mwk for the cellulose-binding domain (CBM). The CD is purple with the cellulose tunnel shown in transparent blue. One glycosylation of the CD is visible in dark red. Further elements marked are the two catalytic amino acids (red and blue stick-and-ball), the linker region (blue with orange mannose glycosylation), the CBM (dark green), and the cellulose sheet (pale green) of which one fibril (yellow-green) is being processed. The sequence and glycosylation is based on UniProt P62694. (**B**) Statistical kinetic modelling considering geometry (size) and binding of the linker to CD shows binding times characteristic of this system. The green area represents typical catalytic times for Cel7A cellulase family (Table S3), whereas the red area marks typical linker region lengths (Figure S2). The four curves correspond to various values of the linker region's partial binding to the CD, which results in it emerging from the CD at different points (see color mark). If we consider the beginning of the CD domain as the origin of the coordinate system and the cellulose filament moves along the X axis, and assume no binding between the linker and the CD, then free end of the linker region reaches −4.2 nm (red in color scale). When the largest portion of the linker is bound to the CD, the starting point of the free linker end is at zero (blue in color scale). Yellow and light blue colors represent intermediate back-binding cases, with −2.5 and −1.5 nm starting points, respectively.

3. Discussion

Processivity is a basic device of enzymes working on (generating, modifying or moving along) polymeric substrates [1]. By its very molecular logic, it increases cellular economy by limiting the production of metabolic by-products and the dissipation of energy, and it enables large-scale molecular changes to occur, thus it is at the heart of many key cellular processes. Due to the all-or-none character of the operation of processive enzymes, however, there have to be very precise and highly controlled cellular mechanisms for turning them on.

As outlined, there are diverse molecular mechanisms underlying processivity, falling into two general categories, structural confinement by well-folded binding elements and spatial confinement by independent binding elements connected through a linker region. This latter mechanism is apparent in dimeric mechanochemical motors and also in monomeric enzymes. The importance of the general kinetic consequence of processivity can be deduced from its convergent appearance in many independent systems. Whereas its mechanistic underpinning is rather well understood in the case of enzymes that rely on structural confinement and is also analyzed rather extensively in the case of mechanochemical motors, it has so far been largely overlooked in the case of monomeric enzymes.

The typical design of such enzymes is embodied by certain bacterial cellulases, which have a modular structure that combines a large CD linked to a smaller CBM by an intrinsically disordered linker [39] that enables a continuum of conformations. A similar feature has been suggested for the matrix metalloproteinase MMP-9 [33,37], which progressively degrades polymeric components of the extracellular matrix, such as collagen. This enzyme also has a modular structure, with an N-terminal unit of a catalytic domain and three fibronectin type II exosite modules, connected by a 54-residues long linker to a C-terminal hemopexin C domain. SAXS and AFM demonstrated that it can assume multiple conformations and that it can crawl in an inchworm-like manner along its substrate [57]. A similar architecture has been suggested and/or theoretically modelled in the case of glycohydrolases, such as Cel7A [58], cellobiohydrolase I [59] and chitinases [60]. The importance of this arrangement is underscored by cellobiohydrolase I, in which the deletion of the linker dramatically reduces the rate of crystalline cellulose degradation [32] and also other glycoside hydrolases, in which the removal of the carbohydrate-binding module results in a significant decrease in their activity [6], without directly affecting their catalytic domain. Apparently, the unifying feature of all these examples is the structural disorder of their linkers, which ensures a high local concentration and relatively restricted conformational search of binding domains around their binding sites.

Here, we used statistical-kinetic modelling of such systems that this structural arrangement can endow such an enzyme with the capacity of processive movements along a polymeric substrate of spatially repeating binding sites. We characterized these types enzymes by the time of (re)binding as a function of linker length, and found that within a certain length range, they have a preference for binding over dissociation, i.e., they show processive kinetic behavior. Geometric features of the domains, direct binding of the linker with the domains themselves and PTMs of the linkers all influence binding kinetics and may thus serve as points of regulatory input. This might be of no negligible importance, as the processive chain of events past the point of activation appears uncontrolled, which may have dire consequences. A proper regulatory input halting the reaction may be a remedy under some circumstances, as suggested by frequent PTMs of processive linkers (Table 2) and their regulated binding to the flanking domains, as shown for MMP-9, for example [33].

These theoretical observations have general relevance and are supported by a collection of 12 such enzymes that all have highly disordered linkers. Notably, despite rapid evolution and sequence variability of IDPs/IDRs in general, and disordered linker regions in particular, the length and flexibility of linkers in the processive enzymes is conserved. Quantitative modelling of the cellulase enzymes is in general agreement with the observed level of processivity and suggests that this functional-kinetic property is manifest in a relatively limited range of linker lengths, which appear to be in co-evolutionary link with the particular step size along their typical substrate. This has been also suggested by the behavior of the related mechanochemical motors kinesin-1 and kinesin-2, the degree of processivity of which sharply changes by changing the length of their linker regions [15]. This feature is also underlined by the observation that short and long linkers are entirely missing in DLD-type processive enzymes.

In a broader functional context, we suggest that this observed behavior is a special case of the entropic chain functions of IDPs/IDRs and appears as a conceptual extension of mechanisms, such as fly casting [27] and monkey-bar mechanism [28]. Processivity appears to draw on all these mechanisms and may represent one of the primary benefits of the flexibility emanating from structural disorder [25,61]. This type of function cannot be supported by a structured protein; thus it is an appealing addition to the functional arsenal of structural disorder, understanding of which may even enable the design and generation of enzymes of improved capacity for the needs of biotechnology.

4. Data and Methods

4.1. Collection of Processive Enzymes and Intrinsically Disordered Proteins

Processive enzymes were collected from the literature by searching for keywords "processive" or "processivity." We aimed for a full coverage of all types of processive enzymes, which resulted in

47 illustrative examples (Table S1), many of which were covered previously [1]. From this collection we selected 12 monomeric enzymes, for further analysis (Table 1). Due to their dominant modular arrangement, we term these monomeric processive enzymes domain-linker-domain (DLD) type. For comparative purposes, we also downloaded 1274 IDP/IDR sequences from the DisProt database (version 7.0) and selected 133 of the IDRs annotated as "linkers" [44].

4.2. Statistical Kinetic Modelling of Linker Regions

To assess the statistical kinetic behavior of DLD proteins we chose the Freely Jointed Chain (FJC) model and simulated it with a Gaussian approximation [36,62]. As shown by details of the model (Supplementary Methods and Figure S1), this only causes minor deviations from the analytical solution at extreme linker lengths.

An important parameter in modelling is the stiffness of the chain that characterizes its nature of spatial distribution. In the FJC model, this is described by Kuhn segments (l_k), whose measure is two times the persistence length. In a freely moving random-coil polypeptide chain this persistence length is 0.44 nm [62], whereas in a stiff polyproline helix it is roughly an order of magnitude longer. To get the number of Kuhn segments, an amino acid chain can be simulated by calculating the contour length of the chain, l_c, divided by l_k.

It is to be noted that the approximation of a kinetic phenomenon of binding and/or dissociation is only tenable if reaching the equilibrium in spatial distribution is much faster than the event of binding and unbinding, i.e., binding/unbinding is not rate-limiting. As diffusion rates of small proteins in water are on the order of 10^{-6} cm^2 s^{-1} [63], which is equivalent to 102 nm$^2 \cdot$s^{-1}, the typical µs time of the unbound ("free," for domain definitions, cf. Figure 1D) domain equilibrating within the boundaries of the model is well below the time scale of processivity steps.

4.3. Assessing Structural Disorder of Linkers

Structural disorder of processive enzymes was predicted by the IUPred algorithm [41], which is based on estimating the total pairwise inter-residue interaction energy gained upon folding of a polypeptide chain. The predictor returns a position-specific disorder score in the range 0.0–1.0, and a residue with score ≥0.5 is considered as locally disordered. To characterize the disorder tendency of domains and linkers, we calculated the ratio of disordered residues within the given region.

4.4. Flexibility of Linker Regions

To quantify the flexibility of linkers, we used DynaMine [45], a backbone dynamics predictor that has been trained on proteins for which NMR-based chemical shifts and experimental amide bond order parameters (S2) were available. Its score falls between 0.0 and 1.0, with a threshold 0.78 separating flexible (below) and rigid (above) regions. Residue-level DynaMine values were averaged for the entire sequence of linkers to calculate an overall measure of flexibility.

4.5. Charge State and Kappa Value Calculation of Linkers

The charge state of linkers was characterized by three parameters [47,48]. The net charge per residue value (NCPR) is defined as |f+ − f−|, where f+ and f− are the fractions of positively- and negatively-charged residues within the linker region, respectively. The total fraction of charged residues (FCR) is defined as (f+) + (f−). The linear distribution of opposite charges is described by the kappa (κ) parameter [48], which is the mean-square deviation of local charge asymmetry from the overall sequence charge asymmetry weighted on the maximal asymmetry allowed for a given amino-acid composition. Kappa can range from 0 (when opposite charges are evenly distributed) to 1 (when opposite charges are segregated into two clusters). Kappa has a basic influence on IDP/IDR conformation, as there appears to be an inverse correlation between the kappa value and the radius of gyration of the polypeptide chain.

4.6. Amino-Acid Composition and Length Distribution of Linkers

The length and amino acid composition of each processive linker (Table 1) and all IDPs/IDRs in DisProt [44] were calculated. For classification purposes, we also determined composition in terms of a reduced set of amino acid types (positive/basic: Arg, Lys; negative/acidic: Asp, Glu; polar: Ser, Thr, Cys, Gln, His, Tyr, Asn; hydrophobic: Ala, Val, Met, Trp, Phe, Leu, Ile; and special: Pro, Gly).

4.7. Variability and Conservation of Linker Regions

The DLD-type processive enzymes studied here contain two globular domains connected by a disordered linker. To analyze their evolutionary relatedness, we applied the MAFFT (Multiple Alignment using Fast Fourier Transform) program to generate multiple alignments [64] of the sequences from several species, anchored by the flanking ordered binding domain(s), which are highly conserved. Evolutionary conservation of a given region (either disordered or folded) was calculated by an algorithm that computes the average of genetic distances between each pair of sequences in the alignment. The details of the applied method are given in [65]. The species used for alignments and conservation analysis are listed for each protein in Table S2.

Author Contributions: Conceptualization, P.T., A.T., L.B.C., K.H.H.; methodology, T.H., E.S., B.S.; software, L.K., T.H.; formal analysis, B.S., E.S., T.H., L.B.C.; data curation, B.S., E.S., N.M., writing—original draft preparation, A.T., L.B.C.; writing—review and editing, P.T., L.B.C.; A.T., P.T.; funding acquisition, P.T., L.B.C., K.H.H.

Abbreviations

AFM	atomic-force microscopy
DLD	domain-linker-domain
FJC	freely jointed chain
ID	intrinsically disordered
IDP	intrinsically disordered protein
IDR	intrinsically disordered region
MMP-9	matrix metalloproteinase-9
PTM	post-translational modification
RNAse-H1	ribonuclease H1
SAXS	small-angle X-ray scattering

References

1. Breyer, W.A.; Matthews, B.W. A structural basis for processivity. *Protein Sci.* **2001**, *10*, 1699–1711. [CrossRef]
2. Bambara, R.A.; Uyemura, D.; Choi, T. On the processive mechanism of *Escherichia coli* DNA polymerase I. Quantitative assessment of processivity. *J. Biol. Chem.* **1978**, *253*, 413–423.
3. Bonderoff, J.M.; Lloyd, R.E. Time-dependent increase in ribosome processivity. *Nucleic Acids Res.* **2010**, *38*, 7054–7067. [CrossRef] [PubMed]
4. Breyer, W.A.; Matthews, B.W. Structure of *Escherichia coli* exonuclease I suggests how processivity is achieved. *Nat. Struct. Biol.* **2000**, *7*, 1125–1128. [PubMed]
5. Gaidamakov, S.A.; Gorshkova, I.I.; Schuck, P.; Steinbach, P.J.; Yamada, H.; Crouch, R.J.; Cerritelli, S.M. Eukaryotic RNases H1 act processively by interactions through the duplex RNA-binding domain. *Nucleic Acids Res.* **2005**, *33*, 2166–2175. [CrossRef]
6. Boraston, A.B.; Bolam, D.N.; Gilbert, H.J.; Davies, G.J. Carbohydrate-binding modules: fine-tuning polysaccharide recognition Carbohydrate-binding modules: Fine-tuning polysaccharide recognition. *Biochem. J.* **2004**, *382*, 769–781. [CrossRef]

7.   Akopian, T.N.; Kisselev, A.F.; Goldberg, A.L. Processive degradation of proteins and other catalytic properties of the proteasome from Thermoplasma acidophilum. *J. Biol. Chem.* **1997**, *272*, 1791–1798. [CrossRef]

8.   Schrader, E.K.; Harstad, K.G.; Matouschek, A. Targeting proteins for degradation. *Nat. Chem. Biol.* **2009**, *5*, 815–822. [CrossRef]

9.   Gyimesi, M.; Sarlos, K.; Kovacs, M. Processive translocation mechanism of the human Bloom's syndrome helicase along single-stranded DNA. *Nucleic Acids Res.* **2010**, *38*, 4404–4414. [CrossRef] [PubMed]

10.   Hochstrasser, M. Lingering mysteries of ubiquitin-chain assembly. *Cell* **2006**, *124*, 27–34. [CrossRef] [PubMed]

11.   Sowa, M.E.; Harper, J.W. From loops to chains: Unraveling the mysteries of polyubiquitin chain specificity and processivity. *ACS Chem. Biol.* **2006**, *1*, 20–24. [CrossRef] [PubMed]

12.   Gyimesi, M.; Sarlós, K.; Derényi, I.; Kovács, M. Streamlined determination of processive run length and mechanochemical coupling of nucleic acid motor activities. *Nucleic Acids Res.* **2010**, *38*, e102. [CrossRef] [PubMed]

13.   Kolomeisky, A.B.; Fisher, M.E. Molecular motors: A theorist's perspective. *Annu. Rev. Phys. Chem.* **2007**, *58*, 675–695. [CrossRef] [PubMed]

14.   Rock, R.S.; Ramamurthy, B.; Dunn, A.R.; Beccafico, S.; Rami, B.R.; Morris, C.; Spink, B.J.; Franzini-Armstrong, C.; Spudich, J.A.; Sweeney, H. A flexible domain is essential for the large step size and processivity of myosin VI. *Mol. Cell* **2005**, *17*, 603–609. [CrossRef] [PubMed]

15.   Shastry, S.; Hancock, W.O. Neck linker length determines the degree of processivity in kinesin-1 and kinesin-2 motors. *Curr. Biol.* **2010**, *20*, 939–943. [CrossRef] [PubMed]

16.   Krishna, T.S.; Fenyö, D.; Kong, X.-P.; Gary, S.; Chait, B.T.; Burgers, P.; Kuriyan, J. Crystallization of proliferating cell nuclear antigen (PCNA) from Saccharomyces cerevisiae. *J. Mol. Biol.* **1994**, *241*, 265–268. [CrossRef]

17.   Krishna, T.S.; Kong, X.-P.; Gary, S.; Burgers, P.M.; Kuriyan, J. Crystal structure of the eukaryotic DNA polymerase processivity factor PCNA. *Cell* **1994**, *79*, 1233–1243. [CrossRef]

18.   Huang, H.; Chopra, R.; Verdine, G.L.; Harrison, S.C. Structure of a covalently trapped catalytic complex of HIV-1 reverse transcriptase: Implications for drug resistance. *Science* **1998**, *282*, 1669–1675. [CrossRef]

19.   Asenjo, A.B.; Weinberg, Y.; Sosa, H. Nucleotide binding and hydrolysis induces a disorder-order transition in the kinesin neck-linker region. *Nat. Struct. Mol. Biol.* **2006**, *13*, 648–654. [CrossRef] [PubMed]

20.   Carter, A.P. Crystal clear insights into how the dynein motor moves. *J. Cell Sci.* **2013**, *126*, 705–713. [CrossRef]

21.   Tompa, P. Unstructural biology coming of age. *Curr. Opin. Struct. Biol.* **2011**, *21*, 419–425. [CrossRef]

22.   Varadi, M.; Guharoy, M.; Zsolyomi, F.; Tompa, P. DisCons: A novel tool to quantify and classify evolutionary conservation of intrinsic protein disorder. *BMC Bioinform.* **2015**, *16*, 153. [CrossRef]

23.   Tompa, P.; Fuxreiter, M.; Oldfield, C.J.; Simon, I.; Dunker, A.K.; Uversky, V.N. Close encounters of the third kind: Disordered domains and the interactions of proteins. *Bioessays* **2009**, *31*, 328–335. [CrossRef] [PubMed]

24.   Wright, P.E.; Dyson, H.J. Linking folding and binding. *Curr. Opin. Struct. Biol.* **2009**, *19*, 31–38. [CrossRef] [PubMed]

25.   Tompa, P. The interplay between structure and function in intrinsically unstructured proteins. *FEBS Lett.* **2005**, *579*, 3346–3354. [CrossRef]

26.   Tompa, P.; Fuxreiter, M. Fuzzy complexes: Polymorphism and structural disorder in protein-protein interactions. *Trends Biochem. Sci.* **2008**, *33*, 2–8. [CrossRef] [PubMed]

27.   Shoemaker, B.A.; Portman, J.J.; Wolynes, P.G. Speeding molecular recognition by using the folding funnel: The fly-casting mechanism. *Proc. Natl. Acad. Sci. USA* **2000**, *97*, 8868–8873. [CrossRef]

28.   Vuzman, D.; Azia, A.; Levy, Y. Searching DNA via a "Monkey Bar" mechanism: The significance of disordered tails. *J. Mol. Biol.* **2010**, *396*, 674–684. [CrossRef]

29.   Mittag, T.; Orlicky, S.; Choy, W.-Y.; Tang, X.; Lin, H.; Sicheri, F.; Kay, L.E.; Tyers, M.; Forman-Kay, J.D. Dynamic equilibrium engagement of a polyvalent ligand with a single-site receptor. *Proc. Natl. Acad. Sci. USA* **2008**, *105*, 17772–17777. [CrossRef]

30.   Song, J.; Ng, S.C.; Tompa, P.; Lee, K.A.; Chan, H.S. Polycation-pi interactions are a driving force for molecular recognition by an intrinsically disordered oncoprotein family. *PLoS Comput. Biol.* **2013**, *9*, e1003239. [CrossRef]

31.   Carrard, G.; Koivula, A.; Söderlund, H.; Béguin, P. Cellulose-binding domains promote hydrolysis of different sites on crystalline cellulose. *Proc. Natl. Acad. Sci. USA* **2000**, *97*, 10342–10347. [CrossRef] [PubMed]

32.   Srisodsuk, M.; Reinikainen, T.; Penttilä, M.; Teeri, T.T. Role of the interdomain linker peptide of Trichoderma reesei cellobiohydrolase I in its interaction with crystalline cellulose. *J. Biol. Chem.* **1993**, *268*, 20756–20761.

33. Rosenblum, G.; Steen, P.E.V.D.; Cohen, S.R.; Grossmann, J.G.; Frenkel, J.; Sertchook, R.; Slack, N.; Strange, R.W.; Opdenakker, G.; Sagi, I. Insights into the structure and domain flexibility of full-length pro-matrix metalloproteinase-9/gelatinase B. *Structure* **2007**, *15*, 1227–1236. [CrossRef]

34. Tompa, P. Intrinsically unstructured proteins. *Trends Biochem. Sci.* **2002**, *27*, 527–533. [CrossRef]

35. Chen, Y.; Jiang, T.; Mao, A.; Xu, J. Esophageal cancer stem cells express PLGF to increase cancer invasion through MMP9 activation. *Tumour Biol.* **2014**, *35*, 12749–12755. [CrossRef]

36. Gao, D.; Chundawat, S.P.S.; Sethi, A.; Balan, V.; Gnanakaran, S.; Dale, B.E. Increased enzyme binding to substrate is not necessary for more efficient cellulose hydrolysis. *Proc. Natl. Acad. Sci. USA* **2013**, *110*, 10922–10927. [CrossRef]

37. Rosenblum, G.; Meroueh, S.; Toth, M.; Fisher, J.F.; Fridman, R.; Mobashery, S.; Sagi, I. Molecular structures and dynamics of the stepwise activation mechanism of a matrix metalloproteinase zymogen: Challenging the cysteine switch dogma. *J. Am. Chem. Soc.* **2007**, *129*, 13566–13574. [CrossRef]

38. Tilbeurgh, H.V.; Tomme, P.; Claeyssens, M.; Bhikhabhai, R.; Pettersson, G. Limited proteolysis of the cellobiohydrolase I from Trichoderma reesei. Separation of functional domains. *FEBS Lett.* **1986**, *204*, 223–227. [CrossRef]

39. Von Ossowski, I.; Eaton, J.T.; Czjzek, M.; Perkins, S.J.; Frandsen, T.P.; Schülein, M.; Panine, P.; Henrissat, B.; Receveur-Bréchot, V. Protein disorder: Conformational distribution of the flexible linker in a chimeric double cellulase. *Biophys. J.* **2005**, *88*, 2823–2832. [CrossRef]

40. Pell, G.; Szabo, L.; Charnock, S.J.; Xie, H.; Gloster, T.M.; Davies, G.J.; Gilbert, H.J. Structural and biochemical analysis of Cellvibrio japonicus xylanase 10C: How variation in substrate-binding cleft influences the catalytic profile of family GH-10 xylanases. *J. Biol. Chem.* **2004**, *279*, 11777–11788. [CrossRef]

41. Dosztanyi, Z.; Csizmok, V.; Tompa, P.; Simon, I. IUPred: Web server for the prediction of intrinsically unstructured regions of proteins based on estimated energy content. *Bioinformatics* **2005**, *21*, 3433–3434. [CrossRef]

42. Noivirt-Brik, O.; Prilusky, J.; Sussman, J.L. Assessment of disorder predictions in CASP8. *Proteins* **2009**, *77* (Suppl. S9), 210–216. [CrossRef]

43. Bellay, J.; Han, S.; Michaut, M.; Kim, T.; Costanzo, M.; Andrews, B.J.; Boone, C.; Bader, G.D.; Myers, C.L.; Kim, P.M. Bringing order to protein disorder through comparative genomics and genetic interactions. *Genome Biol.* **2011**, *12*, R14. [CrossRef] [PubMed]

44. Piovesan, D.; Tabaro, F.; Mičetić, I.; Necci, M.; Quaglia, F.; Oldfield, C.J.; Aspromonte, M.C.; Davey, N.E.; Davidović, R.; Dosztányi, Z.; et al. DisProt 7.0: A major update of the database of disordered proteins. *Nucleic Acids Res.* **2017**, *45*, D219–D227. [CrossRef] [PubMed]

45. Cilia, E.; Pancsa, R.; Tompa, P.; Lenaerts, T. From protein sequence to dynamics and disorder with DynaMine. *Nat. Commun.* **2013**, *4*, 2741. [CrossRef] [PubMed]

46. Daughdrill, G.W.; Narayanaswami, P.; Gilmore, S.H.; Belczyk, A.; Brown, C.J. Dynamic behavior of an intrinsically unstructured linker domain is conserved in the face of negligible amino acid sequence conservation. *J. Mol. Evol.* **2007**, *65*, 277–288. [CrossRef] [PubMed]

47. Mao, A.H.; Crick, S.L.; Vitalis, A.; Chicoine, C.L.; Pappu, R.V. Net charge per residue modulates conformational ensembles of intrinsically disordered proteins. *Proc. Natl. Acad. Sci. USA* **2010**, *107*, 8183–8188. [CrossRef]

48. Das, R.K.; Pappu, R.V. Conformations of intrinsically disordered proteins are influenced by linear sequence distributions of oppositely charged residues. *Proc. Natl. Acad. Sci. USA* **2013**, *110*, 13392–13397. [CrossRef] [PubMed]

49. George, R.A.; Heringa, J. An analysis of protein domain linkers: Their classification and role in protein folding. *Protein Eng.* **2002**, *15*, 871–879. [CrossRef]

50. Fuxreiter, M.; Tompa, P.; Simon, I. Local structural disorder imparts plasticity on linear motifs. *Bioinformatics* **2007**, *23*, 950–956. [CrossRef] [PubMed]

51. The UniProt Consortium. UniProt: The universal protein knowledgebase. *Nucleic Acids Res.* **2017**, *45*, D158–D169. [CrossRef] [PubMed]

52. Pan, C.; Olsen, J.V.; Daub, H.; Mann, M. Global effects of kinase inhibitors on signaling networks revealed by quantitative phosphoproteomics. *Mol. Cell. Proteom.* **2009**, *8*, 2796–2808. [CrossRef] [PubMed]

53. Harrison, M.J.; Nouwens, A.S.; Jardine, D.R.; Zachara, N.E.; Gooley, A.A.; Nevalainen, H.; Packer, N.H. Modified glycosylation of cellobiohydrolase I from a high cellulase-producing mutant strain of Trichoderma reesei. *Eur. J. Biochem.* **1998**, *256*, 119–127. [CrossRef]

54. Chung, J.; Khadka, P.; Chung, I.K. Nuclear import of hTERT requires a bipartite nuclear localization signal and Akt-mediated phosphorylation. *J. Cell Sci.* **2012**, *125*, 2684–2697. [CrossRef] [PubMed]

55. Jeong, S.A.; Kim, K.; Lee, J.H.; Cha, J.S.; Khadka, P.; Cho, H.S.; Chung, I.K. Akt-mediated phosphorylation increases the binding affinity of hTERT for importin alpha to promote nuclear translocation. *J. Cell Sci.* **2015**, *128*, 2287–2301. [CrossRef]

56. Kang, S.S.; Kwon, T.; Kwon, D.Y.; Do, S.I. Akt protein kinase enhances human telomerase activity through phosphorylation of telomerase reverse transcriptase subunit. *J. Biol. Chem.* **1999**, *274*, 13085–13090. [CrossRef]

57. Overall, C.M.; Butler, G.S. Protease yoga: Extreme flexibility of a matrix metalloproteinase. *Structure* **2007**, *15*, 1159–1161. [CrossRef]

58. Zhao, Y.; Wang, Y.; Zhu, J.; Ragauskas, A.; Deng, Y.; Ragauskas, A. Enhanced enzymatic hydrolysis of spruce by alkaline pretreatment at low temperature. *Biotechnol. Bioeng.* **2008**, *99*, 1320–1328. [CrossRef]

59. Igarashi, K.; Koivula, A.; Wada, M.; Kimura, S.; Penttilä, M.; Samejima, M. High speed atomic force microscopy visualizes processive movement of Trichoderma reesei cellobiohydrolase I on crystalline cellulose. *J. Biol. Chem.* **2009**, *284*, 36186–36190. [CrossRef]

60. Seidl, V. Chitinases of filamentous fungi: A large group of diverse proteins with multiple physiological functions. *Fungal Biol. Rev.* **2008**, *22*, 36–42. [CrossRef]

61. Van der Lee, R.; Buljan, M.; Lang, B.; Weatheritt, R.J.; Daughdrill, G.W.; Dunker, A.K.; Fuxreiter, M.; Gough, J.; Gsponer, J.; Jones, D.T.; et al. Classification of intrinsically disordered regions and proteins. *Chem. Rev.* **2014**, *114*, 6589–6631. [CrossRef] [PubMed]

62. Czovek, A.; Szollosi, G.J.; Derenyi, I. The relevance of neck linker docking in the motility of kinesin. *Biosystems* **2008**, *93*, 29–33. [CrossRef] [PubMed]

63. Czovek, A.; Szollosi, G.J.; Derenyi, I. Neck-linker docking coordinates the kinetics of kinesin's heads. *Biophys. J.* **2011**, *100*, 1729–1736. [CrossRef] [PubMed]

64. Katoh, K. MAFFT: A novel method for rapid multiple sequence alignment based on fast Fourier transform. *Nucleic Acids Res.* **2002**, *30*, 3059–3066. [CrossRef] [PubMed]

65. Capra, J.A.; Singh, M. Predicting functionally important residues from sequence conservation. *Bioinformatics* **2007**, *23*, 1875–1882. [CrossRef] [PubMed]

2

Pathogens and Disease Play Havoc on the Host Epiproteome—The "First Line of Response" Role for Proteomic Changes Influenced by Disorder

Erik H. A. Rikkerink

The New Zealand Institute for Plant & Food Research Ltd., 120 Mt. Albert Rd., Private Bag 92169, Auckland 1025, New Zealand; erik.rikkerink@plantandfood.co.nz

Abstract: Organisms face stress from multiple sources simultaneously and require mechanisms to respond to these scenarios if they are to survive in the long term. This overview focuses on a series of key points that illustrate how disorder and post-translational changes can combine to play a critical role in orchestrating the response of organisms to the stress of a changing environment. Increasingly, protein complexes are thought of as dynamic multi-component molecular machines able to adapt through compositional, conformational and/or post-translational modifications to control their largely metabolic outputs. These metabolites then feed into cellular physiological homeostasis or the production of secondary metabolites with novel anti-microbial properties. The control of adaptations to stress operates at multiple levels including the proteome and the dynamic nature of proteomic changes suggests a parallel with the equally dynamic epigenetic changes at the level of nucleic acids. Given their properties, I propose that some disordered protein platforms specifically enable organisms to sense and react rapidly as the first line of response to change. Using examples from the highly dynamic host-pathogen and host-stress response, I illustrate by example how disordered proteins are key to fulfilling the need for multiple levels of integration of response at different time scales to create robust control points.

Keywords: intrinsically disordered proteins; epiproteome; disordered protein platform; molecular recognition feature; post-translational modifications; physiological homeostasis; stress response; RIN4; p53; molecular machines

1. Introduction

Survival of both individuals and a species is predicated, in no small measure, on their ability to respond to a changing environment. Faced with the challenge of drastic changes, organisms have stark options of fight or flight. Flight comes with its own series of challenges (including adapting to a new environment, or competing with others that have already occupied the new niche). Either way there is a strong evolutionary imperative to acquire an ability to adapt. Adaptation is likely to require response at different time scales from immediate (at the level of the individual) to geological scale (at the level of species and genus). Rapid response can be both a benefit and a cost, as a quick change in direction can sometimes prove to be detrimental in the fullness of time and might, therefore, be likely to favor rapid responses that are also readily reversible. Critical decision points used by the organism to drive response in a particular direction need to be robustly integrated into the core physiology of the cell. In this review I argue in favor of the broader interpretation of the term epiproteome to encapsulate the concepts that (1) changes at the post-translational level are ideally placed to respond in real time and that (2) flexible proteins displaying significant disorder are ideal platforms that can be decorated with post-translational changes and used to integrate responses that potentially have competing impacts on cellular resources.

The term epiproteome was first coined by Dai and Rasmussen [1] to refer to proteomic changes directly associated with epigenetic modifications, namely histone acetylation. Some researchers argue that histone modifications are part of epigenetics, although others argue that their lack of heritability means they should not be included in that term. A search for epiproteome/epiproteomics in PubMed Central yields references to post-translational modification (PTM) changes in histones and a small number that use the term in a wider sense to refer to other PTM [2]. Below I argue in favor of the broad interpretation that includes all PTM.

I suggest that an understanding of the epiproteome (i.e., changing alternative post-translational protein states) in combination with the critical nodal positions occupied by disordered proteins, provides a new basis to comprehend the hypervariable PTM theatre. Its features enable integration of multiple post-translational signals to match the demands of a flexible response. The best examples of hypervariable theatres of response to stress are the battle between hosts and their pathogens and/or their changing environment. Epiproteomic changes offer the host an elegant real-time control of its responses. Unfortunately, this also makes the PTM theatre an Achilles heel, able to be exploited by pathogens. Arguably this explains why so much of a pathogens weaponry appears to be enzymatic and focused on the PTM level of host organisation [3].

2. Review

Below I address five key points or questions that address the key demands on a highly integrated cellular stress control point, namely: (1) A broad interpretation of the concept of the epiproteome; (2) How an organism can communicate between (and marshal) sets of proteins that need to respond to stress while also integrating the, on occasion conflicting, demands of distinct but simultaneous stresses; (3) Are there some key exemplars in plants and animals that point towards solutions to meet the demanding challenge of multiple stresses? (4) What are the characteristics required for a node that can successfully integrate response to simultaneous challenges? (5) How can multiple diverse signals be coordinated in real-time to deliver a coherent response?

2.1. Why Use the Broad Interpretation of the Term Epiproteome?

Our concepts of how the molecular machinery in a cell operates have changed radically over the last three decades. One-dimensional models of static proteins acting alone to promote a particular enzymatic step have been superseded by an understanding that proteins typically act as parts of molecular nano-machines in complexes, and are dynamically controlled by a combination of their own intrinsic flexibility [4], their micro-environment, location and their interactions with their partners. We know that robust control of cellular processes needs to occur at multiple levels [5] that can include modifications of chromatin, transcription, post-transcription, translation and PTM. Epigenetic changes and their role in host plasticity have been widely discussed over the last decade [6,7], including their role in responding to challenges such as pathogens [8]. More recently epitranscriptomic changes have become a topic of renewed interest [9]. It is timely therefore to focus on the epiproteome and the key role that PTM could play in coordinated cellular response to pathogens and other stresses. There are of course thousands of papers referring to specific post-translational modifications or similar terms. Initially epiproteomics referred simply to changes in the specific proteome associated with DNA epigenetic changes [1]. And indeed it is still sometimes used in this way now [10]. Only recently have papers used the term to refer to the sum of all post-translational changes in all proteins or a subset such as the redox or cysteine epiproteome [2,11]. The term epiproteomics evokes a parallel with the temporal nature of epigenetics that is entirely appropriate, and perhaps even central, to the importance of the PTM level of control. Therefore below the term epiproteome is applied in a much more general way to any post-translational modification of any protein in any protein complex. PTM can lead to a large ensemble of forms of the components of the proteome existing in a dynamic state within a cell. Unfortunately, the research tools required to analyse PTM states properly are still expensive to run and hence our current picture of the dynamics of these states is inadequate. PTM alternative states are however likely to be more significant than random noise and there are numerous individual cases where this is confirmed.

2.2. Marshalling a Diverse Set of Responding Proteins More or Less in Unison

When combined with the concept of the role of intrinsic disorder in signalling, the significance of epiproteomic changes are placed in a new light. Coherently controlled epiproteomic changes would have the potential to alter the response of many proteins simultaneously, whether they be members of the same protein complex or dispersed complexes that need to be coordinated with each other. Individual PTM changes have been shown to play key roles in a number of different properties including the formation or dissolution of protein interactions [12], the conformation of a protein [13], the membrane localisation of a protein [14] or the inactivation [15] or degradation [16] of a protein. When such changes are driven by a significant change in physiology of the entire cell, such as its redox potential or pH, there is significant scope for matching coordinated changes in the protein complexes within the cell that need to react to the new state. Thus, the epiproteome is uniquely positioned to play a vital role in marshalling multi-protein responses. There are some known examples of this in situations of stress in nature already. A plant pathogen effector was recently shown to acetylate several proteins that interact with each other in a complex [17]. Another example is the cell signalling that results from electrophilic oxidized lipid products, so called reactive lipid species (RLS), that can react with the amino acids cysteine, lysine and histidine because of their nucleophilic nature. RLS effects on signalling events are largely restricted to the modification of cysteine residues in proteins. RLS induced modifications appear to participate in multiple physiological processes including inflammation, induction of antioxidants and even cell death through the modification of signalling proteins [18].

2.3. Animal and Plant Exemplars: p53 and RIN4

The p53 transcription factor in mammals is best known as a target for cancer therapy and understanding the interaction between stress and cancer but is also associated with facets of aging and microbial responses [19]. Additionally, p53 is a target for microbial manipulation by both viruses and bacteria [20]. The p53 protein interacts with a remarkable array of partners and a key characteristic that allows p53 to act as such a key node/hub is believed to be its disordered characteristics [21]. A particularly pertinent facet of disorder in this discussion is its accessibility to PTM events. In highly structured proteins, only a minority of surface exposed residues and short flexible disordered loops are available for PTM. In disordered proteins/regions, the majority of residues are exposed and provide a readily available platform for epiproteomic modification. Disorder is particularly common in regulatory proteins such as transcription factors (TF) in both animals and plants [22]. The TF p53 has a typical platform with a number of disordered regions (often highly charged) attached to a more structured domain that interacts with DNA. The number of proteins that p53 interacts with, the roles played by epiproteome changes in p53 that link with protein–protein interactions, and the types of PTM events have grown into a very complex interacting network [23,24]. The p53 platform epitomizes the plasticity of disordered proteins and the vital importance of epiproteomic changes to their flexible response. PTM changes are clustered in and around Molecular Recognition Features (MoRFs)—short semi-ordered segments within a largely unstructured backbone that drive interactions with multiple protein partners [25] (see Figure 1).

Disorder-associated properties of p53 have likely also played a significant role in the evolution of this protein family. In a recent study of the evolution of p53 and related proteins in metazoans, Joerger et al. [26] suggest that mutations which stabilized formation of tetrameric p53 forms early in the evolution of vertebrates may have freed up the C-terminal region to adopt a disordered structure. Disorder then may have allowed the C-terminus to undergo numerous PTM and evolve the ability to interact with multiple partner regulatory proteins. They suggest that this disorder assisted evolutionary path allowed p53 to acquire many novel somatic functions by rewiring signalling pathways. Different parts of the p53 disordered regions have been shown to possess significant variation in divergence rates [27]. Indeed, it has been argued for some time that disordered regions in general have novel properties and show increased rates of mutation that suggest they are often under diversifying selection pressure and may be important to allow organisms to adapt [28–31].

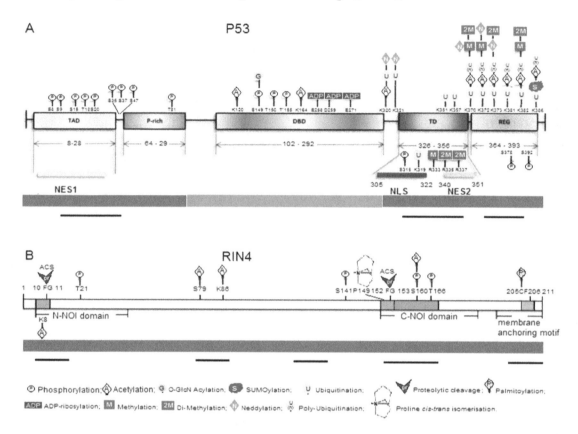

Figure 1. Multiple PTMs cluster in the Molecular Recognition Features within the disordered regions in p53 (**A**) and RIN4 (**B**). Consensus disordered regions are indicated by the red bar at the bottom while ordered regions are indicated by the green bar, putative MoRFs within the disordered regions are indicated by black bars as determined by Uversky (2016; p. 53, [23]) or Sun et al. (2014; RIN4, [32]) by application of disorder prediction programs. PTMs for p53 are a modified form of those identified by Gu and Zhu (2012) [24]. TAD: Transactivation domain; P-rich: proline-rich domain; DBD: DNA binding domain; TD: tetramerization domain; REG: C-terminal regulatory domain; NES1/2: N-terminal (1) and C-terminal (2) nuclear export sequences; NLS: nuclear localization sequence; N-NOI: N-terminal Nitrate induced domain; C-NOI: C-terminal Nitrate induced domain; ACS: AvrRpt2 cleavage site.

The nature of the role played by RPM1-Interacting protein 4 (RIN4) in plant defence is an enigma. That its role is important is hard to question, as RIN4 is targeted directly or indirectly by a number of plant pathogen effectors [32,33]. Moreover the activity of effectors result in several different epiproteomic changes to RIN4 including phosphorylation [34,35], proteolytic cleavage [36], proline isomerisation [37] and acetylation [17]. We have suggested that, like p53, RIN4 is largely intrinsically disordered and is therefore a viable platform for multiple PTM events [32]. As for p53, RIN4 has MoRFs that correlate with conserved motifs and sites of critical importance within RIN4 that are targeted for epiproteomic modification by pathogens and/or the cell itself, or are juxtaposed to modified residues (Figure 1). Two known examples of how RIN4 works are illustrated by recent research [33,37]. Chung and colleagues suggest that two phosphorylation sites within RIN4 are in competition with each other and may be responsible for driving RIN4 in the direction of either innate (molecular pattern triggered) immunity or effector triggered immunity. The authors suggest that RIN4 is a 'phospho-switch' and I note that one of these sites (T166p-phosphorylated) sits in the middle of one of the MoRFs identified by Sun et al. [32], while the other sits near the boundary of this MoRF (S141p). Li and colleagues [37] identify a proline isomerisation site at P149 in the same MoRF that is, in turn, influenced by phosphorylation at T166. The T166p epiproteomic variant has a reduced affinity for the ROC1 enzyme that drives a *cis* to *trans* isomerisation at P149. These first examples of multiple proteomic forms illustrate how RIN4 constitutes the most compelling example yet of a plant protein playing a parallel role to that of p53, as a platform that appears primed to 'collect' epiproteomic signals.

2.4. Characteristics Required for an Integrated Response to Simultaneous Challenges

In order to be able to integrate responses, a hub must be capable of multiple interactions with various partners and collect signatures from various input pathways that can then be coherently interpreted. A high number of flexible and reversible interactions, and ability to make subtle changes to the equilibrium between the various states of control, would constitute a further advantage for such a hub. As these requirements match key characteristics of disordered regions it has been recognised for some time that such hub proteins are highly enriched for disorder [38]. Epiproteomic modification to MoRFs or neighbouring sites in a disordered platform could either block interactions, block other PTM changes at the same site (e.g., acetylation of a serine residue sometimes phosphorylated), or change the charge profile in disordered regions that then changes the dynamic of how (and/or whether) a MoRF interacts with a specific partner. While a degree of subtlety is important sometimes, a hard on-off switch will be important at other times. Disordered proteins can also undergo major conformational switches and even these can be linked to epiproteomic changes by adding larger modifying groups (e.g., glutathionylation, or AMPylation), isomerisation events around critical prolines, or by targeting the entire protein for proteolytic degradation for example.

In a recent analysis of human cells, Chavez and colleagues [39] used novel protein cross-linking methods combined with mass spectrometry to directly identify PTM decorated proteins that are physically associated with each other in complexes. New software advances have also enabled data analysis to focus on cross-linked peptides [40]. In cross-linking analyses distance constraints can be imposed by the type of chemical linker arm used, while addition of biotin groups permits enrichment for cross-linked fragments (e.g., by using avidin-mediated affinity capture technologies). Although the majority of cross-linked peptides identified were derived from homo-dimer interactions, acetylated and methylated peptides from core histone proteins participating in hetero-dimers were particularly common in this analysis. Almost half of the cross-linked histone peptides were found to contain at least one PTM event. Histones are known carry a number of highly significant PTM events. The multiple cases of linkages found between specific peptides increases the likelihood that these have biological relevance in terms of the protein interaction zones between the partners. Interestingly many of the cross-linked peptides with PTM contained modified lysine or arginine residues (residues that are also particularly enriched in disordered regions of proteins). Cross-linking sites were common in the disordered N- and C-termini of histones. In fact, it has been known for more than two decades that the histone tails are the sites where some of the most significant PTM takes place and that these modifications play key roles in the formation or dissolution of chromatin remodelling complexes. These tails serve as recognition sites for chromatin assembly as well as the assembly of the multi-component transcription machinery [41]. The largely positively charged disordered N-terminal tail also contributes to inter-nucleosome binding by contacting an acidic patch in the structured component of histone H2A/H2B dimers to influence histone stacking [42].

The cross-linking analysis allowed Chavez and colleagues [39] to build a significant interactome network map and highlights the importance of the combination of disordered regions and PTM to interactions in such networks, the hub position occupied by histones and the importance of their lysine/arginine rich disordered tails to drive their ability to organize into multi-component complexes. Other biophysical methods of experimentation can also provide indications of how closely associated proteins are in vitro or in vivo. Hydrogen-deuterium exchange (HDX) provides a measure of how exposed different parts of proteins are to PTM [43]. Changes in HDX patterns upon binding with partners can indicate likely interaction zones in protein complexes and were initially used to map antibody binding sites [44]. Other techniques like Förster (fluorescence) resonance energy transfer (FRET) also lend themselves to analysing protein disorder. For example, Vassall et al. [45] used FRET measurements to analyse the order-to-disorder transition of the myelin basic protein (MBP). MBP is largely disordered in aqueous conditions but forms alpha helical recognition fragments upon binding to membranes and its protein partners. The MBP FRET studies, when combined with other tools used to probe structural transition in largely disordered proteins (such as circular dichroism and the membrane-mimetic solvent trifluoroethanol), yielded some surprising results. The data suggested that an intermediate conformational form between disorder and alpha helical state is in fact more compact than the alpha helical form (the latter would normally be expected to have more compactness). This longer form may provide a better bridge across to its complexing protein partners as well as facilitating faster binding to the membrane. Disorder-associated characteristics possessed by histone hubs allow them to integrate epigenetic marks with downstream modifications in mRNA expression response and transfer signals between the epigenetics and transcriptomics levels of response. The disorder properties of MBP on the other hand allow MBP to peripherally attach itself to the cytoplasmic membrane as well as interact with both cytoskeletal proteins like actin and signalling proteins that respond to Ca^{2+}-triggered protein cascades.

One of the ways that cells coordinate their response to changing situation such as stress is to form recognizable sub-cellular organelles. Examples include stress granules (SG), processing bodies (P-bodies) and nuclear stress bodies. Such organelles do not contain membranes, a factor that differentiates them from permanent cellular compartments like the ER, nucleus and mitochondria. Functional organelles must be able to keep interacting with their surrounding liquid environment and yet they must have an ability to form an interphase boundary with this environment. In a recent review Uversky [46] suggests that disorder can provide a crucial component required for forming this liquid-to-liquid interphase. Examples of this are the role that the RNA-binding protein TIA-1 plays to promote assembly of SG through its disordered domains and the disordered regions of a number of the RNA-binding proteins found in human and yeast stress granules. The latter were found to be able to undergo liquid-liquid phase transition in vitro on their own, or when combined with RNA [47]. The phase separated droplets promoted by this organisation can then also recruit other proteins with disordered regions. Furthermore mutations in the key disordered regions or PTM sites involved in regulation can then lead to aberrant fibers or granules that may then contribute to neurodegenerative conditions.

2.5. How Can Multiple Diverse Signals Be Coordinated in Real-Time?

Responses need to be organised at both the temporal and spatial levels. An important biological question is how can organisms create control points that match such elaborate requirements? Significant PTM changes can be very rapid with response times measured in minutes as opposed to hours or even days for many other types of regulation responses [48]. Rapid response makes this level of regulation ideal for responding in real time to challenges perceived by the organisms. A successful reaction to stress is dynamic and requires both sequential, temporal and spatial separation of components and the ability to be nimble in response. The high degree of sophistication required by a successful response is elegantly matched with the opportunities offered by disordered platforms to rapidly integrate PTM signals through multiple MoRFs, multiple targeted PTM sites and reversible as well as competing PTM changes at particular sites. Moreover, PTM changes can also be spatially compartmentalised by limiting where matching substrates and enzymatic functions are co-expressed. As discussed above compartmentalisation can even be aided or driven by the ability of disordered proteins to contribute to phase transition in examples like stress bodies.

Importantly many PTM changes are reversible involving balancing modifications such as phosphorylation/dephosphorylation or acetylation/de-acetylation. Pathogens in turn interfere with PTM processes by developing modifications that can compete with these changes, e.g., phospholyase reactions that break a unique phospho-threonine bond in a protein kinase activation site and make this site un-available for re-phosphorylation [15]. The very properties that make PTM changes so dynamic also make this level of response technically very demanding to illustrate. In order to capture such dynamic potential, sampling time needs to be adjusted to a much finer timescale than commonly used. In addition, techniques that can capture protein associations in real time and are not affected by their readily reversible nature (such as cross-linking techniques) will be required. This will need to be matched with detection techniques sensitive enough to identify any PTM, yet robust enough to be able to scan across complex proteomes. Physical and software enrichment strategies that can overcome the challenge of these limitations in concert with much more sensitive mass spectrometry instrumentation have recently become available. I suggest that disordered protein regions in particular have properties that indicate they are likely to feature prominently in these novel analyses in the near future. Their dynamic ability to change their binding partnerships and to be decorated by multiple PTM events, as illustrated by the examples of p53, RIN4 and histones presented above, suggest that this is one of the major reasons that disorder has become such a common feature of proteins in complex multi-cellular organisms. Indeed this fits with the proposal that disorder was a key enabler on the road to multi-cellular lifestyles [31]. The ability of disordered regions to sense the physiological milieu in which they find themselves by a combination of PTM events, charge profiles and electrostatic interactions suggests that sensing change in this milieu is a specific biological niche that disordered proteins occupy.

3. Conclusions

The animal and plant exemplars, p53 and RIN4, show some key similarities. I suggest that their disordered platform is specifically designed to integrate diverse signals that arrive via alternate post-translational changes inside (or sometimes in close proximity to) MoRFs. PTM changes have a great deal of flexibility and can be very rapid and reversible (e.g., phosphorylation and de-phosphorylation) and the term epiproteomics evokes the dynamic nature of these changes. In addition to reversibility, PTM sites can be; locked in competitive battles (e.g., phosphorylation and acetylation [24]), display competition between sites (as illustrated by the RIN4 phospho-switch concept [33]), result in more subtle shifts in equilibrium (e.g., by changing the charge profile and flexibility of the environment around a MoRF), or result in drastic conformational changes suited to acting as a molecular on/off switch (e.g., by proline isomerisation or multiple phosphorylations). I suggest that the main role of the p53 and RIN4 (and probably many other) proteins containing large disordered domains is to act as sensors and integrators of stress signals from multiple distinct sources via changes to the epiproteome. Moreover, this could explain why examples such as RIN4 and p53 play such important roles in plant and animal disease respectively.

Acknowledgments: This research was funded by Discovery Science Grants to EHAR on intrinsically disordered proteins (DS-1166 and DS-2002) from The New Zealand Institute for Plant & Food Research Ltd. (Auckland, New Zealand). The author thanks Joanna Bowen and Xiaolin Sun for editorial suggestions.

References

1. Dai, B.; Rasmussen, T.P. Global epiproteomic signatures distinguish embryonic stem cells from differentiated cells. *Stem Cells* **2007**, *25*, 1567–2574. [CrossRef] [PubMed]
2. Go, Y.M.; Jones, D.P. The redox proteome. *J. Biol. Chem.* **2013**, *288*, 26512–26520. [CrossRef] [PubMed]
3. Block, A.; Alfano, J.R. Plant targets for *Pseudomonas syringae* type III effectors: Virulence targets or guarded decoys? *Curr. Opin. Microbiol.* **2011**, *14*, 39–46. [CrossRef] [PubMed]
4. Uversky, V.N.; Dunker, A.K. Controlled chaos. *Science* **2008**, *322*, 1340–1341. [CrossRef] [PubMed]
5. Payne, J.L.; Wagner, A. Mechanisms of mutational robustness in transcriptional regulation. *Front. Genet.* **2015**, *6*, 322. [CrossRef] [PubMed]
6. Riddihough, G.; Zahn, L.M. What is epigenetics? *Science* **2010**, *330*, 611. [CrossRef] [PubMed]

7. Cortijo, S.; Wardenaar, R.; Colomé-Tatché, M.; Gilly, A.; Etcheverry, M.; Labadie, K.; Caillieux, E.; Hospital, F.; Aury, J.M.; Wincker, P.; et al. Mapping the epigenetic basis of complex traits. *Science* **2014**, *343*, 1145–1148. [CrossRef] [PubMed]

8. Gómez-Díaz, E.; Jordà, M.; Peinado, M.A.; Rivero, A. Epigenetics of host–pathogen interactions: The road ahead and the road behind. *PLoS Pathog.* **2012**, *8*, e1003007. [CrossRef] [PubMed]

9. Gokhale, N.S.; Horner, S.M. RNA modifications go viral. *PLoS Pathog.* **2017**, *13*, e1006188. [CrossRef] [PubMed]

10. Zheng, Y.; Huang, X.; Kelleher, N.L. Epiproteomics: Quantitative analysis of histone marks and codes by mass spectrometry. *Curr. Opin. Chem. Biol.* **2016**, *33*, 142–150. [CrossRef] [PubMed]

11. Go, Y.-M.; Chandler, J.D.; Jones, D.P. The cysteine proteome. *Free Radic. Biol. Med.* **2015**, *84*, 227–245. [CrossRef] [PubMed]

12. Nishi, H.; Hashimoto, K.; Panchenko, A.R. Phosphorylation in protein-protein binding: Effect on stability and function. *Structure* **2011**, *19*, 1807–1815. [CrossRef] [PubMed]

13. Andreotti, A.H. Native state proline isomerisation: An intrinsic molecular switch. *Biochemistry* **2003**, *42*, 9515–9524. [CrossRef] [PubMed]

14. Resh, M.D. Covalent lipid modifications of proteins. *Curr. Biol.* **2013**, *23*, R431–R435. [CrossRef] [PubMed]

15. Li, X.; Lin, H.; Zou, Y.; Zhang, J.; Long, C.; Li, S.; Chen, S.; Zhou, J.M.; Shao, F. The phosphothreonine lyase activity of a bacterial type III effector family. *Science* **2007**, *315*, 1000–1003. [CrossRef] [PubMed]

16. Rosebrock, T.R.; Zeng, L.R.; Brady, J.J.; Abramovitch, R.B.; Xiao, F.M.; Martin, G.B. A bacterial E3 ubiquitin ligase targets a host protein kinase to disrupt plant immunity. *Nature* **2007**, *448*, 370–374. [CrossRef] [PubMed]

17. Lee, J.; Manning, A.J.; Wolfgeher, D.; Jelenska, J.; Cavanaugh Keri, A.; Xu, H.; Fernandez, S.M.; Michelmore, R.W.; Kron, S.J.; Greenberg, J.T. Acetylation of an NB-LRR plant immune-effector complex suppresses immunity. *Cell Rep.* **2015**, *13*, 1670–1682. [CrossRef] [PubMed]

18. Higdon, A.; Diers, A.R.; Oh, J.Y.; Landar, A.; Darley-Usmar, V.M. Cell signalling by reactive lipid species: New concepts and molecular mechanisms. *Biochem. J.* **2012**, *442*, 453–464. [CrossRef] [PubMed]

19. Maclaine, N.J.; Hupp, T.R. The regulation of p53 by phosphorylation: A model for how distinct signals integrate into the p53 pathway. *Aging* **2009**, *1*, 490–502. [CrossRef] [PubMed]

20. Zaika, A.I.; Wei, J.; Noto, J.M.; Peek, R.M. Microbial regulation of p53 tumor suppressor. *PLoS Pathog.* **2015**, *11*, e1005099. [CrossRef] [PubMed]

21. Oldfield, C.J.; Meng, J.; Yang, J.Y.; Yang, M.Q.; Uversky, V.N.; Dunker, A.K. Flexible nets: Disorder and induced fit in the associations of p53 and 14-3-3 with their partners. *BMC Genom.* **2008**, *9* (Suppl. 1), S1. [CrossRef] [PubMed]

22. Sun, X.; Rikkerink, E.H.A.; Jones, W.T.; Uversky, V.N. Multifarious roles of intrinsic disorder in proteins illustrate its broad impact on plant biology. *Plant Cell* **2013**, *25*, 38–55. [CrossRef] [PubMed]

23. Uversky, V.N. P53 proteoforms and intrinsic disorder: An illustration of the protein structure–function continuum concept. *Int. J. Mol. Sci.* **2016**, *17*, 1874. [CrossRef] [PubMed]

24. Gu, B.; Zhu, W.-G. Surf the post-translational modification network of p53 regulation. *Int. J. Biol. Sci.* **2012**, *8*, 672–684. [CrossRef] [PubMed]

25. Cheng, Y.G.; Oldfield, C.J.; Meng, J.W.; Romero, P.; Uversky, V.N.; Dunker, A.K. Mining alpha-helix-forming molecular recognition features with cross species sequence alignments. *Biochemistry* **2007**, *46*, 13468–13477. [CrossRef] [PubMed]

26. Joerger, A.C.; Wilcken, R.; Andreeva, A. Tracing the evolution of the p53 tetramerization domain. *Structure* **2014**, *22*, 1301–1310. [CrossRef] [PubMed]

27. Xue, B.; Brown, C.J.; Dunker, A.K.; Uversky, V.N. Intrinsically disordered regions of p53 family are highly diversified in evolution. *Biochim. Biophys. Acta* **2013**, *1834*, 725–738. [CrossRef] [PubMed]

28. Dunker, A.K.; Garner, E.; Guilliot, S.; Romero, P.; Albrecht, K.; Hart, J.; Obradovic, Z.; Kissinger, C.; Villafranca, J.E. Protein disorder and the evolution of molecular recognition: Theory, predictions and observations. *Pac. Symp. Biocomput.* **1998**, *3*, 473–484.

29. Brown, C.J.; Johnson, A.K.; Dunker, A.K.; Daughdrill, G.W. Evolution and disorder. *Curr. Opin. Struct. Biol.* **2011**, *21*, 441–446. [CrossRef] [PubMed]

30. Nilsson, J.; Grahn, M.; Wright, A.P. Proteome-wide evidence for enhanced positive Darwinian selection within intrinsically disordered regions in proteins. *Genome Biol.* **2011**, *12*, R65. [CrossRef] [PubMed]

31. Romero, P.R.; Zaidi, S.; Fang, Y.Y.; Uversky, V.N.; Radivojac, P.; Oldfield, C.J.; Cortese, M.S.; Sickmeier, M.; LeGall, T.; Obradovic, Z.; et al. Alternative splicing in concert with protein intrinsic disorder enables increased functional diversity in multicellular organisms. *Proc. Natl. Acad. Sci. USA* **2006**, *103*, 8390–8395. [CrossRef] [PubMed]

32. Sun, X.; Greenwood, D.R.; Templeton, M.D.; Libich, D.S.; McGhie, T.K.; Xue, B.; Yoon, M.; Cui, W.; Kirk, C.A.; Jones, W.T.; et al. The intrinsically disordered structural platform of the plant defence hub protein RIN4 provides insights into its mode of action in the host-pathogen interface and evolution of the NOI protein family. *FEBS J.* **2014**, *281*, 3955–3979. [CrossRef] [PubMed]

33. Chung, E.H.; El-Kasmi, F.; He, Y.; Loehr, A.; Dangl, J.L. A plant phosphoswitch platform repeatedly targeted by type III effector proteins regulates the output of both tiers of plant immune receptors. *Cell Host Microbe* **2014**, *16*, 484–494. [CrossRef] [PubMed]

34. Mackey, D.; Holt, B.F.; Wiig, A.; Dangl, J.L. RIN4 interacts with *Pseudomonas syringae* type III effector molecules and is required for RPM1-mediated resistance in *Arabidopsis*. *Cell* **2002**, *108*, 743–754. [CrossRef]

35. Liu, J.; Elmore, J.M.; Lin, Z.-J.D.; Coaker, G. A receptor-like cytoplasmic kinase phosphorylates the host target RIN4, leading to the activation of a plant innate immune receptor. *Cell Host Microbe* **2011**, *9*, 137–146. [CrossRef] [PubMed]

36. Axtell, M.J.; Chisholm, S.T.; Dahlbeck, D.; Staskawicz, B.J. Genetic and molecular evidence that the *Pseudomonas syringae* type III effector protein AvrRpt2 is a cysteine protease. *Mol. Microbiol.* **2003**, *49*, 1537–1546. [CrossRef] [PubMed]

37. Li, M.; Ma, X.; Chiang, Y.H.; Yadeta, K.A.; Ding, P.; Dong, L.; Zhao, Y.; Li, X.; Yu, Y.; Zhang, L.; et al. Proline isomerization of the immune receptor-interacting protein RIN4 by a cyclophilin inhibits effector-triggered immunity in *Arabidopsis*. *Cell Host Microbe* **2014**, *16*, 473–483. [CrossRef] [PubMed]

38. Haynes, C.; Oldfield, C.J.; Ji, F.; Klitgord, N.; Cusick, M.E.; Radivojac, P.; Uversky, V.N.; Vidal, M.; Iakoucheva, L.M. Intrinsic disorder is a common feature of hub proteins from four eukaryotic interactomes. *PLoS Comput. Biol.* **2006**, *2*, e100. [CrossRef] [PubMed]

39. Chavez, J.D.; Weisbrod, C.R.; Zheng, C.; Eng, J.K.; Bruce, J.E. Protein interactions, post-translational modifications and topologies in human cells. *Mol. Cell. Proteom.* **2013**, *12*, 1451–1467. [CrossRef] [PubMed]

40. Weisbrod, C.R.; Chavez, J.D.; Eng, J.K.; Yang, L.; Zheng, C.; Bruce, J.E. In vivo protein interaction network identified with novel real-time chemical cross-linked peptide identification strategy. *J. Proteome Res.* **2012**, *12*, 1569–1579. [CrossRef] [PubMed]

41. Luger, K.; Richmond, T.J. The histone tails of the nucleosome. *Curr. Opin. Genet. Dev.* **1998**, *8*, 140–146. [CrossRef]

42. Chen, Q.; Yang, R.; Korolev, N.; Liu, C.F.; Nordenskiöld, L. Regulation of nucleosome stacking and chromatin compaction by the histone H4 N-terminal tail–H2A acidic patch interaction. *J. Mol. Biol.* **2017**, *429*, 2075–2092. [CrossRef] [PubMed]

43. Sheerin, D.J.; Buchanan, J.; Kirk, C.; Harvey, D.; Sun, X.; Spagnuolo, J.; Li, S.; Liu, T.; Woods, V.A.; Foster, T.; et al. Inter- and intra-molecular interactions of *Arabidopsis thaliana* DELLA protein RGL1. *Biochem. J.* **2011**, *435*, 629–639. [CrossRef] [PubMed]

44. Paterson, Y.; Englander, S.W.; Roder, H. An antibody binding site on cytochrome c defined by hydrogen exchange and two-dimensional NMR. *Science* **1990**, *249*, 755–759. [CrossRef] [PubMed]

45. Vassall, K.A.; Jenkins, A.D.; Bamm, W.; Haruaz, G. Thermodynamic analysis of the disorder-to-α-helical transition of 18.5-kDa myelin basic protein reveals an equilibrium intermediate representing the most compact conformation. *J. Mol. Biol.* **2015**, *427*, 1977–1992. [CrossRef] [PubMed]

46. Uversky, V.N. Protein intrinsic disorder-base liquid-liquid phase transitions in biological systems: Complex coacervates and membrane-less organelles. *Adv. Colloid Interface Sci.* **2017**, *239*, 97–114. [CrossRef] [PubMed]

47. Lin, Y.; Protter, D.S.; Rosen, M.K.; Parker, R. Formation and maturation of phase-separated liquid droplets by RNA-binding proteins. *Mol. Cell* **2015**, *60*, 208–219. [CrossRef] [PubMed]

48. Yosef, N.; Regev, A. Impulse control: Temporal dynamics in gene transcription. *Cell* **2011**, *144*, 886–896. [CrossRef] [PubMed]

3

Depicting Conformational Ensembles of α-Synuclein by Single Molecule Force Spectroscopy and Native Mass Spectroscopy

Roberta Corti [1,2], **Claudia A. Marrano** [1], **Domenico Salerno** [1], **Stefania Brocca** [3], **Antonino Natalello** [3], **Carlo Santambrogio** [3], **Giuseppe Legname** [4], **Francesco Mantegazza** [1], **Rita Grandori** [3,*] **and Valeria Cassina** [1,*]

[1] School of Medicine and Surgery, Nanomedicine Center NANOMIB, University of Milan-Bicocca, 20900 Monza, Italy; r.corti9@campus.unimib.it (R.C.); claudia.marrano@unimib.it (C.A.M.); domenico.salerno@unimib.it (D.S.); francesco.mantegazza@unimib.it (F.M.)

[2] Department of Materials Science, University of Milan-Bicocca, 20125 Milan, Italy

[3] Department of Biotechnology and Biosciences, University of Milan-Bicocca, 20126 Milan, Italy; stefania.brocca@unimib.it (S.B.); antonino.natalello@unimib.it (A.N.); carlo.santambrogio@unimib.it (C.S.)

[4] Scuola Internazionale Superiore di Studi Avanzati, SISSA, 34136 Trieste, Italy; giuseppe.legname@sissa.it

[*] Correspondence: rita.grandori@unimib.it (R.G.); valeria.cassina@unimib.it (V.C.)

Abstract: Description of heterogeneous molecular ensembles, such as intrinsically disordered proteins, represents a challenge in structural biology and an urgent question posed by biochemistry to interpret many physiologically important, regulatory mechanisms. Single-molecule techniques can provide a unique contribution to this field. This work applies single molecule force spectroscopy to probe conformational properties of α-synuclein in solution and its conformational changes induced by ligand binding. The goal is to compare data from such an approach with those obtained by native mass spectrometry. These two orthogonal, biophysical methods are found to deliver a complex picture, in which monomeric α-synuclein in solution spontaneously populates compact and partially compacted states, which are differently stabilized by binding to aggregation inhibitors, such as dopamine and epigallocatechin-3-gallate. Analyses by circular dichroism and Fourier-transform infrared spectroscopy show that these transitions do not involve formation of secondary structure. This comparative analysis provides support to structural interpretation of charge-state distributions obtained by native mass spectrometry and helps, in turn, defining the conformational components detected by single molecule force spectroscopy.

Keywords: α-synuclein; single molecule force spectroscopy; intrinsically disordered proteins; native mass spectrometry

1. Introduction

Intrinsically disordered proteins (IDPs) play crucial regulatory roles in biological systems and lack a specific tertiary structure under physiological conditions [1–4]. Molecular characterization of IDPs requires description of the conformational ensembles populated by the disordered polymers in solution. Single-molecule approaches offer information on dynamic and heterogeneous ensembles, capturing distinct and less populated states, overcoming the limitations of average parameter assessment, intrinsic to bulk methods [5–8].

Usually employed in imaging mode [9,10], atomic force microscopy (AFM) can be used in single-molecule force spectroscopy (SMFS) to characterize the statistical distribution of distinct protein conformers in solution. Indeed, protein unfolding under the action of a pulling force has been demonstrated to characterize the molecular structure of tens of distinct proteins and to distinguish among different conformations induced by ligand binding or mutations [11–13]. In the case of the human, amyloidogenic IDP α-synuclein (AS), at least three major conformational states can be recognized [14–16]: random coil (RC), collapsed states stabilized by weak interactions (WI), and compact conformations stabilized by strong interactions (SI). The SMFS technique has been applied to explore the conformational space populated by the different structures of the protein, revealing distinct conformers of the molecular ensemble and structural effects of point mutations linked to familial Parkinson's disease [4,14–17].

Pure AS in vitro, in the absence of interactors, is largely unstructured at neutral pH, with a small fraction of the population in collapsed states of different compactness, as revealed by NMR spectroscopy [18] and small angle X-ray scattering [19]. A particularly compact, globular state is populated in vivo, as indicated by in-cell NMR in neuronal and non-neuronal mammalian cell types [20]. Dopamine (DA) and epigallocatechin-3-gallate (EGCG) are known to bind AS and redirect the aggregation pathway toward soluble oligomers with different structure and toxicity [21,22].

Native mass spectrometry (native MS) has developed into a central tool for structural biology [23–26]. The analysis of charge states populated by globular and disordered proteins by native MS has shown effects of denaturants [27], stabilizers [28], metal binding [29], and protein–protein interactions [30], just to mention some examples. The application of native MS to free AS in solution reveals multimodal charge-state distributions (CSDs), which are suggestive of a conformational ensemble populated by different conformers, in line with the above-mentioned, in vitro and in vivo evidence [29,31–33]. The charge states obtained by proteins in electrospray have long been recognized as affected by protein compactness at the moment of transfer from solution to gas phase [27,34]. This effect can be rationalized by an influence of protein structure on solvent-accessible surface area [35–37] and apparent gas-phase basicity [38].

A large amount of evidence suggests that the ionization patterns of globular and disordered proteins are similarly affected by conformational properties [23,39–41]. Native MS has described conformational responses of AS to alcohols, pH, and copper binding consistent with NMR and other solution methods [29,33]. Native MS has also suggested that binding of DA and EGCG have distinct structural effects on AS soluble monomers [42,43]. While DA preferentially binds and stabilizes an intermediate form, EGCG promotes accumulation of the most compact AS conformer [42]. This different conformational selectivity could help rationalizing the different structure and toxicity of the resulting oligomers, although the two ligands have similar fibrillation-inhibition effects [42]. Nonetheless, the difficulty to capture IDP compact states by small-angle X-ray scattering and ensemble-optimization method has led to the hypothesis that IDP bimodal CSDs are artifacts resulting from a bifurcated ESI mechanism, rather than distinct components reflecting structural heterogeneity of the original protein sample [44]. The aim of this work is to describe AS conformational ensemble and its response to ligands by orthogonal and highly sensitive biophysical techniques, such as SMFS, in order to test the effect of ligand binding in solution and help interpretation of the available native-MS data on AS and IDPs in general. It is found that, while spectroscopic methods sensitive to secondary structure do not capture these conformational transitions, SMFS and native MS reveal rearrangements of the conformational ensembles, consistent with a loss of structural disorder induced by the ligands.

2. Results

2.1. Single Molecule Force Spectroscopy (SMSF)

The SMFS experiments have been performed on a polyprotein construct containing eight repeats of titin immunoglobulin-like domain (I27) and one grafted AS domain [45–47]. A schematic representation of the polyprotein construct and typical unfolding curves in the absence of ligands are reported in Figure 1A,B.

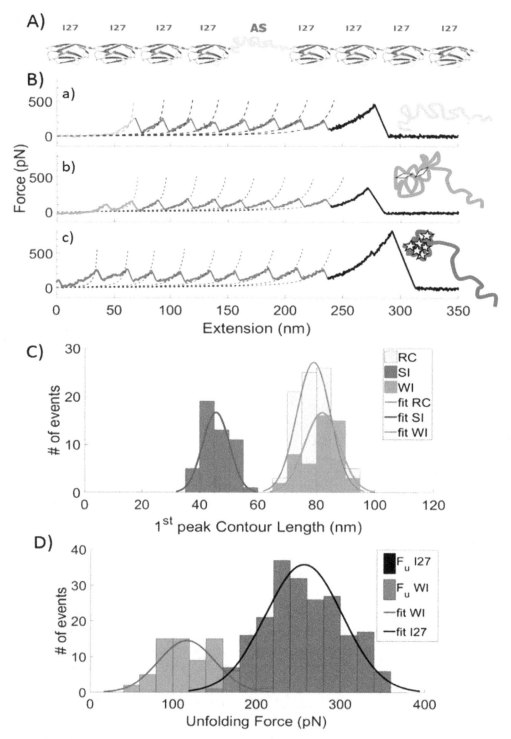

Figure 1. Representative single molecule force spectroscopy (SMFS) recording of α-synuclein (AS) polyprotein and relative statistical analysis. (**A**) Polyprotein construct encompassing the AS full-length polypeptide chain for SMFS experiments. (**B**) Representative force curves of the mechanical unfolding of the polyprotein in distinct conformations stabilized by RC (a), WI (b), and SI (c). Dotted lines are worm-like-chain (WLC) fits to the force-extension curves with free contour length L_C and a fixed persistence length $L_p = 0.36$ nm (see Figure S2 for raw data). Sketches of AS conformations are shown on the right. Diamonds represent weak interactions stabilizing the AS protein, while stars represent strong interactions. (**C**) Statistical distribution of the contour length of the first peak for RC ($L_C = 79 \pm 6$ nm), WI ($L_C = 82 \pm 6$ nm), and SI ($L_C = 46 \pm 5$ nm) conformations. Solid lines represent the Gaussian fits of the histograms. (**D**) Unfolding force statistical distribution of WI ($F_{WI} = 117 \pm 34$ pN) and I27 modules ($F_{I27} = 257 \pm 46$ pN).

As apparent from Figure 1B, the observed SMFS curves show the typical "sawtooth" pattern, in which the initial part is related to the presence of AS and it is characterized by different mechanical resistances to the unfolding. Each following regular peak is due to the unfolding of an individual I27 domain. Every curve was fitted by means of the worm-like-chain (WLC) model to extract the contour length L_C of each peak (both for I27 and AS) [48]. Consistent with the presence of a heterogeneous conformational ensemble, three distinct patterns can be recognized by analyzing the L_C of the first peak (Figure 1B,C). A first class of curves displays $L_C = 79 \pm 6$ nm (light blue curve, first line of Figure 1B); a second class is characterized by $L_C = 82 \pm 6$ nm and by the presence of at least one small peak before the first regular peak (green curve, second line of Figure 1B); a third class displays $L_C = 46 \pm 5$ nm and it is characterized by the presence of an additional peak whose height is comparable to the one related to an I27 unfolding event (red curve, third line in Figure 1B). In detail, the light blue curves are ascribed to unstructured conformations of AS and classified as random coil (RC), since no additional peak is detected in the first ~80 nm. The green curves display small (one or more) peaks corresponding to an unfolding force ($F_{WI} = 117 \pm 34$ pN) sensibly lower than I27 ($F_{I27} = 257 \pm 46$ pN, Figure 1D, Figure S1 and Table S1). These curves are interpreted as representative of a collapsed state of AS mainly stabilized by weak interactions (WI), characterized by an energy barrier to overcome smaller than the one involved in the I27 unfolding. The third and the latter type of curves, characterized by a shorter L_C is assigned to a collapsed state of AS, mainly stabilized by strong interactions (SI), which presents resistance to unfolding similar to the one shown by the highly mechanostable protein I27. The extension of the first peak (in the curves assigned to the RC conformation) and that of the first of the higher peaks (in the curves assigned to the WI) are all around 80 nm. These peaks occur when AS is completely extended and flanked by eight I27 folded modules. The measured length is due to the contribution of the eight folded I27 modules (a folded module of I27 is 3 nm long, i.e., 3 nm × 8 = 24 nm), the length of eight linkers between each protein module (a linker is 2 aa, i.e., 8 × 0.36 nm × 2 = 5.76 nm) [47], and the length of the completely extended AS (i.e., 140 aa × 0.36 nm = 50.4 nm). By summing all the contributions, one obtains a total extension of 80.16 nm, which is coherent with the measured values. The subset of curves presenting a L_C of the first peak higher than 95 nm, which could be associated with an undesired misfolding event of a I27 module [49], was discarded.

2.2. Effect of DA and EGCG on the Conformational Ensemble

The SMFS measurements were repeated in the presence of either 200 µM DA or 25 µM EGCG (see Figures S3–S6 and Tables S2–S6 for more details). These concentrations were chosen to compare SMFS results with native-MS data [42]. The statistical distributions of AS conformations obtained by SMFS in the presence or absence of ligands are reported in Figure 2A. In solution, at neutral pH and without ligands, AS behaves partially as RC (62% of the molecules) and partially populates collapsed states (~30%, mainly stabilized by WI and ~8%, mainly stabilized by SI), consistent with previously reported SMFS data [14–16]. The addition of either ligand leads to a loss of the RC conformation in favor of the SI conformation, with the most pronounced effect of EGCG (drop of RC from 62% to 36%). The same conditions had been investigated by native MS, showing the presence of intermediate states (I1 and I2), together with random coil (RC) and compact (C) conformations (Figure 2B).

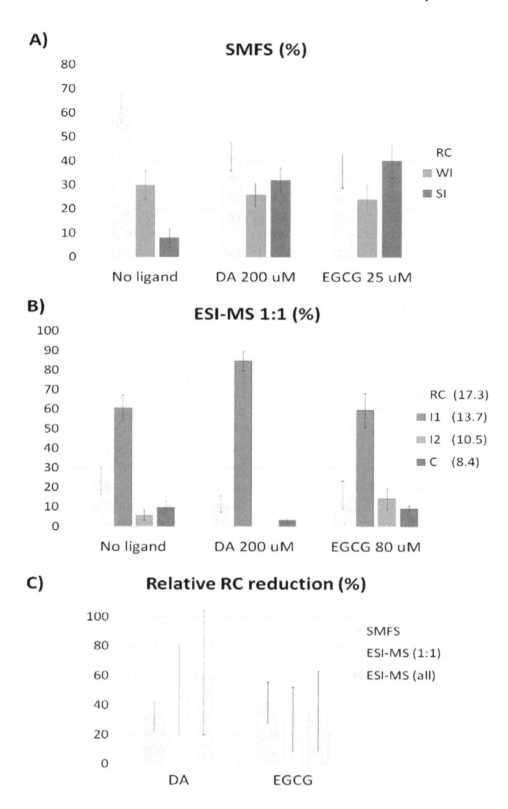

Figure 2. Species distributions as obtained by (**A**) SMFS and (**B**) native MS. The intensity-weighted average charge of the peak envelopes is reported in brackets (i.e., RC = 17.3; I1 = 13.7; I2 = 10.5; C = 8.4). Error bars in panel (A) represent the standard deviation calculated for the normal distribution. Error bars in panel (B) represent the standard deviations from three independent experiments. (**C**) RC reduction in response to ligand binding, relative to the free protein, as obtained by SMFS and native MS, considering the 1:1 protein:ligand complexes (1:1) or the cumulative MS data (all). Error bars in panel (C) represent the propagated standard deviation.

A quantitative comparison between the species distributions obtained by SMFS and native-MS data is shown in Figure 2C. An intrinsic difference between SMFS and MS concerns the discrimination between free and ligand-bound protein molecules, which is possible only by the latter technique. Thus, native-MS data in Figure 2C have been processed by two alternative ways. In one case, only signals of the 1:1 protein:ligand complexes have been considered. This procedure yields more reliable information on the conformational changes induced by ligand binding but is, at the same time, not exactly comparable to the blind molecular selection performed by SMFS. Thus, "cumulative" MS data are also shown (labeled as ESI-MS(all) in Figure 2C), derived by Gaussian fitting of the artificial CSD obtained by the summation of the species-specific CSDs corresponding to the different binding stoichiometries, including the free protein. In either way, the aggregated data for the unstructured (RC) component, represented as relative change from the reference condition of the protein in the absence of ligands, indicate a remarkable loss of the most disordered conformation induced by ligand binding, as assessed by both techniques.

2.3. Comparison to CD and FTIR

For comparison with complementary spectroscopic methods, sensitive to protein secondary structure, far-UV circular dichroism (CD) and Fourier-transform infrared spectroscopy (FTIR) analyses were performed. Representative results are reported in Figure 3. It can be noted that AS spectra in the presence or absence of the ligands, acquired by either technique under the same conditions employed for SMFS experiments, are almost superimposable. Thus, bulk methods probing secondary structure do not capture the conformational changes induced by ligand binding in monomeric AS in solution.

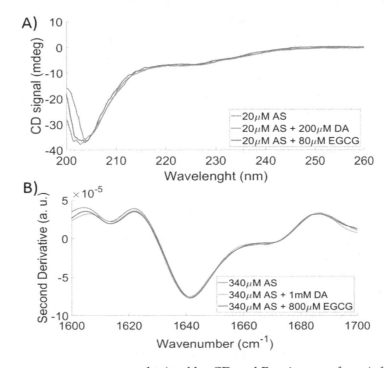

Figure 3. Secondary-structure content as obtained by CD and Fourier-transform infrared spectroscopy (FTIR) techniques. (**A**) Far-UV CD spectra of 20 μM AS in PBS buffer in the absence of ligands (gray), in the presence of 200 μM DA (orange) or 80 μM epigallocatechin-3-gallate (EGCG) (purple). (**B**) Second derivatives in the Amide I region of the FTIR absorption spectra of 340 μM AS in deuterated PBS buffer in the absence of ligands (gray) and in the presence of 1 mM DA (orange) or 800 μM EGCG (purple).

3. Discussion

The results reported here provide direct evidence of the different conformers populated by AS in solution and the structural effects elicited by ligand binding, resulting in a rearrangement of the conformational ensemble [50]. The structural heterogeneity of free AS in solution captured by SMFS is consistent with previous reports by the same approach [14,15], as well as with results from native MS [23,29,33,42], computational simulations [51,52], and chemical crosslinking [53], indicating the presence of at least three different conformational states characterized by different degrees of intramolecular interactions. Furthermore, SMFS is applied here for the first time to probe the effects of the fibrillation inhibitors DA and EGCG on AS conformational properties in solution.

Since the reliability of CSD analysis in the investigation of IDP conformational ensembles by native MS has been questioned [44], the SMFS results obtained in this work are compared to native-MS data. In analogy with SMFS, the CSD analysis of nano-ESI-MS spectra identifies, in addition to the RC component, the presence of three non-RC components, namely the intermediate species I1, I2, and the compact conformation C. Furthermore, both techniques indicate a loss of the most disordered component in response to ligand binding, resulting in the accumulation of the more structured species. Thus, not only the presence of multimodal profiles is confirmed by both techniques, but also a reorganization of the conformational ensemble in the same direction is consistently indicated in the presence of ligands.

Nonetheless, the structural intermediates detected by SMFS and native MS cannot be related in a straightforward way. These discrepancies can be due to the fact that the physical properties detected by the two techniques are different. While SMFS discriminates protein structures according to their mechanical stability under an external tension (quantified by the unfolding force), native MS is affected by structural compactness (quantified by the acquired net charge). Different compaction levels can correspond to similar unfolding force and vice-versa. Accordingly, the WI state, as detected by SMFS, is characterized by a number of variable peaks ranging from 1 to 3 different species, which could be compatible with different AS compaction states. Furthermore, the SMFS instrumental noise, related to the minimum measurable force (around 20pN) limits the minimal detectable unfolding force, below which the less stable AS compact states are counted as RC molecules.

It should also be noted that the conformations with lower unfolding force, as detected by SMFS, could include some components with higher charge-state detected by ESI-MS. This hypothesis can be verified by comparing the two techniques in terms of the response of the RC component to the binding of the ligand. Indeed, upon binding of either ligand there is a compatible trend of loss in such a component, as observed by both techniques, in favor of more compact structures (native MS) or stronger interactions (SMFS). Therefore, interpreting the low-charge components of CSDs as collapsed and partially structured conformational states leads to compatible pictures delivered by SMFS and native MS. Both techniques reveal the presence of partially structured conformers, thus suggesting that the bimodal or multimodal CSDs detected by native MS do not simply reflect artefacts of the ESI mechanism. It is worth pointing out that the conditions employed in this work do not lead neither to AS oxidation (Figure S7) nor to AS oligomerization, which requires incubation at 37 °C, shaking and higher protein concentrations [54], as also indicated by the lack of higher-order aggregates in native-MS spectra [42].

It cannot be ruled out that different ionization and/or transmission efficiency of compact and extended protein ions in native MS might lead to distortions of the apparent molecular ensemble, adding to the difficulties of direct comparison with SMFS data. Indeed, it has been suggested that folded and unfolded molecules could undergo different ESI mechanisms, resulting in different signal yields [55]. However, this effect seems to be protein-specific, since quantitative agreement with solution methods has been observed describing, for instance, the pH-dependent unfolding transition of cytochrome c [56]. The underlying mechanism has been identified in the different hydropathy of the exposed regions of normally folded proteins in different conformational states, which could affect their surface activity inside ESI droplets [55]. Such an effect is expected to be much more modest for IDPs, which lack a structured hydrophobic core and whose collapsed conformations are mostly promoted by electrostatic interactions [57]. More systematic, quantitative comparison between native MS and solution methods will be required to further elucidate this point. This first comparative study between a single-molecule technique and CSD analysis by native MS supports the feasibility of combined approaches to describe IDP molecular ensembles.

Based on this study, it seems safe to conclude that SMFS and structural interpretation of CSDs consistently indicate the simultaneous presence of collapsed and partially structured conformers of AS monomer in solution and, most importantly, reveal induced-folding transitions elicited by ligand binding. Furthermore, this study shows that single-molecule protein unfolding can capture changes in AS conformational landscape, induced by variable solution conditions, with remarkable sensitivity and reproducibility. These results indicate that the conformational ensemble depicted by two orthogonal biophysical principles is heterogeneous and reshaped in the same direction by ligand binding.

Another implication of this study is that the AS conformational transitions detected by SMFS under these conditions should not be interpreted in terms of secondary structure formation [14,15]. Indeed, the measured WI and SI components cannot be simply seen as the distinct contributions of van-der-Waals interactions and ordered secondary structure, respectively. In fact, an increase of almost 30% in the SI component, as observed here, would be detected by CD and FTIR spectroscopies, if ascribable to secondary structure. It is conceivable that the AS conformational components detected by SMFS under these conditions differ by contact order, type, and number of interactions, within a picture of similar secondary-structure content. Hence, a new structural interpretation of SMFS data is proposed, in particular for the SI population, differing from the one reported in the literature [14,15], where the SI component was directly associated with the presence of secondary structure.

This comparison points out that the ion-sorting mechanism inherent to MS analyses makes the MS methods more comparable to single-molecules approaches, rather than to bulk spectroscopic techniques, and underscores the importance of multi-technological approach to ensemble characterization. Nevertheless, the WI population detected by SMFS and the intermediate species (I1 and I2) detected by native MS do not necessarily coincide. Actually, two intermediates are detected by MS and only one by SMFS and the WI species found by SMFS does not respond to ligands, while the MS-detected intermediates do. These results indicate that both techniques capture the decrease in structural disorder induced by the ligands, but they describe the partially structured species of the conformational ensemble in different ways. In particular, it seems that the collapsed and partially structured species detected by MS contribute cumulatively to the SI component by SMFS, while the WI component by SMFS does not find correspondence in the MS spectra. These interactions could be too weak to survive the ionization/desolvation step.

4. Materials and Methods

4.1. Cloning, Expression, and Purification of the (I27)4_AS_(I27)4 Polyprotein

In order to obtain a (I27)4_AS_(I27)4 polyprotein, consisting of a single AS molecule, flanked by four repetitions of titin immunoglobulin-like domain (I27) at the N-terminus and at the C-terminus, the cDNA of the human AS (NP_000336) was cloned in the pRSet.A(I27)8 expression vector [47], taking advantage of the NheI restriction site placed in the middle of (I27)8 encoding sequence. A mutagenic PCR was performed on the pEGFP_AS vector [58] to delete the start and stop codons and to insert a NheI restriction site at both extremities of the AS gene. The PCR was carried out using the Q5®High-Fidelity DNA Polymerase (NEB, cat. #M0491) with the following primers: forward primer 5′ AAAAGCTAGCGATGTATTCATGAAAGGAC 3′, reverse primer 5′ AATTGCTAGCGGCTTCAGGTTCGTAG 3′, (in bold, the NheI restriction site). After sequencing, the pRSet.A (I27)4_AS_(I27)4 vector was used to transform BL21(DE3) *Escherichia coli* cells. Transformed cells were grown in Luria-Bertani medium at 37 °C until they reached an OD600 of 0.4–0.6 and the expression of the polyprotein was induced overnight at 22 °C by the addition of 1 mM IPTG. Cells were subsequently harvested by centrifugation and resuspended in lysis buffer (50 mM Na_2HPO_4, 300 mM NaCl, 10 mM imidazole, 4% Triton™ X-100, and 0.5 mM phenylmethylsulfonyl fluoride) before sonication on ice. The purification was performed by gravity flow column ion metal affinity chromatography (IMAC), taking advantage of the 6× His-tag present at the N-terminus of the polyprotein. The soluble fraction of cell lysate was incubated on Ni+-NTA resin (Roche, cat. #05893682001) for 1 h at 4 °C with gentle agitation. The washing step was carried out in 50 mM Na_2HPO_4, 300 mM NaCl added with 20 mM imidazole, elution was achieved in the same buffer, added with 250 mM imidazole. The presence of the protein in the eluted fractions was verified by SDS-PAGE on a 4–12% polyacrylamide gel (InvitrogenTM, ThermoFisher Scientific, cat. #NW04120BOX) stained with Coomassie Brilliant Blue.

4.2. AFM—Single Molecule Force Spectroscopy

SMFS experiments were carried out on a Nanowizard II (JPK Instruments, Berlin, Germany) at room temperature. Prior to each experiment, every cantilever (Si_3N_4, Bruker MLCT-BIO, Cantilever D, Nominal spring constant k = 0.03 N/m) was individually calibrated using the Equipartition Theorem in the JPK software. Approximately 20 µL of protein (at a concentration of ~2 µM) were deposited onto an evaporated gold coverslip and allowed to adsorb for about 15 min. After this time, 1.8 mL of PBS buffer (pH 7.4, 150 mM) were added to reach an overall protein final concentration of ~20 nM. Constant-velocity, single-molecule pulling experiments were performed at 1 µm/s, with a recorded rate of 4096 Hz. Each experiment was carried out in fresh PBS buffer, to which EGCG (stock diluted in PBS) and DA (stock diluted in acidic MilliQ, pH 4) (stored at 4 °C protected from light) were added to reach the desired final concentration. Each solution was filtered on a filter screen with a porosity of 0.2 µm before each experiment.

4.3. AFM Data Analysis

The resulting force curves were then processed by means of both the JPK-Data Processed software (JPK Instruments, Berlin, Germany) and MATLAB custom-written software. The contour length (L_C) of each peak (both I27 and AS) was calculated by means of WLC fit as a single parameter, while the persistence length (L_P) was kept constant (0.36 nm) [59]. Only curves with a single clear detachment peak, at least seven I27 peaks, and traces with a spurious signal below 45pN in the first 25 nm of the force-extension were considered.

4.4. Native-MS Experiments

Nano-ESI-MS data were taken from Konijnenberg [42]. In particular, nano-ESI-MS spectra were collected after 10-minute incubation of protein-ligand mixtures in 10 mM ammonium acetate, pH 7.4, at a final AS concentration of 20 µM. Quantification from native-MS data was based on Gaussian fitting of CSDs, upon transformation to x = z abscissa axis. The reported values refer to the area of the components obtained for the protein in the absence of ligand and for the 1:1 AS:ligand complexes, from three independent experiments.

4.5. CD and FTIR Experiments

CD and FTIR analyses were performed as previously described [43]. In particular, Far-UV CD spectra of 20 µM AS in PBS buffer were acquired on a J-815 spectropolarimeter (JASCO Corp., Tokyo, Japan) under the following instrumental settings: data pitch, 0.1 nm; scan speed, 20 nm/min; bandwidth, 1 nm; accumulation spectra, 2. A 1-mm path length quartz cuvette was employed. FTIR spectra of 340 µM AS in deuterated PBS buffer were acquired on a Varian 670-IR spectrometer (Varian Australia Pty. Ltd., Mulgrave, VIC, Australia) under the following instrumental settings: resolution, 2 cm^{-1}; scan speed, 25 kHz; scan coadditions, 1000; apodization, triangular; nitrogen-cooled mercury cadmium telluride detector. A temperature-controlled transmission cell with two BaF_2 windows separated by a 100-µm Teflon spacer was employed. Representative spectra from three independent experiments are shown.

5. Conclusions

Single-molecule description of AS conformational ensemble in solution detects differently structured components that are overseen by bulk spectroscopic methods, which probe secondary structure, but are consistent with the different degrees of compactness suggested by CSD analysis. Thus, although ion-mobility studies and molecular-dynamics simulations have shown that IDPs rearrange in the gas phase in a charge-dependent fashion [40], the extent of ionization at the moment of transfer from solution to gas phase, i.e., CSDs, seems to reflect structural heterogeneity in solution rather than ESI artifacts. This correspondence is experimentally established here, independently of assumptions on the underlying ESI mechanism. Combined description by orthogonal biophysical methods can provide valuable constraints for computational simulations of IDP conformational ensembles in the presence or absence of interactors [51].

Author Contributions: Conceptualization, R.G., V.C., D.S., and F.M.; methodology, R.C., V.C., C.A.M., A.N., and C.S.; software, R.C.; investigation, R.C., V.C., C.A.M., A.N., and C.S.; resources, R.G., F.M.; data curation, R.C., V.C., D.S., A.N., and C.S.; writing—original draft preparation, R.C., V.C., C.A.M., R.G., and F.M.; writing—review and editing, V.C., S.B., G.L., R.G., and F.M.; supervision, V.C., D.S., S.B., G.L., R.G., and F.M.

Acknowledgments: We thank G. Cappelletti and J. Clarke for the kind gift of the DNA plasmids used in this work.

Abbreviations

AS	α-synuclein
C	compact structure detected in native MS
CD	circular dichroism
CSDs	charge state distributions
DA	dopamine
EGCG	epigallocatechin-3-gallate
ESI-MS	electrospray ionization mass spectrometry
F	unfolding force
FTIR	Fourier-transform infrared spectroscopy
I1, I2	Intermediate 1 and 2 detected in native MS
I27	27th titin immunoglobulin-like domain
IDP	intrinsically disordered protein
IDPs	intrinsically disordered proteins
L_C	contour length
L_P	persistence length
native MS	native mass spectrometry
NMR	nuclear magnetic resonance
RC	random coil
SAXS-EOM	small-angle X-ray scattering and ensemble-optimization method
SI	strong interactions
SMFS	single molecule force spectroscopy
WI	weak interactions
WLC	worm-like-chain

References

1. Rezaei-Ghaleh, N.; Parigi, G.; Soranno, A.; Holla, A.; Becker, S.; Schuler, B.; Luchinat, C.; Zweckstetter, M. Local and Global Dynamics in Intrinsically Disordered Synuclein. *Angew. Chem. Int. Ed. Engl.* **2018**, *57*, 15262–15266. [CrossRef] [PubMed]

2. Borgia, A.; Kemplen, K.R.; Borgia, M.B.; Soranno, A.; Shammas, S.; Wunderlich, B.; Nettels, D.; Best, R.B.; Clarke, J.; Schuler, B. Transient misfolding dominates multidomain protein folding. *Nat. Commun.* **2015**, *6*, 8861. [CrossRef] [PubMed]

3.   Gruebele, M.; Dave, K.; Sukenik, S. Globular Protein Folding In Vitro and In Vivo. *Annu. Rev. Biophys.* **2016**, *45*, 233–251. [CrossRef] [PubMed]

4.   Wright, P.E.; Dyson, H.J. Intrinsically disordered proteins in cellular signalling and regulation. *Nat. Rev. Mol. Cell. Biol.* **2015**, *16*, 18–29. [CrossRef] [PubMed]

5.   Oberhauser, A.F.; Marszalek, P.E.; Carrion-Vazquez, M.; Fernandez, J.M. Single protein misfolding events captured by atomic force microscopy. *Nat. Struct. Biol.* **1999**, *6*, 102510–102528. [CrossRef]

6.   Rounsevell, R.; Forman, J.R.; Clarke, J. Atomic force microscopy: Mechanical unfolding of proteins. *J. Methods* **2004**, *34*, 100–111. [CrossRef] [PubMed]

7.   Ferreon, A.C.M.; Deniz, A.A. Protein folding at single-molecule resolution. *Biochim. Biophys. Acta* **2011**, *1814*, 1021–1029. [CrossRef]

8.   Junker, J.P.; Rief, M. Single-molecule force spectroscopy distinguishes target binding modes of calmodulin. *Proc. Natl. Acad. Sci. USA* **2009**, *106*, 14361–14366. [CrossRef]

9.   Cassina, V.; Manghi, M.; Salerno, D.; Tempestini, A.; Iadarola, V.; Nardo, L.; Brioschi, S.; Mantegazza, F. Effects of cytosine methylation on DNA morphology: An atomic force microscopy study. *Biochim. Biophys. Acta Gen. Subj.* **2016**, *1860*, 1–7. [CrossRef]

10. Cassina, V.; Seruggia, D.; Beretta, G.L.; Salerno, D.; Brogioli, D.; Manzini, S.; Zunino, F.; Mantegazza, F. Atomic force microscopy study of DNA conformation in the presence of drugs. *Eur. Biophys. J.* **2011**, *40*, 59–68. [CrossRef]

11. Beedle, A.E.M.; Lezamiz, A.; Stirnemann, G.; Garcia-Manyes, S. The mechanochemistry of copper reports on the directionality of unfolding in model cupredoxin proteins. *Nat. Commun.* **2015**, *6*, 7894. [CrossRef] [PubMed]

12. Walder, R.; LeBlanc, M.A.; Van Patten, W.J.; Edwards, D.T.; Greenberg, J.A.; Adhikari, A.; Okoniewski, S.R.; Sullan, R.M.A.; Rabuka, D.; Sousa, M.C.; et al. Rapid Characterization of a Mechanically Labile α-Helical Protein Enabled by Efficient Site-Specific Bioconjugation. *J. Am. Chem. Soc.* **2017**, *39*, 9867–9875. [CrossRef] [PubMed]

13. Garcia-Manyes, S.; Kuo, T.L.; Fernández, J.M. Contrasting the individual reactive pathways in protein unfolding and disulfide bond reduction observed within a single protein. *J. Am. Chem. Soc.* **2011**, *133*, 3104–3113. [CrossRef] [PubMed]

14. Sandal, M.; Valle, F.; Tessari, I.; Mammi, S.; Bergantino, E.; Musiani, F.; Brucale, M.; Bubacco, L.; Samorì, B. Conformational equilibria in monomeric alpha-synuclein at the single-molecule level. *PLoS Biol.* **2008**, *6*, 99–108. [CrossRef]

15. Brucale, M.; Sandal, M.; Di Maio, S.; Rampion, A.; Tessari, I.; Tosatto, L.; Bisaglia, M.; Bubacco, L.; Samorì, B. Pathogenic mutations shift the equilibria of alpha-synuclein single molecules towards structured conformers. *ChemBioChem* **2009**, *10*, 176–183. [CrossRef]

16. Hervàs, R.; Oroz, J.; Galera-Prat, A.; Goñi, O.; Valbuena, A.; Vera, A.M.; Gòmez-Sicilia, A.; Losada-Urzáiz, F.; Uversky, V.N.; Menéndez, M.; et al. Common features at the start of the neurodegeneration cascade. *PLoS Biol.* **2012**, *10*, 1001335. [CrossRef]

17. Zhang, Y.; Hashemi, M.; Lv, Z.; Williams, B.; Popov, K.I.; Dokholyan, N.V.; Lyubchenko, Y.L. High-speed atomic force microscopy reveals structural dynamics of α-synuclein monomers and dimers. *J. Chem. Phys.* **2018**, *148*, 123322. [CrossRef]

18. Stephens, A.D.; Zacharopoulou, M.; Kaminski Schierle, G.S. The Cellular Environment Affects Monomeric α-Synuclein Structure. *Trends Biochem. Sci.* **2019**, *44*, 453–466. [CrossRef]

19. Curtain, C.C.; Kirby, N.M.; Mertens, H.D.; Barnham, K.J.; Knott, R.B.; Masters, C.L.; Cappai, R.; Rekas, A.; Kenche, V.B.; Ryan, T. α-synuclein oligomers and fibrils originate in two distinct conformer pools: A small angle X-ray scattering and ensemble optimisation modelling study. *Mol. Biosyst.* **2015**, *11*, 190–196. [CrossRef]

20. Theillet, F.X.; Binolfi, A.; Bekei, B.; Martorana, A.; Rose, H.M.; Stuiver, M.; Verzini, S.; Lorenz, D.; van Rossum, M.; Goldfarb, D.; et al. Structural disorder of monomeric α-synuclein persists in mammalian cells. *Nature* **2016**, *530*, 45–50. [CrossRef]

21. Zhao, J.; Liang, Q.; Sun, Q.; Chen, C.; Xu, L.; Ding, Y.; Zhou, P. (-)-Epigallocatechin-3-gallate (EGCG) inhibits fibrillation, disaggregates amyloid fibrils of α-synuclein, and protects PC12 cells against alpha-synuclein-induced toxicity. *RSC Adv.* **2017**, *7*, 32508–32517. [CrossRef]

22. Lee, H.J.; Baek, S.M.; Ho, D.H.; Suk, J.E.; Cho, E.D.; Lee, S.J. Dopamine promotes formation and secretion of non-fibrillar alpha-synuclein oligomers. *Exp. Mol. Med.* **2011**, *43*, 216–222. [CrossRef] [PubMed]

23. Santambrogio, C.; Natalello, A.; Brocca, S.; Ponzini, E.; Grandori, R. Conformational Characterization and Classification of Intrinsically Disordered Proteins by Native Mass Spectrometry and Charge-State Distribution. *Proteomics* **2019**, *19*, 1800060. [CrossRef] [PubMed]

24. Lössl, P.; Van de Waterbeemd, M.; Heck, A.J. The diverse and expanding role of mass spectrometry in structural and molecular biology. *EMBO J.* **2016**, *35*, 2634–2657. [CrossRef]

25. Konijnenberg, A.; Butterer, A.; Sobott, F. Native ion mobility-mass spectrometry and related methods in structural biology. *Biochim. Biophys. Acta* **2013**, *1834*, 1239–1256. [CrossRef]

26. Loo, R.R.; Loo, J.A. Salt Bridge Rearrangement (SaBRe) Explains the Dissociation Behavior of Noncovalent Complexes. *J. Am. Soc. Mass Spectrom.* **2016**, *27*, 975–990. [CrossRef]

27. Chowdhury, S.K.; Katta, V.; Chait, B.T. Probing conformational changes in proteins by mass spectrometry. *J. Am. Chem. Soc.* **1990**, *112*, 9012–9013. [CrossRef]

28. Grandori, R.; Matecko, I.; Mayr, P.; Müller, N. Probing protein stabilization by glycerol using electrospray mass spectrometry. *J. Mass Spectrom.* **2001**, *36*, 918–922. [CrossRef]

29. Natalello, A.; Benetti, F.; Doglia, S.M.; Legname, G.; Grandori, R. Compact conformations of α-synuclein induced by alcohols and copper. *Proteins* **2011**, *79*, 611–621. [CrossRef]

30. D'Urzo, A.; Konijnenberg, A.; Rossetti, G.; Habchi, J.; Li, J.; Carloni, P.; Sobott, F.; Longhi, S.; Grandori, R. Molecular basis for structural heterogeneity of an intrinsically disordered protein bound to a partner by combined ESI-IM-MS and modeling. *J. Am. Soc. Mass Spectrom.* **2015**, *26*, 472–481. [CrossRef]

31. Wongkongkathep, P.; Han, J.Y.; Choi, T.S.; Yin, S.; Kim, H.I.; Loo, J.A. Native Top-Down Mass Spectrometry and Ion Mobility MS for Characterizing the Cobalt and Manganese Metal Binding of α-Synuclein Protein. *J. Am. Soc. Mass Spectrom.* **2018**, *29*, 1870–1880. [CrossRef] [PubMed]

32. Testa, L.; Brocca, S.; Santambrogio, C.; D'Urzo, A.; Habchi, J.; Longhi, S.; Uversky, V.N.; Grandori, R. Extracting structural information from charge-state distributions of intrinsically disordered proteins by non-denaturing electrospray-ionization mass spectrometry. *Intrinsic. Disord. Proteins* **2013**, *1*, 25068. [CrossRef] [PubMed]

33. Frimpong, A.K.; Abzalimov, R.R.; Uversky, V.N.; Kaltashov, I.A. Characterization of intrinsically disordered proteins with electrospray ionization mass spectrometry: Conformational heterogeneity of α-synuclein. *Proteins* **2010**, *78*, 714–722. [CrossRef] [PubMed]

34. Verkerk, U.H.; Kebarle, P. Ion-ion and ion-molecule reactions at the surface of proteins produced by nanospray. Information on the number of acidic residues and control of the number of ionized acidic and basic residues. *J. Am. Soc. Mass Spectrom.* **2005**, *16*, 1325–1341. [CrossRef]

35. Testa, L.; Brocca, S.; Grandori, R. Charge-surface correlation in electrospray ionization of folded and unfolded proteins. *Anal. Chem.* **2011**, *83*, 6459–6463. [CrossRef]

36. Kaltashov, I.A.; Mohimen, A. Estimates of protein surface areas in solution by electrospray ionization mass spectrometry. *Anal. Chem.* **2005**, *77*, 5370–5379. [CrossRef]

37. Hall, Z.; Robinson, C.V. Do charge state signatures guarantee protein conformations? *J. Am. Soc. Mass Spectrom.* **2012**, *23*, 1161–1168. [CrossRef]

38. Li, J.; Santambrogio, C.; Brocca, S.; Rossetti, G.; Carloni, P.; Grandori, R. Conformational effects in protein electrospray-ionization mass spectrometry. *Mass Spectrom. Rev.* **2016**, *35*, 111–122. [CrossRef]

39. Natalello, A.; Santambrogio, C.; Grandori, R. Are Charge-State Distributions a Reliable Tool Describing Molecular Ensembles of Intrinsically Disordered Proteins by Native MS? *J. Am. Soc. Mass Spectrom.* **2017**, *28*, 21–28. [CrossRef]

40. Beveridge, R.; Migas, L.G.; Das, R.K.; Pappu, R.V.; Kriwacki, R.W.; Barran, P.E. Ion Mobility Mass Spectrometry Uncovers the Impact of the Patterning of Oppositely Charged Residues on the Conformational Distributions of Intrinsically Disordered Proteins. *J. Am. Chem. Soc.* **2019**, *141*, 4908–4918. [CrossRef]

41. Stuchfield, D.; Barran, P. Unique insights to intrinsically disordered proteins provided by ion mobility mass spectrometry. *Curr. Opin. Chem. Biol.* **2018**, *42*, 177–185. [CrossRef] [PubMed]

42. Konijnenberg, A.; Ranica, S.; Narkiewicz, J.; Legname, G.; Grandori, R.; Sobott, F.; Natalello, A. Opposite Structural Effects of Epigallocatechin-3-gallate and Dopamine Binding to α-Synuclein. *Anal. Chem.* **2016**, *88*, 8468–8475. [CrossRef] [PubMed]

43. Ponzini, E.; De Palma, A.; Cerboni, L.; Natalello, A.; Rossi, R.; Moons, R.; Konijnenberg, A.; Narkiewicz, J.; Legname, G.; Sobott, F.; et al. Methionine oxidation in α-synuclein inhibits its propensity for ordered secondary structure. *J. Biol. Chem.* **2019**, *294*, 5657–5665. [CrossRef] [PubMed]

44. Borysik, A.J.; Kovacs, D.; Guharoy, M.; Tompa, P. Ensemble Methods Enable a New Definition for the

Solution to Gas-Phase Transfer of Intrinsically Disordered Proteins. *J. Am. Chem. Soc.* **2015**, *137*, 13807–13817. [CrossRef]

45. Steward, A.; Toca-Herrera, J.L.; Clarke, J. Versatile cloning system for construction of multimeric proteins for use in atomic force microscopy. *J. Protein Sci.* **2002**, *11*, 2179–2183. [CrossRef]

46. Hoffman, T.; Dougan, L. Single molecule force spectroscopy using polyproteins. *Chem. Soc. Rev.* **2012**, *41*, 4781–4796. [CrossRef]

47. Best, R.B.; Brockwell, D.J.; Toca-Herrera, J.L.; Blake, A.W.; Smith, A.; Radford, S.E.; Clarke, J. Force mode atomic force microscopy as a tool for protein folding studies. *J. Anal. Chim. Acta* **2003**, *479*, 87–105. [CrossRef]

48. Bustamante, C.; Marko, J.F.; Siggia, E.D.; Smith, S. Entropic elasticity of lambda-phage DNA. *Science* **1994**, *265*, 1599–1600. [CrossRef]

49. Borgia, M.B.; Borgia, A.; Best, R.B.; Steward, A.; Nettels, D.; Wunderlich, B.; Schuler, B.; Clarke, J. Single-molecule fluorescence reveals sequence-specific misfolding in multidomain proteins. *Nature* **2011**, *474*, 662–665. [CrossRef]

50. Heller, G.T.; Bonomi, M.; Vendruscolo, M. Structural Ensemble Modulation upon Small-Molecule Binding to Disordered Proteins. *J. Mol. Biol.* **2018**, *430*, 2288–2292. [CrossRef]

51. Rossetti, G.; Musiani, F.; Abad, E.; Dibenedetto, D.; Mouhib, H.; Fernandez, C.O.; Carloni, P. Conformational ensemble of human α-synuclein physiological form predicted by molecular simulations. *Phys. Chem. Chem. Phys.* **2016**, *18*, 5702–5706. [CrossRef] [PubMed]

52. Balupuri, A.; Choi, K.E.; Kang, N.S. Computational insights into the role of α-strand/sheet in aggregation of α-synuclein. *Sci. Rep.* **2019**, *9*, 59. [CrossRef] [PubMed]

53. Brodie, N.I.; Popov, K.I.; Petrotchenko, E.V.; Dokholyan, N.V.; Borchers, C.H. Conformational ensemble of native α-synuclein in solution as determined by short-distance crosslinking constraint-guided discrete molecular dynamics simulations. *PLoS Comput. Biol.* **2019**, *15*, e1006859. [CrossRef] [PubMed]

54. Ehrnhoefer, D.E.; Bieschke, J.; Boeddrich, A.; Herbst, M.; Masino, L.; Lurz, R.; Engemann, S.; Pastore, A.; Wanker, E.E. EGCG redirects amyloidogenic polypeptides into unstructured, off-pathway oligomers. *Nat. Struct. Mol. Biol.* **2008**, *15*, 558–566. [CrossRef]

55. Kuprowski, M.C.; Konermann, L. Signal response of coexisting protein conformers in electrospray mass spectrometry. *Anal. Chem.* **2007**, *79*, 2499–2506. [CrossRef]

56. Samalikova, M.; Matecko, I.; Müller, N.; Grandori, R. Interpreting conformational effects in protein nano-ESI-MS spectra. *Anal. Bioanal. Chem.* **2004**, *378*, 1112–1123. [CrossRef]

57. Marsh, J.A.; Forman-Kay, J.D. Sequence determinants of compaction in intrinsically disordered proteins. *Biophys. J.* **2010**, *98*, 2383–2390. [CrossRef]

58. Cartelli, D.; Aliverti, A.; Barbiroli, C.; Santambrogio, C.; Raggi, E.M.; Casagrande, F.V.M.; Cantele, F.; Beltramone, S.; Marangon, J.; De Gregorio, C.; et al. α-Synuclein is a Novel Microtubule Dynamase. *Sci. Rep.* **2016**, *6*, 33289. [CrossRef]

59. Marszalek, P.E.; Lu, H.; Li, H.; Carrion-Vazquez, M.; Oberhauser, A.F.; Schulten, K.; Fernandez, J.M. Mechanical unfolding intermediates in titin modules. *Nature* **1999**, *402*, 100–103. [CrossRef]

4

The Role of Post-Translational Modifications in the Phase Transitions of Intrinsically Disordered Proteins

Izzy Owen and Frank Shewmaker *

Department of Biochemistry, Uniformed Services University of the Health Sciences, Bethesda, MD 20814, USA;
izzy.owen@usuhs.edu
* Correspondence: fshewmaker@usuhs.edu

Abstract: Advances in genomics and proteomics have revealed eukaryotic proteomes to be highly abundant in intrinsically disordered proteins that are susceptible to diverse post-translational modifications. Intrinsically disordered regions are critical to the liquid–liquid phase separation that facilitates specialized cellular functions. Here, we discuss how post-translational modifications of intrinsically disordered protein segments can regulate the molecular condensation of macromolecules into functional phase-separated complexes.

Keywords: liquid–liquid phase separation; intrinsically disordered regions; post-translational modifications; membraneless organelles

1. Introduction to Liquid–Liquid Phase Separation and Membraneless Organelles

Cells contain crowded molecular environments hosting discrete functions that must be separated within time and space. Membrane-less compartments resulting from liquid–liquid phase separation (LLPS) are increasingly being recognized as mechanisms for organizing cellular activities. These distinct regions may be referred to as biomolecular condensates or membrane-less organelles (MLOs). As the names suggest, these organelles are not encapsulated in a membrane, yet contain enriched sets of specific macromolecules. Thus, LLPS is the biologically regulated process by which specific macromolecular components are concentrated into a specific MLO.

MLOs contain proteins, and frequently nucleic acids, and are dynamic in size (generally submicrometer), formation, and composition [1]. They behave like liquid droplets, capable of fusing, deforming, and rearranging [2]—all while being solvated in the larger aqueous environment of the cell. The macromolecular components of MLOs have a higher affinity for each other than for surrounding molecules, allowing for separation from the bulk solution by demixing, thus forming two co-existing liquid states with differing concentrations of particular solutes [3].

The network of multivalent interactions within an MLO is not ordered like a conventional protein complex [4–6]. The interactions are typically characterized as non-static and more dynamic, with less specificity and weaker binding than the forces that hold macromolecular complexes—such as the proteasome or ribosome—into rigid stoichiometric structures [2]. For example, a ribosome consists of large and small subunits with more-or-less specific quaternary arrangement of components that together form a large macromolecular machine. Interactions in MLOs are thought to be less specific, with greater fluctuation of molecular contacts and stoichiometry. The plasticity of interactions may permit these organelles to react more dynamically to specific cellular conditions.

Numerous distinct functional MLOs have been characterized, and their many unique protein constituents have been previously reviewed [7,8]. A recently developed database of nearly 3000 non-redundant LLPS-associated proteins suggests that many MLOs have yet to be fully characterized [9]. Of the MLOs that have been characterized, their diversity and ubiquity is remarkable. MLOs have been observed in cytoplasm and nucleoplasm, and also in canonical membrane-enclosed organelles like mitochondria or chloroplasts [10]. Most commonly, MLOs are linked to specific functions involving ribonucleic acid, such as germ granules [11]. Pathological examples have also been proposed, such as the cytoplasmic inclusion bodies (IBs) within which measles viral RNA is replicated [12]. MLOs may exist transiently, like stress granules (SGs), which are stalled translation complexes that form upon cellular stress [13]. Alternatively, MLOs can have a more persistent presence, like the nucleolus, which is a constant site of ribosome production in the nucleus [7,14]. MLOs may also form in response to spatial necessity, such as neuronal RNA granules, which function in transport of mRNAs from dendrite bodies to distant synapses [15,16].

MLOs may also have roles in the pathogenesis of many diseases, particularly neurodegenerative disorders [17]. For example, many proteins linked to amyotrophic lateral sclerosis (ALS) and frontotemporal dementia (FTD) can undergo LLPS and accumulate within MLOs [18]. Mutations in these proteins not only cause disease but can alter LLPS and the physical properties of the phase-separated state [19,20]. It is hypothesized that aberrant irreversible phase transitions may result in proteinaceous neuronal inclusions that lead directly to cellular dysfunction [2].

2. Intrinsically Disordered Regions Facilitate LLPS

An interesting feature of proteins that undergo LLPS is they frequently contain long segments that lack well-defined three-dimensional structure. These segments are typically termed intrinsically disordered regions (IDRs), or intrinsically disordered proteins (IDPs), because they have no single equilibrium structure; instead, they exist as broad structural (or population) ensembles or they exchange between multiple conformations rapidly. IDRs are usually defined as being approximately 30 amino acids or longer [21], and their distinguishing characteristic is a relative paucity of hydrophobic amino acids to drive folding into a narrow conformational landscape.

The sequence composition of IDRs can vary, but is commonly disproportionately represented by only a few amino acids (i.e., low-complexity). Some low-complexity sequences are called yeast prion-like because they are compositionally very similar to the domains that enable certain yeast proteins to form self-propagating amyloid fibers. Yeast prion domains (and prion-like domains (PrLDs)) are usually very rich in hydrophilic amino acids (e.g., asparagine, glutamine, serine, and tyrosine). Other low-complexity sequences may disproportionately contain charged amino acids, such as the arginine/glycine repeats (RGG or GRG), which occur in several IDRs within liquid phase-separating proteins. Repeating (or spatially distributed) motifs of a subset of amino acids are also common to IDRs [6]. IDRs of MLO-forming proteins are also enriched in amino acids that can form π–π interactions, in which induced electrostatic interactions occur between sp^2 hybridized atoms [22]. These interactions can also involve the backbone amide bonds, which are accessible due to the non-folded arrangement of IDRs.

The significance of IDRs in liquid phase-separating proteins is they enable diverse networks of transient interactions with moderate affinities (i.e., reversible, due partly to entropic penalties of IDRs adopting binding conformations). Relative to folded domains, IDRs have greater accessible conformational space and flexibility for forming molecular contacts. The frequent presence of repetitive motifs can enable numerous low-affinity interactions with the potential for high-avidity binding [23]. IDRs can therefore support multivalent interactions, meaning they can form multiple molecular contacts with a potential variety of binding partners. Thus, an MLO may emerge from a continuous network of IDRs forming inter-protein (or RNA-protein) multivalent contacts.

3. Post-Translational Modifications of Intrinsically Disordered Domains Can Govern LLPS

The lack of secondary structure makes IDRs especially susceptible to post-translational modifications (PTMs) [24]. In fact, IDRs are disproportionately modified post-translationally relative to the entire proteome [25–27]. A variety of PTMs can alter IDR charge, hydrophobicity, size, and structure. These changes may occur through additions of functional groups (e.g., phosphoryl, methyl, acyl, glycosyl, alkyl, etc.), or subtler chemical changes such as oxidation, deimidation, and deamidation [28].

There are many biological examples of PTMs serving as on/off switches, where they regulate a cellular event, such as protein signaling, localization, and degradation. In the case of IDRs and phase separation, PTMs can similarly have on/off functions by altering the nature of intermolecular contacts that support MLO formation or dissolution [29] (Table 1). Here, we discuss examples of PTMs and IDRs in proteins that undergo functional phase separation in cells. We evaluate the hypothesis that the combination of IDR multivalency and the capacity to be extensively modified results in reversible networks of interactions that can be regulated by specific cellular cues (Figure 1).

Table 1. Examples of post-translational modifications (PTMs) of intrinsically disordered regions (IDRs) altering the liquid–liquid phase separation (LLPS) of proteins. The underlined proteins have multiple PTMs that affect the phase separation propensity. Arrows indicate if PTMs promote (↑) or inhibit (↓) LLPS. C-terminal domain (CTD), N-terminal domain (NTD), prion-like domain (PrLD), arginine-glycine-glycine (RGG), stress granules (SGs), amyotrophic lateral sclerosis (ALS), and frontal temporal dementia (FTD).

PTM	Protein Example	Region Modified	Proposed Effects of PTM on LLPS (↓↑)		Type of MLO	Disease Link
Serine/Threonine Phosphorylation	FMRP	CTD IDR	↑	Increases electrostatic interactions	Neuronal granules	Fragile X syndrome
	TIAR-2	CTD PrLD	↑		SGs	
	Phosphoprotein	Internal IDR	↑		Inclusion bodies	Measles
	tau	Internal IDR	↑		SGs	Alzheimer's disease
	MEG-3	Internal IDR	↓	Introduces electrostatic repulsion	P granule	
	FUS	NTD PrLD	↓		SGs, nuclear paraspeckles	ALS, FTD
	TDP-43	NTD domain & CTD IDR	↓		SGs	ALS, FTD
Arginine Methylation	(SDMA) LSM4	49 aa, NTD RGG domain	↑	Changes hydrophobicity and H-bonding	Processing bodies	
	(ADMA) hnRNPA2	CTD IDR at RGG sites	↓		SGs	ALS, FTD
	RAP55A	36 aa, NTD RGG domain	↓	Changes hydrophobicity and H-bonding	SGs, Processing bodies	Primary biliary cirrhosis
	FUS	41 aa, CTD RGG domain	↓		SGs, nuclear paraspeckles	ALS, FTD
Arginine Citrullination	FUS	RGG domain	↓	Disrupts charge-charge interactions	SGs, nuclear paraspeckles	ALS, FTD
Lysine Acetylation	DDX3X	NTD IDR	↓	Disrupts cation-π interactions	SGs	Intellectual disability
	tau	Internal IDR	↓		SGs	Alzheimer's disease
Lysine Ribosylation	hnRNPA1	Glycine-rich region	↑	Increases multivalency	SGs	ALS, FTD

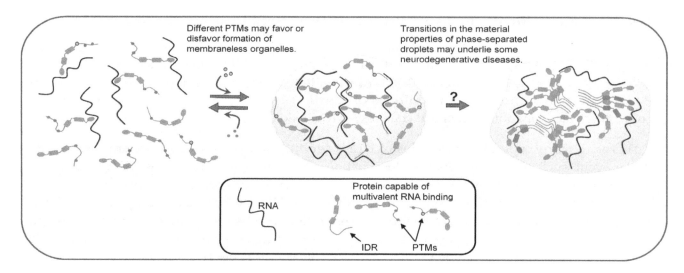

Figure 1. Liquid–liquid phase separation (LLPS) of biopolymers, such as proteins and RNA, is a mechanism by which cells organize their contents into specific functional structures called membraneless organelles (MLOs). Post-translational modifications (PTMs) of intrinsically disordered proteins can influence LLPS and thus regulate the formation and dissolution of MLOs. The figure depicts different patterns of PTMs favoring dispersed or condensed states. Changes in the material properties of liquid-phase separated granules are hypothesized to cause some neurodegenerative diseases. According to this hypothesis, droplets lose their liquid (reversible) properties and adopt more rigid (less reversible) internal structures, which may be glass-like, or in some cases, may have solid amyloid-like structures. These irreversible phase states may have gain-of-function toxicity to neurons.

3.1. Serine/Threonine/Tyrosine Phosphorylation

Phosphorylation is the covalent attachment of a phosphoryl group to an amino acid hydroxyl group. The phosphoryl group is negatively charged, so its addition changes a polar, uncharged residue to a negatively charged amino acid. Serine is the most commonly phosphorylated residue, followed by threonine and tyrosine [30]. The addition of charges to macromolecules may promote certain charge–charge interactions that drive complex coacervation (phase separation of oppositely charged polymers) [31]. Alternatively, addition of phosphates may cause charge repulsion or steric hindrance, thus inhibiting phase separation [2]. Depending on the protein context, the phosphate modification of amino acids can either favor or disfavor phase separation.

Serine/threonine phosphorylation has been shown to promote phase separation of IDPs such as fragile X mental retardation protein (FMRP) [32], TIA-1/TIAL RNA binding protein homolog (TIAR-2) [33], and microtubule-binding protein tau [34]. FMRP has 12 serine residues within its C-terminal IDR (aa 445–632) that have been identified as targets of casein kinase II (CKII). In vitro phosphorylation by CKII results in an increase in the negative charge densities throughout this IDR, increasing the propensity for multivalent electrostatic interactions and promoting phase separation [32]. TIAR-2 also contains a C-terminal intrinsically disordered PrLD that facilitates its LLPS into cytosolic granules [33]. The PrLD of TIAR-2 was shown to be serine phosphorylated when expressed in mechanosensory neurons. Ten serine residues in the intrinsically disordered PrLD were predicted as phospho-sites using NetPhos3.1 [35]. Expression of a non-phosphorylatable (S→A) TIAR-2 mutant (at 10, 8, or 2 serine residues) showed significantly less granule formation in the axons of neurons when compared to wild type. Alternatively, phosphomimetic (S→E) TIAR-2 mutants showed similar levels of granule formation when compared to wild type. These data suggest phosphorylation of serine residues promotes LLPS and formation of TIAR-2 positive granules in neurons of *C. elegans* [33].

PTMs also affect virally encoded proteins and their LLPS capabilities in cells. Measles virus phosphoprotein is a 507-amino acid virally encoded protein composed of multiple IDRs. Measles virus phosphoprotein and nucleoprotein undergo LLPS to form IBs. Phosphoprotein is phosphorylated at multiple sites, but Serine 86 and Serine 151—both of which are in IDRs—have been identified as regulatory sites for IB formation. Mutation or inhibition of phosphorylation at these two sites results in irregular and small IBs [12].

Examples of serine/threonine phosphorylation that disrupt LLPS include maternal-effect germline proteins (MEGs) in P granules of *C. elegans* [36], fused in sarcoma (FUS) [37–39] and TAR DNA-binding protein 43 (TDP-43) [40]. Proper segregation of P granules in zygotes of *C. elegans* requires the expression of MEG proteins [41]. Interestingly, two of the MEG family proteins (MEG-1 and MEG-3) are phosphorylated within their IDRs by a regulatory kinase (MBK-2) [36]. MBK-2 activity and counteractive phosphatase (PPTR-1) activity on MEGs is required for P granule disassembly and formation, respectively [36].

FUS and TDP-43 are frequently studied proteins because their phase separation in vivo has been linked to amyotrophic lateral sclerosis (ALS) [42]. A current hypothesis is that MLOs containing these proteins may promote their stochastic conversion into solid, pathological aggregates [43]. FUS contains an intrinsically disordered N-terminal PrLD, which is necessary and sufficient to drive LLPS [5]. The ~160 amino acid PrLD has 32 putative phosphorylation sites, 12 of which have been identified as PIKK family kinase consensus sites [44]. Phosphomimetic substitution (S/T→E) at 6 or 12 PIKK consensus sites diminishes FUS's ability to phase separate and form fibrillar aggregates in vitro [37]. In cells, a decrease in cytoplasmic aggregation is also observed upon increase in phosphomimetic substitution, suggesting a potential therapeutic target for disrupting pathological aggregate formation [37]. TDP-43 has a C-terminal PrLD, which is multiphosphorylated and aggregated in ALS motor neurons [45]. Two phosphorylation sites in the PrLD, Serine 409 and 410, identified in samples from frontotemporal lobar dementia patients were shown to regulate TDP-43 cytoplasmic granule formation. Phosphomimetic substitution (S→D) at Serine 409 and 410 showed a significant reduction in the number of cells containing TDP-43 puncta [46]. Interestingly, TDP-43 phase separation is regulated by PTMs in both an IDR and a structured domain. A single phosphorylation event in its N-terminal structured domain at Serine 48 is sufficient to suppress its LLPS in vitro and in cells [40]. Serine 48 is conserved in most species evaluated, including flies, mice, and humans [40].

There are two kinases, SKY1 and DYRK3, that have the ability to phase separate into SGs and, in the stages of recovery following a stress response, phosphorylate proteins containing IDRs, resulting in dissolution of the granules [47,48]. SKY1 is a yeast protein kinase with a PrLD that enables its recruitment into SGs. In SGs, SKY1 phosphorylates NLP3 at Serine 441, which is located in its serine-arginine rich C-terminal IDR [49]. This phosphorylation event promotes SG dissolution [48]. Similarly, DYRK3 (human homolog of MBK-2) was shown to phase separate into SGs via its intrinsically disordered N-terminal domain. Aside from regulation of SGs, DYRK3 was identified as a factor that controls phase separation and dissolution of several condensates containing IDPs during mitosis [50]. This kinase is interesting because it has broad-specificity and is generally proline-directed. Some proteins sensitive to DYRK3 inhibition, all of which contain IDRs, include splicing-speckle marker SC35, SG marker PABP, and pericentriolar-material protein PCM1. DYRK3 expression results in dissolution of these granules during mitosis, whereas a kinase-dead mutant or inhibition of DYRK3 results in the persistence of granules [50].

For some proteins, phosphorylation can be a driver or inhibitor of condensate formation, but there are instances where it is not clearly binary. Tau441 contains numerous IDRs and putative serine/threonine phosphorylation sites throughout the protein [51]. In experiments performed with bacterially produced recombinant full-length tau441 and molecular crowding agents, LLPS was driven mostly by electrostatic intermolecular interactions. There was no requirement for phosphorylation [52]. However, in a different study, phosphorylation was found to be required to initiate tau441 LLPS in vitro [34]. Of importance, there are 22 phospho-sites analyzed in this study, 15 of the sites are located in IDRs. Interestingly, LLPS of p-tau was dependent on hydrophobic interactions [34], whereas unphosphorylated tau LLPS was more dependent on ionic interactions [52]. These examples suggest the biophysical mechanism driving LLPS can be altered by the post-translational state of the protein.

3.2. Arginine Methylation

Methylation of arginine residues is important for regulating phase separation and recruitment of proteins into MLOs. The side chain of arginine contains a positively charged guanidinium head group that can be multiply methylated. This reaction is catalyzed by protein methyltransferases (PRMTs) using a donor methyl group from S-adenosyl methionine (SAM). Arginine methylation does not change the charge of the side group, but instead alters its volume, charge distribution, hydrophobicity, and potential for hydrogen bonding [53].

Including unmethylated arginine, there are four differential arginine methylation patterns. Arginines can be monomethylated (MMA) or dimethylated (DMA); arginine dimethylation can exist as symmetrical dimethylation (SDMA) or asymmetrical dimethylation (ADMA). Symmetric dimethylation occurs when two methyl groups are added to the two different nitrogen atoms within the guanidino group, whereas asymmetric dimethylation is when two methyl groups are added to the same nitrogen [54]. These reactions are catalyzed by different methyltransferases. PRMTs are grouped into three subtypes based on their catalytic activity. Type I PRMTs include PRMT1, 2, 3, 4, 6, and 8, while type II PRMTs include PRMT5 and 9 and there is one type III methyltransferase, PRMT7 [55,56]. All three classes of PRMTs have the ability to catalyze MMA reactions, but the DMA reactions that occur subsequent to this reaction are specific to type I and II enzymes. Type I PRMT enzymes catalyze the reaction of MMA to ADMA and type II enzymes catalyze MMA to SDMA. PRMT enzymes target arginine residues within glycine-arginine-glycine (GRG) or arginine-glycine-glycine (RGG) sequences [57], which are preferred sites of arginine methylation [58], and are frequently encoded as multiple repeats within low-complexity regions [6].

The hydrogen bond potential of the head group is the same for SDMA and ADMA, but the location of the methyl group can alter the orientation of hydrogen bonding. Additionally, SDMA and ADMA have different electrostatic surface potentials to the head group, resulting in shifting of charge [59]. The hydrophobicity of the head group is also modified upon methylation. Arginine hydrophobicity incrementally increases following the addition of MMA, ADMA, and SDMA, respectively [53]. Hydrophobic residues are generally located within the folded core of proteins, so within the context of an IDR, these changes may have profound effects on the propensity to fold or bind other macromolecules.

The guanidinium electrons are delocalized into π orbitals, enabling interactions via π-stacking [54]. Arginines can form cation-π interactions, which occur between the positively charged guanidinium group and the available electrons of the π orbital of aromatic rings [60]. These cation-π interactions have been shown to drive protein condensate formation [61]. Methylation of arginine has been shown to both favor and disfavor phase separation in certain contexts. ADMA in RGG motifs has been shown to disrupt favorable interactions and thus disrupt phase separation of IDPs such as dead-box helicase 4 (DDX4) [6], heterogeneous nuclear ribonucleoproteins A2 (hnRNPA2) [62], RAP55A [63,64], and FUS [65]. SDMA within RGG motifs, however, has been shown to drive phase separation. In the case of U6 snRNA-associated Sm-like protein (LSM4), SDMA is necessary for phase separation and processing body formation [66]. PRMT5 catalyzes SDMA of LSM4 at multiple arginine residues in its intrinsically disordered C-terminal RGG-containing domain. Mutation of the arginine residues in this domain or knockdown of PRMT5 diminish SDMA and processing body formation in cells [66].

YTH domain-containing family (YTHDF) of proteins contain numerous IDRs, which allows them to undergo LLPS. These proteins are found in cytoplasmic phase-separated SGs, P-bodies, and neuronal RNA granules [67]. In vitro experiments show that enzymatic modifications drive phase separation, but interestingly, the modifications occur to mRNA, not the protein. Methylation of adenosine in RNA, specifically the formation of N^6-methyladenosine (m^6A), is the most commonly modified nucleotide in mRNA. This modification increases mRNA's multivalency and seeds phase separation of YTHDF proteins in vitro [67]. The protein has the ability to form droplets in solution without the presence of m^6A, but only at much higher concentrations. Importantly, the mRNA that seeded droplet formation of YTHDF was multimethylated; unmethylated or singly methylated mRNA did not show droplet formation [67].

3.3. Arginine Citrullination

Citrullination is another PTM that occurs to arginine residues. Instead of the addition of a functional group, the arginine side chain undergoes an oxidation (or deimination) reaction. In this reaction, peptidylarginine deiminases (PADs) catalyze the oxidation of an imine group (=NH), forming a ketone group (=O) [60,68]. This modification removes the positive charge, leaving a neutrally charged amino acid. Interestingly, the consensus site for PADs are the same RG/RGG motifs that are common to many RNA-binding and phase separating proteins [69]. Citrullination of FUS by PAD4 was shown to diminish FUS recruitment to SGs [69]. PAD4 knockout in mouse embryonic fibroblasts showed a greater amount of FUS sequestration into SGs than when PAD4 was overexpressed in these cells, suggesting citrullination hinders FUS phase separation. Cation-π interactions between arginine residues and the π orbitals of tyrosine residues modulate FUS phase separation [60], but when citrullination occurs, the positive charge of the arginine side chain is removed, disrupting cation-π interactions and disrupting FUS phase separation in vitro [60].

3.4. Lysine Acetylation

Similar to citrullination, acetylation neutralizes the positive charge of an amino acid. Lysine residues contain a positively charged amino head group that can be neutralized by addition of an acetyl group; this not only changes lysine's charge state but also increases its hydrophobicity [70]. Acetyl groups are enzymatically added via acetyltransferases and removed by deacetylases [71]. Acetylation has been shown to disrupt simple coacervation of DDX3X (dead box RNA helicase 3) in vitro [72]. DDX3X has two IDRs: at the C-terminus and the N-terminus. Analysis of an acetylome dataset identified several acetylated lysines in the N-terminal IDR that play a role in DDX3X incorporation into SGs [72]. DDX3X is a substrate of acetyltransferase CREB-binding protein (CBP) and histone deacetylase 6 (HDAC6). To better understand the role of acetylation of DDX3X and SG incorporation, acetyl mimetic (K→Q) constructs and acetyl-dead (K→R) were constructed and expressed in DDX3X knock-out cell lines. Expression of acetyl-dead DDX3X (or inhibition of CBP) increased SG volume, whereas expression of acetyl-mimetic mutant (or inhibition of HDAC6) decreased SG volume.

The formation of SG has been proposed as a two-step process. First a stable core structure is formed, which is followed by the recruitment of IDPs into an outer shell structure [73]. The increase in volume is an important step in SG maturation. Of significance to this growth mechanism, the interaction partners of the acetyl-dead and acetyl mimetic DDX3X mutants were different. The non-acetylated DDX3X interacts with numerous SG components, whereas the acetyl mimetic loses its capacity for interactions with SG proteins, thus showcasing how lysine acetylation can be used to regulate MLO maturation [72].

Another protein that is lysine acetylated is tau, which is of particular interest since its solid-phase aggregation in neurons is linked to Alzheimer's disease [74]. Tau is an IDP, and like DDX3X, its ability to phase separate is disrupted by lysine acetylation [75]. Ferreon et al. found that recombinant tau, when incubated with enzymatically-active p300 histone acetyltransferase (HAT), becomes hyperacetylated (ac-tau) [75]. This acetylation (removal of positive charges) was observed to disfavor LLPS, which is consistent with the previous observations of Wegmann et al. that phosphorylation (addition of negative charges) promotes tau phase separation [34]. Using mass spectrometry analysis, 15 acetylation sites were identified, 8 of which are located in IDRs. Tau readily undergoes LLPS in vitro in low-salt conditions, but ac-tau was unable to form droplets under the same conditions. The neutralization of charged residues was concluded to disrupt electrostatic interactions required for tau LLPS [52,75]. Interestingly, tau also phase separates into SGs [76]. Ukmar-Godec et al. showed that tau association into SGs is altered by the acetylation state of the lysines. Unmodified full length tau441 readily associated with SGs following proteasome inhibition by MG132. Consistent with the in vitro findings above, acetylation of tau strongly reduced the association of the protein with SGs in HeLa cells [77]. Lastly, Ac-tau also showed decreased solid-phase aggregation propensity and reduced thioflavin-t reactivity, which indicates less propensity to form amyloid-like solid aggregates.

This suggests acetyltransferases and deacetylases are potential therapeutic targets for prevention of pathological tau aggregation [75].

3.5. Poly(ADP-Ribosylation)

Poly(ADP-ribosyl)ation or PARylation is a reversible covalent addition of multiple NAD-derived ADP-ribose (ADPr) molecules to a protein [78]. ADPr units can be added to glutamate, aspartate, lysine, arginine, or serine residues by poly(ADP-ribose) polymerases (PARPs) and removed by PAR glycohydrolases (PARGs) [78]. Aside from the physical addition of ADPr units, polyADP-ribose (PAR) molecules are freely synthesized polymers that can modify phase separation of some IDPs [79]. PAR is a multivalent, anionic, nucleic acid-mimicking (similar to RNA) biopolymer that can be bound by phase separating proteins [2]. Cellular stress conditions and DNA damage have been shown to cause an upregulation of PAR synthesis [79]. PAR, PARPs, and PARGs have all been shown to play a regulatory role in SG dynamics [80].

The SG component hnRNPA1 contains both a PAR-binding domain and a PARylation site at Lysine 298 within a glycine rich IDR. PARylation at hnRNPA1 Lysine 298 is important for hnRNPA1 nucleocytoplasmic shuttling, a necessary step for localization to SGs following cellular stress [81]. Interestingly, like numerous other proteins, hnRNPA1 contains a PAR-binding motif (PBM). In vitro experiments showed hnRNPA1 phase separation increasing in response to increased PAR concentration in solution. Mutating the hnRNPA1 PBM resulted in no phase separation in the presence of PAR, implying this interaction is domain specific. TDP-43 also contains a PBM and co-phase separates with hnRNPA1 in vitro and in SGs [81,82]. In vitro, PAR binding via hnRNPA1 PBM is necessary for the co-phase separation of TDP-43 and hnRNPA1 in low-salt concentrations. In cells, both TDP-43 and hnRNPA1 need functional PBMs to localize to SGs, highlighting the role of PAR in protein–protein interaction and phase transition [81,82].

4. Membraneless Organelles and Neurodegenerative Diseases

A connection between MLOs and neurodegenerative disease has been widely observed [83]. Specifically, many proteins that are genetically or histopathologically linked to neurodegeneration are also found in neuronal MLOs [18]. Likewise, proteins with intrinsically disordered PrLDs are notoriously disproportionately linked to neurodegenerative disease [84], and many of these proteins are both capable of undergoing LLPS and frequently found within inclusions of diseased neurons [85].

Why do the same proteins that functionally undergo LLPS appear to adopt pathological meso-scale aggregates in cells? A leading hypothesis is proteins within MLOs may undergo additional transitions into oligomeric species, solid-phase aggregates [86], or droplet-like structures with dramatically different material properties [17] (Figure 1). For example, expression of an ALS-linked TDP-43 mutant results in an MLO that is more viscous and resistant to solvation, suggesting it has a stabilized internal structure [19]. Similarly, in vitro, ALS-mutant FUS can transition from a droplet state into a solid aggregate more rapidly than wild-type FUS [42]. Once formed, such aggregates are thought to be detrimental to cell function and contribute to neuronal degeneration. Possibly, the high concentration of specific proteins within MLOs may potentiate these stochastic, irreversible phase transitions. Since disease-linked proteins like tau and TDP-43 are hyper- and multi-phosphorylated, respectively, within neuronal cytoplasmic inclusions, it is possible the PTMs are facilitating solid-phase transitions; alternatively, the PTMs may simply mark failed attempts at solubilization.

In the case of many IDRs, the abundance of hydrophilic amino acids and lack of stable tertiary structures may facilitate solid-phase transitions into highly ordered amyloid conformations. Amyloid is a well-ordered, filamentous polymeric state composed of a single protein species, much like a one-dimensional crystal [87]. It usually consists of polypeptides aligning in parallel in-register beta sheets [88] and is notoriously difficult to solubilize. MLOs may provide an environment in which some enriched IDR-containing proteins can stochastically adopt amyloid-like conformations, thus explaining the presence of certain MLO-linked proteins in pathological neuronal inclusions. Examples include tau

(Alzheimer's disease), TDP-43 (ALS), and FUS (frontotemporal dementia). Importantly, crystal-like arrangements would be disrupted by PTMs occurring within the structural core of amyloid [89]; thus, targeting specific modifying enzymes could offer a viable therapeutic strategy for neurodegenerative disorders that feature solid-phase inclusions.

5. Future Directions

Experimentation with MLOs is frequently focused on a few protein species. However, in vivo, the entire repertoire of macromolecules within individual biocondensates remains largely unknown. Additionally, for any given MLO, specific protein components can exhibit a broad array of PTMs, thus making it difficult to dissect which modifications are altering LLPS or perhaps serving other non-structural functional roles. Going forward, a major challenge will be to determine the precise relationship between MLOs and disease processes. For example, many human viruses encode proteins with PrLDs [90]. Given what we know about this type of protein domain, it is possible many viruses exploit LLPS during replication and infection, yet antiviral drugs do not specifically target LLPS mechanisms. In the case of neurodegenerative diseases, the pathological connections between aberrant phase transitions and neuronal death are not fully understood, and there are many non-unifying hypotheses. No drugs specifically target phase-separation processes in any neurodegenerative disease. Yet, given the almost complete lack of drugs for treating these diseases, manipulating the enzymes that regulate biocondensation may provide a new target paradigm.

Author Contributions: This manuscript was written by I.O. and F.S.

Acknowledgments: We thank Debra Yee and Hala Wyne for reviewing and editing our text.

Abbreviations

ADMA	Asymmetric dimethyl arginine
ADPr	NAD-derived ADP-ribose
IBs	Inclusion bodies
IDP	Intrinsically disordered protein
IDR	Intrinsically disordered region
LCD	Low-complexity domain
LLPS	Liquid–liquid phase separation
MMA	Monomethyl arginine
MLO	Membrane-less organelle
PTM	Post-translational modification
PrLD	Prion-like domain
SDMA	Symmetric dimethyl arginine
SG	Stress granule
RGG/RGR	Arginine/glycine repeats

References

1. Darling, A.L.; Liu, Y.; Oldfield, C.J.; Uversky, V.N. Intrinsically Disordered Proteome of Human Membrane-Less Organelles. *Proteomics* **2018**, *18*, e1700193. [CrossRef]
2. Shin, Y.; Brangwynne, C.P. Liquid phase condensation in cell physiology and disease. *Science* **2017**, *357*, eaaf4382. [CrossRef]
3. Hyman, A.A.; Weber, C.A.; Jülicher, F. Liquid-Liquid Phase Separation in Biology. *Annu. Rev. Cell Dev. Biol.* **2014**, *30*, 39–58. [CrossRef]

4.   Murthy, A.C.; Dignon, G.L.; Kan, Y.; Zerze, G.H.; Parekh, S.H.; Mittal, J.; Fawzi, N.L. Molecular interactions underlying liquid-liquid phase separation of the FUS low-complexity domain. *Nat. Struct. Mol. Biol.* **2019**, *26*, 637–648. [CrossRef]

5.   Burke, K.A.; Janke, A.M.; Rhine, C.L.; Fawzi, N.L. Residue-by-Residue View of In Vitro FUS Granules that Bind the C-Terminal Domain of RNA Polymerase II. *Mol. Cell* **2015**, *60*, 231–241. [CrossRef]

6.   Nott, T.J.; Petsalaki, E.; Farber, P.; Jervis, D.; Fussner, E.; Plochowietz, A.; Craggs, T.D.; Bazett-Jones, D.P.; Pawson, T.; Forman-Kay, J.D.; et al. Phase transition of a disordered nuage protein generates environmentally responsive membraneless organelles. *Mol. Cell* **2015**, *57*, 936–947. [CrossRef]

7.   Uversky, V.N. Intrinsically disordered proteins in overcrowded milieu: Membrane-less organelles, phase separation, and intrinsic disorder. *Curr. Opin. Struct. Biol.* **2017**, *44*, 18–30. [CrossRef]

8.   Courchaine, E.M.; Lu, A.; Neugebauer, K.M. Droplet organelles? *EMBO J.* **2016**, *35*, 1603–1612. [CrossRef]

9.   You, K.; Huang, Q.; Yu, C.; Shen, B.; Sevilla, C.; Shi, M.; Hermjakob, H.; Chen, Y.; Li, T. PhaSepDB: A database of liquid-liquid phase separation related proteins. *Nucleic Acids Res.* **2019**. [CrossRef]

10.   Antonicka, H.; Shoubridge, E.A. Mitochondrial RNA Granules Are Centers for Posttranscriptional RNA Processing and Ribosome Biogenesis. *Cell Rep.* **2015**, *10*, 920–932. [CrossRef]

11.   Marnik, E.A.; Updike, D.L. Membraneless organelles: P granules in Caenorhabditis elegans. *Traffic* **2019**, *20*, 373–379. [CrossRef]

12.   Zhou, Y.; Su, J.M.; Samuel, C.E.; Ma, D. Measles Virus Forms Inclusion Bodies with Properties of Liquid Organelles. *J. Virol.* **2019**, *93*, e00948-19. [CrossRef]

13.   Molliex, A.; Temirov, J.; Lee, J.; Coughlin, M.; Kanagaraj, A.P.; Kim, H.J.; Mittag, T.; Taylor, J.P. Phase separation by low complexity domains promotes stress granule assembly and drives pathological fibrillization. *Cell* **2015**, *163*, 123–133. [CrossRef]

14.   Drino, A.; Schaefer, M.R. RNAs, Phase Separation, and Membrane-Less Organelles: Are Post-Transcriptional Modifications Modulating Organelle Dynamics? *BioEssays* **2018**, *40*, 1800085. [CrossRef]

15.   Sephton, C.F.; Yu, G. The function of RNA-binding proteins at the synapse: Implications for neurodegeneration. *Cell. Mol. Life Sci.* **2015**, *72*, 3621–3635. [CrossRef]

16.   Krichevsky, A.M.; Kosik, K.S. Neuronal RNA granules: A link between RNA localization and stimulation-dependent translation. *Neuron* **2001**, *32*, 683–696. [CrossRef]

17.   Nedelsky, N.B.; Taylor, J.P. Bridging biophysics and neurology: Aberrant phase transitions in neurodegenerative disease. *Nat. Rev. Neurol.* **2019**, *15*, 272–286. [CrossRef]

18.   Ryan, V.H.; Fawzi, N.L. Physiological, Pathological, and Targetable Membraneless Organelles in Neurons. *Trends Neurosci.* **2019**, *42*, 693–708. [CrossRef]

19.   Gopal, P.P.; Nirschl, J.J.; Klinman, E.; Holzbaur, E.L.F. Amyotrophic lateral sclerosis-linked mutations increase the viscosity of liquid-like TDP-43 RNP granules in neurons. *Proc. Natl. Acad. Sci. USA* **2017**, *114*, E2466–E2475. [CrossRef]

20.   Boeynaems, S.; Bogaert, E.; Kovacs, D.; Konijnenberg, A.; Timmerman, E.; Volkov, A.; Guharoy, M.; De Decker, M.; Jaspers, T.; Ryan, V.H.; et al. Phase Separation of C9orf72 Dipeptide Repeats Perturbs Stress Granule Dynamics. *Mol. Cell* **2017**, *65*, 1044–1055.e5. [CrossRef]

21.   Van der Lee, R.; Buljan, M.; Lang, B.; Weatheritt, R.J.; Daughdrill, G.W.; Dunker, A.K.; Fuxreiter, M.; Gough, J.; Gsponer, J.; Jones, D.T.; et al. Classification of Intrinsically Disordered Regions and Proteins. *Chem. Rev.* **2014**, *114*, 6589–6631. [CrossRef] [PubMed]

22.   Vernon, R.M.; Chong, P.A.; Tsang, B.; Kim, T.H.; Bah, A.; Farber, P.; Lin, H.; Forman-Kay, J.D. Pi-Pi contacts are an overlooked protein feature relevant to phase separation. *eLife* **2018**, *7*, e31486. [CrossRef] [PubMed]

23.   Tompa, P.; Davey, N.E.; Gibson, T.J.; Babu, M.M. A Million Peptide Motifs for the Molecular Biologist. *Mol. Cell* **2014**, *55*, 161–169. [CrossRef] [PubMed]

24.   Dyson, H.J. Expanding the proteome: Disordered and alternatively-folded proteins. *Q. Rev. Biophys.* **2011**, *44*, 467–518. [CrossRef] [PubMed]

25.   Holt, L.J.; Tuch, B.B.; Villen, J.; Johnson, A.D.; Gygi, S.P.; Morgan, D.O. Global analysis of Cdk1 substrate phosphorylation sites provides insights into evolution. *Science* **2009**, *325*, 1682–1686. [CrossRef]

26.   Iakoucheva, L.M.; Radivojac, P.; Brown, C.J.; O'Connor, T.R.; Sikes, J.G.; Obradovic, Z.; Dunker, A.K. The importance of intrinsic disorder for protein phosphorylation. *Nucleic Acids Res.* **2004**, *32*, 1037–1049. [CrossRef]

27. Collins, M.O.; Yu, L.; Campuzano, I.; Grant, S.G.N.; Choudhary, J.S. Phosphoproteomic Analysis of the Mouse Brain Cytosol Reveals a Predominance of Protein Phosphorylation in Regions of Intrinsic Sequence Disorder. *Mol. Cell. Proteom.* **2008**, *7*, 1331–1348. [CrossRef]

28. Li, Q.; Shortreed, M.R.; Wenger, C.D.; Frey, B.L.; Schaffer, L.V.; Scalf, M.; Smith, L.M. Global Post-Translational Modification Discovery. *J. Proteome Res.* **2017**, *16*, 1383–1390. [CrossRef]

29. Bah, A.; Forman-Kay, J.D. Modulation of Intrinsically Disordered Protein Function by Post-translational Modifications. *J. Biol. Chem.* **2016**, *291*, 6696–6705. [CrossRef]

30. Hofweber, M.; Dormann, D. Friend or foe-Post-translational modifications as regulators of phase separation and RNP granule dynamics. *J. Biol. Chem.* **2019**, *294*, 7137–7150. [CrossRef]

31. Pak, C.W.; Kosno, M.; Holehouse, A.S.; Padrick, S.B.; Mittal, A.; Ali, R.; Yunus, A.A.; Liu, D.R.; Pappu, R.V.; Rosen, M.K. Sequence Determinants of Intracellular Phase Separation by Complex Coacervation of a Disordered Protein. *Mol. Cell* **2016**, *63*, 72–85. [CrossRef] [PubMed]

32. Tsang, B.; Arsenault, J.; Vernon, R.M.; Lin, H.; Sonenberg, N.; Wang, L.Y.; Bah, A.; Forman-Kay, J.D. Phosphoregulated FMRP phase separation models activity-dependent translation through bidirectional control of mRNA granule formation. *Proc. Natl. Acad. Sci. USA* **2019**, *116*, 4218–4227. [CrossRef] [PubMed]

33. Andrusiak, M.G.; Sharifnia, P.; Lyu, X.; Wang, Z.; Dickey, A.M.; Wu, Z.; Chisholm, A.D.; Jin, Y. Inhibition of Axon Regeneration by Liquid-like TIAR-2 Granules. *Neuron* **2019**, *104*, 290–304. [CrossRef] [PubMed]

34. Wegmann, S.; Eftekharzadeh, B.; Tepper, K.; Zoltowska, K.M.; Bennett, R.E.; Dujardin, S.; Laskowski, P.R.; MacKenzie, D.; Kamath, T.; Commins, C.; et al. Tau protein liquid–liquid phase separation can initiate tau aggregation. *EMBO J.* **2018**, *37*, e98049. [CrossRef] [PubMed]

35. Blom, N.S.; Gammeltoft, S.; Brunak, S. Sequence and structure-based prediction of eukaryotic protein phosphorylation sites. *J. Mol. Biol.* **1999**, *294*, 1351–1362. [CrossRef]

36. Wang, J.T.; Smith, J.; Chen, B.-C.; Schmidt, H.; Rasoloson, D.; Paix, A.; Lambrus, B.G.; Calidas, D.; Betzig, E.; Seydoux, G. Regulation of RNA granule dynamics by phosphorylation of serine-rich, intrinsically disordered proteins in *C. elegans*. *eLife* **2014**, *3*, e04591. [CrossRef]

37. Monahan, Z.; Ryan, V.H.; Janke, A.M.; Burke, K.A.; Rhoads, S.N.; Zerze, G.H.; O'Meally, R.; Dignon, G.L.; Conicella, A.E.; Zheng, W.; et al. Phosphorylation of the FUS low-complexity domain disrupts phase separation, aggregation, and toxicity. *EMBO J.* **2017**, *36*, 2951–2967. [CrossRef]

38. Lin, Y.; Currie, S.L.; Rosen, M.K. Intrinsically disordered sequences enable modulation of protein phase separation through distributed tyrosine motifs. *J. Biol. Chem.* **2017**, *292*, 19110–19120. [CrossRef]

39. Rhoads, S.N.; Monahan, Z.T.; Yee, D.S.; Leung, A.Y.; Newcombe, C.G.; O'Meally, R.N.; Cole, R.N.; Shewmaker, F.P. The prionlike domain of FUS is multiphosphorylated following DNA damage without altering nuclear localization. *Mol. Biol. Cell* **2018**, *29*, 1786–1797. [CrossRef]

40. Wang, A.; Conicella, A.E.; Schmidt, H.B.; Martin, E.W.; Rhoads, S.N.; Reeb, A.N.; Nourse, A.; Montero, D.R.; Ryan, V.H.; Rohatgi, R.; et al. A single N-terminal phosphomimic disrupts TDP-43 polymerization, phase separation, and RNA splicing. *EMBO J.* **2018**, *37*, e97452. [CrossRef]

41. Seydoux, G. The P Granules of *C. elegans*: A Genetic Model for the Study of RNA–Protein Condensates. *J. Mol. Biol.* **2018**, *430*, 4702–4710. [CrossRef] [PubMed]

42. Patel, A.; Lee, H.O.; Jawerth, L.; Maharana, S.; Jahnel, M.; Hein, M.Y.; Stoynov, S.; Mahamid, J.; Saha, S.; Franzmann, T.M.; et al. A Liquid-to-Solid Phase Transition of the ALS Protein FUS Accelerated by Disease Mutation. *Cell* **2015**, *162*, 1066–1077. [CrossRef] [PubMed]

43. March, Z.M.; King, O.D.; Shorter, J. Prion-like domains as epigenetic regulators, scaffolds for subcellular organization, and drivers of neurodegenerative disease. *Brain Res.* **2016**, *1647*, 9–18. [CrossRef] [PubMed]

44. Rhoads, S.N.; Monahan, Z.T.; Yee, D.S.; Shewmaker, F.P. The Role of Post-Translational Modifications on Prion-Like Aggregation and Liquid-Phase Separation of FUS. *Int. J. Mol. Sci.* **2018**, *19*, 886. [CrossRef] [PubMed]

45. Neumann, M.; Kwong, L.K.; Lee, E.B.; Kremmer, E.; Flatley, A.; Xu, Y.; Forman, M.S.; Troost, D.; Kretzschmar, H.A.; Trojanowski, J.Q.; et al. Phosphorylation of S409/410 of TDP-43 is a consistent feature in all sporadic and familial forms of TDP-43 proteinopathies. *Acta Neuropathol.* **2009**, *117*, 137–149. [CrossRef] [PubMed]

46. Brady, O.A.; Meng, P.; Zheng, Y.; Mao, Y.; Hu, F. Regulation of TDP-43 aggregation by phosphorylation and p62/SQSTM1. *J. Neurochem.* **2011**, *116*, 248–259. [CrossRef]

47. Wippich, F.; Bodenmiller, B.; Trajkovska, M.G.; Wanka, S.; Aebersold, R.; Pelkmans, L. Dual Specificity Kinase DYRK3 Couples Stress Granule Condensation/Dissolution to mTORC1 Signaling. *Cell* **2013**, *152*, 791–805. [CrossRef]

48. Shattuck, J.E.; Paul, K.R.; Cascarina, S.M.; Ross, E.D. The prion-like protein kinase Sky1 is required for efficient stress granule disassembly. *Nat. Commun.* **2019**, *10*, 3614. [CrossRef]

49. Gilbert, W.; Siebel, C.W.; Guthrie, C. Phosphorylation by Sky1p promotes Npl3p shuttling and mRNA dissociation. *RNA* **2001**, *7*, 302–313. [CrossRef]

50. Rai, A.K.; Chen, J.-X.; Selbach, M.; Pelkmans, L. Kinase-controlled phase transition of membraneless organelles in mitosis. *Nature* **2018**, *559*, 211–216. [CrossRef]

51. Mair, W.; Muntel, J.; Tepper, K.; Tang, S.; Biernat, J.; Seeley, W.W.; Kosik, K.S.; Mandelkow, E.; Steen, H.; Steen, J.A. FLEXITau: Quantifying Post-translational Modifications of Tau Protein in Vitro and in Human Disease. *Anal. Chem.* **2016**, *88*, 3704–3714. [CrossRef] [PubMed]

52. Boyko, S.; Qi, X.; Chen, T.-H.; Surewicz, K.; Surewicz, W.K. Liquid-liquid phase separation of tau protein: The crucial role of electrostatic interactions. *J. Biol. Chem.* **2019**, *294*, 11054–11059. [CrossRef] [PubMed]

53. Evich, M.; Stroeva, E.; Zheng, Y.G.; Germann, M.W. Effect of methylation on the side-chain pKa value of arginine. *Protein Sci.* **2016**, *25*, 479–486. [CrossRef] [PubMed]

54. Lorton, B.M.; Shechter, D. Cellular consequences of arginine methylation. *Cell Mol. Life Sci.* **2019**, *76*, 2933–2956. [CrossRef] [PubMed]

55. Morales, Y.; Cáceres, T.; May, K.; Hevel, J.M. Biochemistry and regulation of the protein arginine methyltransferases (PRMTs). *Arch. Biochem. Biophys.* **2016**, *590*, 138–152. [CrossRef]

56. Fulton, M.D.; Brown, T.; Zheng, Y.G. The Biological Axis of Protein Arginine Methylation and Asymmetric Dimethylarginine. *Int. J. Mol. Sci.* **2019**, *20*, 3322. [CrossRef]

57. Thandapani, P.; O'Connor, T.R.; Bailey, T.L.; Richard, S. Defining the RGG/RG Motif. *Mol. Cell* **2013**, *50*, 613–623. [CrossRef]

58. Boisvert, F.-M.; Chenard, C.A.; Richard, S. Protein Interfaces in Signaling Regulated by Arginine Methylation. *Sci. Signal.* **2005**, *2005*, 2. [CrossRef]

59. Fuhrmann, J.; Clancy, K.W.; Thompson, P.R. Chemical Biology of Protein Arginine Modifications in Epigenetic Regulation. *Chem. Rev.* **2015**, *115*, 5413–5461. [CrossRef]

60. Qamar, S.; Wang, G.; Randle, S.J.; Ruggeri, F.S.; Varela, J.A.; Lin, J.Q.; Phillips, E.C.; Miyashita, A.; Williams, D.; Ströhl, F.; et al. FUS Phase Separation Is Modulated by a Molecular Chaperone and Methylation of Arginine Cation-π Interactions. *Cell* **2018**, *173*, 720–734. [CrossRef]

61. Wang, J.; Choi, J.-M.; Holehouse, A.S.; Lee, H.O.; Zhang, X.; Jahnel, M.; Maharana, S.; Lemaitre, R.; Pozniakovsky, A.; Drechsel, D.; et al. A Molecular Grammar Governing the Driving Forces for Phase Separation of Prion-like RNA Binding Proteins. *Cell* **2018**, *174*, 688–699. [CrossRef] [PubMed]

62. Ryan, V.H.; Dignon, G.L.; Zerze, G.H.; Chabata, C.V.; Silva, R.; Conicella, A.E.; Amaya, J.; Burke, K.A.; Mittal, J.; Fawzi, N.L. Mechanistic View of hnRNPA2 Low-Complexity Domain Structure, Interactions, and Phase Separation Altered by Mutation and Arginine Methylation. *Mol. Cell* **2018**, *69*, 465–479. [CrossRef] [PubMed]

63. Yang, W.-H.; Yu, J.H.; Gulick, T.; Bloch, K.D.; Bloch, D.B. RNA-associated protein 55 (RAP55) localizes to mRNA processing bodies and stress granules. *RNA* **2006**, *12*, 547–554. [CrossRef] [PubMed]

64. Matsumoto, K.; Nakayama, H.; Yoshimura, M.; Masuda, A.; Dohmae, N.; Matsumoto, S.; Tsujimoto, M. PRMT1 is required for RAP55 to localize to processing bodies. *RNA Biol.* **2012**, *9*, 610–623. [CrossRef]

65. Hofweber, M.; Hutten, S.; Bourgeois, B.; Spreitzer, E.; Niedner-Boblenz, A.; Schifferer, M.; Ruepp, M.-D.; Simons, M.; Niessing, D.; Madl, T.; et al. Phase Separation of FUS Is Suppressed by Its Nuclear Import Receptor and Arginine Methylation. *Cell* **2018**, *173*, 706–719. [CrossRef]

66. Arribas-Layton, M.; Dennis, J.; Bennett, E.J.; Damgaard, C.K.; Lykke-Andersen, J. The C-Terminal RGG Domain of Human Lsm4 Promotes Processing Body Formation Stimulated by Arginine Dimethylation. *Mol. Cell. Biol.* **2016**, *36*, 2226–2235. [CrossRef]

67. Ries, R.J.; Zaccara, S.; Klein, P.; Olarerin-George, A.; Namkoong, S.; Pickering, B.F.; Patil, D.P.; Kwak, H.; Lee, J.H.; Jaffrey, S.R. m6A enhances the phase separation potential of mRNA. *Nature* **2019**, *571*, 424–428. [CrossRef]

68. Anzilotti, C.; Pratesi, F.; Tommasi, C.; Migliorini, P. Peptidylarginine deiminase 4 and citrullination in health and disease. *Autoimmun. Rev.* **2010**, *9*, 158–160. [CrossRef]

69. Tanikawa, C.; Ueda, K.; Suzuki, A.; Iida, A.; Nakamura, R.; Atsuta, N.; Tohnai, G.; Sobue, G.; Saichi, N.; Momozawa, Y.; et al. Citrullination of RGG Motifs in FET Proteins by PAD4 Regulates Protein Aggregation and ALS Susceptibility. *Cell Rep.* **2018**, *22*, 1473–1483. [CrossRef]

70. Patel, J.; Pathak, R.R.; Mujtaba, S. The biology of lysine acetylation integrates transcriptional programming and metabolism. *Nutr. Metab.* **2011**, *8*, 12. [CrossRef]

71. Drazic, A.; Myklebust, L.M.; Ree, R.; Arnesen, T. The world of protein acetylation. *Biochim. Biophys. Acta Proteins Proteom.* **2016**, *1864*, 1372–1401. [CrossRef] [PubMed]

72. Saito, M.; Hess, D.; Eglinger, J.; Fritsch, A.W.; Kreysing, M.; Weinert, B.T.; Choudhary, C.; Matthias, P. Acetylation of intrinsically disordered regions regulates phase separation. *Nat. Chem. Biol.* **2019**, *15*, 51–61. [CrossRef] [PubMed]

73. Jain, S.; Wheeler, J.R.; Walters, R.W.; Agrawal, A.; Barsic, A.; Parker, R. ATPase-Modulated Stress Granules Contain a Diverse Proteome and Substructure. *Cell* **2016**, *164*, 487–498. [CrossRef] [PubMed]

74. Barghorn, S.; Davies, P.; Mandelkow, E. Tau Paired Helical Filaments from Alzheimer's Disease Brain and Assembled in Vitro Are Based on β-Structure in the Core Domain. *Biochemistry* **2004**, *43*, 1694–1703. [CrossRef]

75. Ferreon, J.C.; Jain, A.; Choi, K.-J.; Tsoi, P.S.; MacKenzie, K.R.; Jung, S.Y.; Ferreon, A.C. Acetylation Disfavors Tau Phase Separation. *Int. J. Mol. Sci.* **2018**, *19*, 1360. [CrossRef]

76. Brunello, C.A.; Yan, X.; Huttunen, H.J. Internalized Tau sensitizes cells to stress by promoting formation and stability of stress granules. *Sci. Rep.* **2016**, *6*, 30498. [CrossRef]

77. Ukmar-Godec, T.; Hutten, S.; Grieshop, M.P.; Rezaei-Ghaleh, N.; Cima-Omori, M.-S.; Biernat, J.; Mandelkow, E.; Söding, J.; Dormann, D.; Zweckstetter, M. Lysine/RNA-interactions drive and regulate biomolecular condensation. *Nat. Commun.* **2019**, *10*, 2909. [CrossRef]

78. Alemasova, E.E.; Lavrik, O.I. Poly(ADP-ribosyl)ation by PARP1: Reaction mechanism and regulatory proteins. *Nucleic Acids Res.* **2019**, *47*, 3811–3827. [CrossRef]

79. Altmeyer, M.; Neelsen, K.J.; Teloni, F.; Pozdnyakova, I.; Pellegrino, S.; Grøfte, M.; Rask, M.-B.D.; Streicher, W.; Jungmichel, S.; Nielsen, M.L.; et al. Liquid demixing of intrinsically disordered proteins is seeded by poly(ADP-ribose). *Nat. Commun.* **2015**, *6*, 8088. [CrossRef]

80. Leung, A.K.L.; Vyas, S.; Rood, J.E.; Bhutkar, A.; Sharp, P.A.; Chang, P. Poly(ADP-ribose) regulates stress responses and microRNA activity in the cytoplasm. *Mol. Cell* **2011**, *42*, 489–499. [CrossRef]

81. Duan, Y.; Du, A.; Gu, J.; Duan, G.; Wang, C.; Gui, X.; Ma, Z.; Qian, B.; Deng, X.; Zhang, K.; et al. PARylation regulates stress granule dynamics, phase separation, and neurotoxicity of disease-related RNA-binding proteins. *Cell Res.* **2019**, *29*, 233–247. [CrossRef] [PubMed]

82. McGurk, L.; Gomes, E.; Guo, L.; Mojsilovic-Petrovic, J.; Tran, V.; Kalb, R.G.; Shorter, J.; Bonini, N.M. Poly(ADP-Ribose) Prevents Pathological Phase Separation of TDP-43 by Promoting Liquid Demixing and Stress Granule Localization. *Mol. Cell* **2018**, *71*, 703–717. [CrossRef] [PubMed]

83. Aguzzi, A.; Altmeyer, M. Phase Separation: Linking Cellular Compartmentalization to Disease. *Trends Cell Biol.* **2016**, *26*, 547–558. [CrossRef] [PubMed]

84. An, L.; Harrison, P.M. The evolutionary scope and neurological disease linkage of yeast-prion-like proteins in humans. *Biol. Direct* **2016**, *11*, 32. [CrossRef]

85. Harrison, A.F.; Shorter, J. RNA-binding proteins with prion-like domains in health and disease. *Biochem. J.* **2017**, *474*, 1417–1438. [CrossRef]

86. Li, Y.R.; King, O.D.; Shorter, J.; Gitler, A.D. Stress granules as crucibles of ALS pathogenesis. *J. Cell Biol.* **2013**, *201*, 361–372. [CrossRef]

87. Tycko, R. Amyloid polymorphism: Structural basis and neurobiological relevance. *Neuron* **2015**, *86*, 632–645. [CrossRef]

88. Shewmaker, F.; McGlinchey, R.P.; Wickner, R.B. Structural Insights into Functional and Pathological Amyloid. *J. Biol. Chem.* **2011**, *286*, 16533–16540. [CrossRef]

89. Hu, Z.-W.; Ma, M.-R.; Chen, Y.-X.; Zhao, Y.-F.; Qiang, W.; Li, Y.-M. Phosphorylation at Ser8 as an intrinsic regulatory switch to regulate the morphologies and structures of Alzheimer's 40-residue β-amyloid (Aβ40) fibrils. *J. Biol. Chem.* **2017**, *292*, 8846. [CrossRef]

90. Tetz, G.; Tetz, V. Prion-like Domains in Eukaryotic Viruses. *Sci. Rep.* **2018**, *8*, 8931. [CrossRef]

5

Raman Evidence of p53-DBD Disorder Decrease upon Interaction with the Anticancer Protein Azurin

Sara Signorelli, Salvatore Cannistraro * and Anna Rita Bizzarri

Biophysics & Nanoscience Centre, DEB, Università della Tuscia, 01100 Viterbo, Italy; signorellis@unitus.it (S.S.);
bizzarri@unitus.it (A.R.B.)
* Correspondence: cannistr@unitus.it

Abstract: Raman spectroscopy, which is a suitable tool to elucidate the structural properties of intrinsically disordered proteins, was applied to investigate the changes in both the structure and the conformational heterogeneity of the DNA-binding domain (DBD) belonging to the intrinsically disordered protein p53 upon its binding to Azurin, an electron-transfer anticancer protein from *Pseudomonas aeruginosa*. The Raman spectra of the DBD and Azurin, isolated in solution or forming a complex, were analyzed by a combined analysis based on peak inspection, band convolution, and principal component analysis (PCA). In particular, our attention was focused on the Raman peaks of Tyrosine and Tryptophan residues, which are diagnostic markers of protein side chain environment, and on the Amide I band, of which the deconvolution allows us to extract information about α-helix, β-sheet, and random coil contents. The results show an increase of the secondary structure content of DBD concomitantly with a decrease of its conformational heterogeneity upon its binding to Azurin. These findings suggest an Azurin-induced conformational change of DBD structure with possible implications for p53 functionality.

Keywords: Raman spectroscopy; p53; intrinsically disordered protein; blue copper protein Azurin; protein–protein interaction; Amide I band deconvolution; principal component analysis

1. Introduction

p53 is an important tumor suppressor protein working as a central hub in a complex interaction network in which it regulates numerous cellular processes, including cell cycle progression, apoptosis induction, and DNA repair [1,2]. p53 is a member of the important class of intrinsically disordered proteins (IDPs), possessing both structured and disordered domains under physiological conditions and different conformations coexisting in solution [3]. Such a structural plasticity confers to IDP an extremely high conformational adaptability, allowing them to act according to functional modes not achievable by ordered proteins, with these properties having been recently exploited to develop engineered protein and peptide drugs [4–6].

p53 is a tetrameric protein composed of four identical subunits and acts as a transcription factor. Each monomer of p53 consists of an N-terminal transactivation domain (NTD), a C-terminal domain (CTD), and a core DNA-binding domain (DBD) [7–10] The presence of unstructured portions allows p53 to adopt widely different conformations, which are at the basis of a vast repertoire of available interactions to different biological partners [11]. Among them, Azurin (AZ), a copper-containing electron-transfer anticancer protein secreted by *Pseudomonas aeruginosa* bacteria, has demonstrated the ability to specifically bind to p53, leading both to its stabilization and to an intracellular level increase both in vitro and in vivo [12–17]. Therefore, the formation of the p53-AZ complex has opened new perspectives in cancer treatment, such as the development of an AZ-derived anticancer peptide [18].

Keeping in mind the crucial role of AZ in assisting the oncosuppressive function of p53, in our group, we investigated the interaction between p53 and AZ at the single molecule level by Atomic Force

Microscopy (AFM) and Atomic Force Spectroscopy (AFS) and by computational approaches [12,19–21]. These studies have provided information about the interaction kinetics between p53 or its DBD and AZ, obtaining also some relevant insights on the possible binding sites [21]. However, no experimental evidences on possible structural alterations of p53 upon its binding to AZ are so far available [3]. In this respect, Raman spectroscopy represents a suitable approach to extract information about the secondary structure of proteins as well as to probe their conformational heterogeneity, including IDPs [22]. Indeed, we have previously applied such a technique to investigate the structure and the conformational heterogeneity of wild-type and mutants p53 and, also, of the AZ-derived anticancer p28 peptide, even in different environmental conditions [18,23,24].

In the present work, we have employed a Raman-based approach to investigate if and how the native conformation of DBD is modified by its interaction with AZ. To such an aim, we have focused on an accurate inspection of the Fermi doublets relative to Tyrosine (830 and 850 cm^{-1}; Tyr) and of Tryptophan peaks (1340 and 1360 cm^{-1}), with these Raman signals having been recognized as suitable diagnostic markers of protein side chain environment [25,26]. Additionally, we have investigated the Amide I Raman band (1600–1700 cm^{-1}), of which the deconvolution has demonstrated to be particularly effective in both extracting conformational information (α-helix, β-sheet, and random coil motifs) and which is a reliable reporter on the structural heterogeneity of proteins [22,27–32]. The Raman spectra have also been analyzed by applying principal component analysis (PCA), which performs a dimensionality reduction of the spectra, allowing a revelation of the differences between the complex Raman spectra of the samples and helping to understand the principal factors affecting the spectral variation [33].

The combination of these approaches has put into evidence the occurrence of structural changes within p53DBD upon its interaction with AZ. In particular, passing from isolated DBD to DBD bound to AZ, we found a variation in Tyrosine (Tyr) and Tryptophan (Trp) residues hydrophobicity and an increase of the DBD secondary structure concomitantly with a significant reduction of the conformational heterogeneity. The observed changes in both the structure and conformational heterogeneity of DBD strongly support the ability of AZ to modulate the DBD structure, and this, in turn, may result in a stabilization of the oncosuppressive function of p53.

2. Results and Discussion

2.1. Raman Analysis of AZ and DBD

Figure 1 shows the Raman spectra of AZ and DBD in the 600–1725 cm^{-1} frequency range. The spectra display a complex set of bands arising from the modes of the aromatic amino acids (Tyr, Trp, and Phenylalanine (Phe) and of the peptide backbone, consistent with the typical Raman spectra of proteins [27,34]. The assignments of the main peaks are summarized in Table 1 [27].

Table 1. Typical proteins' Raman vibrational modes (Raman cm^{-1}) and related assignments.

Raman (cm^{-1})	Assignment
643	Tyr
805	Tyr
830,850	Tyr
870	Trp
902	ν_{CC}
930,980	ν_{CCN}
1001	Phe
1103	$\nu_{CC}, \nu_{CN}, \nu_{CO}$
1127	ν_{CC}

Table 1. *Cont.*

Raman (cm^{-1})	Assignment
1174	Tyr
1180	Phe
1210	Tyr
1230–1240	Amide III (α-helices)
1250–1255	Amide III (β-sheets)
1270–1300	Amide III (Random coils)
1320	CH$_2$ deformation
1340,1360	Trp
1403	Symmetric ν_{co2}^{-}
1424	CH$_2$, CH$_3$ deformation
1451	CH$_2$, CH$_3$ deformation
1552	Trp
1604	Phe
1615	Tyr
1650–1680	Amide I

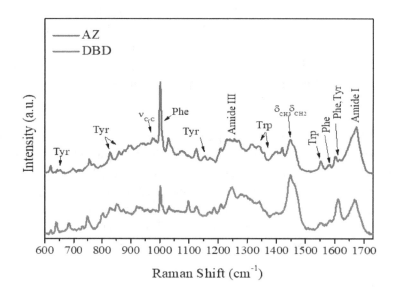

Figure 1. Raman spectra (600–1730 cm^{-1}) with excitation at 532 nm of Azurin (AZ; blue) and DNA-binding domain (DBD; magenta) in Phosphate Buffer Solution (PBS): The principal proteins' vibrational modes are marked. Spectra were normalized in the all spectrum frequency region and baseline corrected for a better visualization.

Among the main Raman markers, we focused our attention on the Raman peaks of Tyr and Trp residues, which allow the extraction of information on protein side-chain local environment and on the Raman band of Amide I, which provides a diagnostic of the protein secondary structure.

Concerning the Tyr residues, the ratio $I_Y = I_{850}/I_{830}$ between the intensity of doublet peaks at 850 and 830 cm^{-1} is related to the donor or acceptor role of the Tyr phenoxyl group. Specifically, a low I_Y value (around 0.3) indicates the phenolic hydroxyl (OH) group acting as a strong hydrogen bond donor, as occurring for buried tyrosine residues. As the I_Y value increases (until 2.5), the phenolic oxygen becomes a stronger hydrogen bond acceptor, while a largely enhanced value ($I_Y > 6.7$) represents a non-hydrogen-bonded state [25,26]. Experimental results on isolated AZ reveal an I_Y value of 0.38 ± 0.07, representative of a buried environment for its two Tyr residues (Tyr[72] and Tyr[108]) in agreement with the X-ray structure of AZ, in which Tyr[72] belongs to the peripheral α-helix region with a moderate solvent accessibility and Tyr[108] is practically inaccessible to solvent (see Figure 2A) [35].

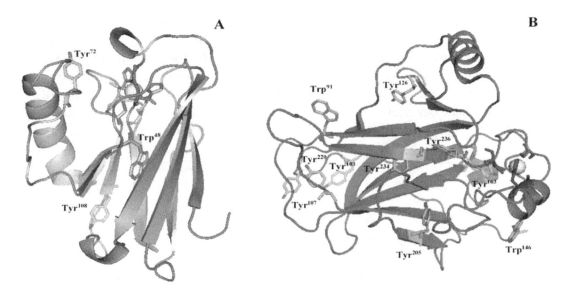

Figure 2. Three-dimensional structures of (**A**) AZ (PDB code: 4AZU) and (**B**) DBD (PDB code: 2XWR): The active site of AZ and the zinc-finger of the DBD are shown as yellow ball and stick models. The aromatic residues are Tyr (green) and Trp (orange). The OH groups in Tyr residues are marked in red.

Isolated DBD exhibits an I_Y ratio of 1.37 ± 0.16, indicating a predominant exposition to the solvent surfaces of the eight Tyr residues. From X-ray structure, the phenolic OH groups of Tyr^{103} and Tyr^{107} are highly oriented towards the solvent (see Figure 2B) [7]. Moreover, Tyr^{126} and Tyr^{205}, located at a crucial protein region interfacing with the DNA, show moderate accessibility, similar to that of Tyr^{220} located on the surface of the protein [7,8]. Finally, the remaining Tyr^{163}, Tyr^{234}, and Tyr^{236} are almost inaccessible to the solvent [8]. Therefore, our results are consistent with the X-ray data endorsing the DBD–Tyr high solvent exposition.

Further information about the side chains can be achieved by analyzing the Fermi doublet bands of Trp residues at 1340 and 1360 cm^{-1}, which are reporters of the hydrophobicity/hydrophilicity neighboring the Trp indole ring [32]. In particular, an intensity ratio $I_W = I_{1360}/I_{1340}$ smaller than 1.0 reflects a hydrophilic environment, while a ratio greater than 1.0 indicates a hydrophobic one [9,32].

For AZ, we found an I_W ratio of 1.54 ± 0.10, indicative of a buried and solvent inaccessible environment for the lone Trp residue. This is in accordance with the AZ X-ray data, showing that the Trp^{48} is deeply embedded in a highly hydrophobic core and surrounded by a closely packed β barrel structure (Figure 2A) [35].

We found for DBD an I_W ratio of 0.68 ± 0.10, which implies, on average, a moderate hydrophilic environment for its Trp^{91} and Trp^{146}. The latter is positioned in a hydrophobic side chain and oriented towards the solvent, while the former is located at the N-terminus of DBD and displays a high solvent accessibility, as it comes out from the X-ray data (Figure 2B) [36]. However, Trp^{91} has been shown to be crucially involved in the packing process of DBD through interaction with the Arg^{174} residue, which reduces its solvent exposure [36]. Therefore, our data suggest that both Trp residues in DBD globally experience a hydrophilic environment.

Information on protein secondary structure can be extracted by the Amide I band (1600–1700 cm^{-1}), mainly arising from C=O stretching and the combination of the C–N stretching, the Cα-C–N bending, and the N–H in-plane bending modes of peptide group. Such a band is usually used as a marker for secondary structure components. In particular, when Amide I band is centered at 1655 cm^{-1}, it indicates a prevailing α-helix conformational arrangement, while a shift of this band peak toward 1670 cm^{-1} is indicative of β-sheet conformation [22]. On the other hand, an analysis of the Amide I shape means an appropriate deconvolution strategy allows for the quantification of the percentage content of secondary structure components present in the protein [22,28].

Specifically, the Amide I band of AZ emerges at about 1670 cm^{-1} (Figure 3A), suggesting a predominant β-sheet conformation [37]. The curve-fitting procedure points out β-sheet conformations predominant for 60%, while the α-helices and random coils account for about 22% and 18%, respectively. The obtained AZ secondary structure is agreement with that determined by X-ray diffraction for the crystals of AZ. Indeed, the major AZ components are β strands and turns (≈69%), which form two sheets arranged in a Greek key motif and with a minor contribution from a rigid α-helix (about 31%), conferring to AZ a low level of flexibility and structural disorder (see Figure 3A) [21,35,38].

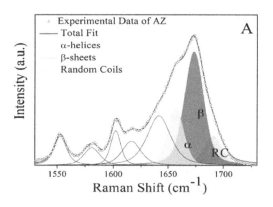 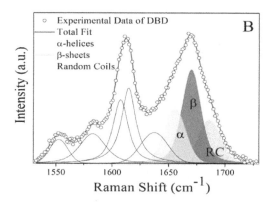

Figure 3. The Amide I band of AZ (**A**) and DBD (**B**) in PBS: The percentage of secondary structure for these proteins has been estimated from the relative area of deconvoluted bands of this spectral region of which the fitting parameters are reported in Table 2.

Table 2. Assignments, relative central frequency (Raman cm^{-1}), and integrated intensities (Area %) of the main Amide I band components (α-helix, β-sheet, or random coil) for AZ, DBD, and DBD:AZ complex obtained by a fitting procedure. $\chi2 = 0.002$ for all curve fitting analysis.

Sample In PBS	Secondary Structure	Raman Shift cm^{-1}	Area (%)
	α-helix	1659	22
AZ	β-sheet	1674	60
	random coil	1688	18
	α-helix	1655	25
DBD	β-sheet	1670	46
	random coil	1686	29
	α-helix	1655	26
DBD:AZ	β-sheet	1669	51
	random coil	1687	23

Concerning DBD (Figure 3B), the band corresponding to the β-structures provides 46% of the total, while those related to α-helix and to random coils have 25% and 29%, respectively. These results indicate that DBD is characterized by a partially ordered structure, combined with the presence of significant disordered regions. Additionally, the results confirm those reported in our recent study on different sample batches of DBD (aminoacids 81–300), from which a content of 27% and 50% for α-helical and β conformations, respectively, have been estimated [23]. Moreover, these data are in agreement with X-ray data indicating a 30% of β-arrangement with an 18% of α-structures (see Figure 3B) [36]. The DBD propensity to adopt a predominant β-conformation is actually related to the large presence in its sequence of hydrophobic residues, such as Cysteine (Cys), Trp, and Leucine (Leu), generally promoting an ordered structure [39].

2.2. Raman Analysis of the DBD:AZ Complex

The previous analysis on the Raman spectra of AZ and DBD proteins, isolated in solution, has provided information on their structural properties paving the way to investigate possible structural change when they are involved in the formation of a complex. The spectrum of DBD:AZ solution, obtained by mixing equimolar amounts of DBD and AZ in the 600–1725 cm^{-1} frequency region is shown in Figure 4. We note almost the same general features displayed by the isolated protein spectra with no significant shifts in frequency for the main vibrational modes. From the Tyr peaks visible at 828 and 854 cm^{-1}, the Fermi doublet ratio I_Y is 0.58 ± 0.08, which is indicative of a predominant hydrophobic environment. Such a value is closer to that of AZ ($I_Y = 0.38 \pm 0.07$) with respect to that of DBD ($I_Y = 1.37 \pm 0.16$), suggesting some changes in the Tyr microenvironment resulting from the interaction between the two biomolecules. To further support such a hypothesis, we have analyzed the Raman spectrum obtained by directly summing the spectra of isolated DBD and AZ molecules acquired at the same concentration used to form the complex (Figure 4), with the resulting spectrum being called added spectrum (AS) in the following. The analysis of the Tyr peaks in the AS spectrum reveals an I_Y of 1.11 ± 0.18, which is indicative of an average hydrophilic environment for all the Tyr residues in the system, as expected for isolated proteins. Therefore, the marked differences between the I_Y values from the complex and AS spectra can be ascribed to changes due to the interaction between the molecules. Although the spectroscopic results alone cannot allow us to identify the Tyrs that are involved in the structural changes, they support literature data that point out the involvement of the S$_7$–S$_8$ loops, comprising Tyr^{220}, Tyr^{234}, and Tyr^{236} and also Tyr^{126} at DBD binding sites with AZ, which, in turn, is engaged through its a.a 50–77 fragment, including Tyr^{72} [40].

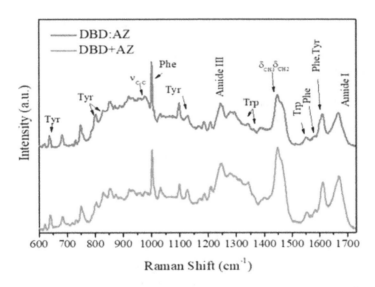

Figure 4. Comparison among the 532-nm-excited Raman spectra (600–1730 cm^{-1}) of DBD:AZ complex (green) and of added spectrum AS (orange) in PBS: The principal proteins' vibrational modes are marked. Spectra were normalized in the all spectrum frequency region and baseline corrected for a better visualization.

The spectrum of DBD:AZ shows that the Trp Raman peaks are located at the same frequencies as in the isolated proteins, with an I_W ratio of 1.15 ± 0.17, indicating a high hydrophobicity for the three Trps residues (AZ–Trp^{48} and DBD–Trp^{91}/Trp^{146}). The found value of I_W for DBD:AZ, slightly lower than that for AZ ($I_W = 1.54 \pm 0.10$) and higher than both isolated DBD ($I_W = 0.68 \pm 0.10$) and AS ($I_W = 0.53 \pm 0.10$), suggests some modifications in the environment experienced by these residues upon complex formation. The AZ–Trp^{48} is well-known to be strongly buried in the central hydrophobic core of the AZ; therefore, these changes can be due to variations in the DBD–Trp neighboring. Additionally, since the DBD–Trp^{91} has been shown to be engaged with the Arg^{174} [36], we suggest that the observed modifications of DBD as due to AZ interactions occurring within the DBD–Trp^{146} environment, with this being in agreement with Docking and Molecular Dynamics (MD) data showing the involvement of Trp^{146} in AZ-binding site [40].

Figure 5A,B shows the fitted curves of the Amide I band for the DBD:AZ and AS spectra, respectively. The results of the best fit for both experimental and AS Amide I bands, obtained by applying the same method used for isolated molecules, are reported in Table 2. DBD:AZ shows a predominant contribution from β-sheet structures (51%) and an α-helix amount of about 26%, while the random coil conformations contributes to 23% of the total Amide I band area. Best fit of AS reveals a predominant β structure (41%) with α-helices and random coils percentages of 31% and 28%, respectively. The observed changes in the secondary structure composition in DBD:AZ with respect to those of DBD and AZ can be attributed to the interaction between these proteins. Furthermore, since AZ is characterized by a highly structured conformation, the decrease of random coil structures can be mainly attributed to DBD. Such a result is supported by previously reported molecular dynamics simulations and docking studies showing that the binding of AZ at the peripheral, unstable, L_1 and S_7–S_8 loops of DBD can enhance their stability upon restraining their flexibility [21], with this being in agreement with a reduction of the DBD disordered regions upon binding to AZ. Accordingly, it could be hypothesized that the increase of structural stability of DBD could be at the basis of the anticancer effect exerted by AZ. Since the structural dynamics and the interactions between proteins are strictly connected, a deeper characterization of the structural–functional relations is of fundamental interest for developing AZ-based drugs, of which effective action in vivo requires, however, further validations [41].

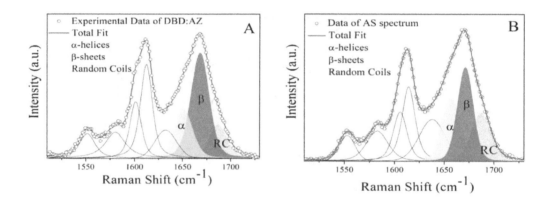

Figure 5. Amide I Raman band (open circles) excited at 532 nm of (**A**) DBD:AZ and (**B**) AS spectra, fitted through the Levenberg–Marquardt minimization algorithm (LMA; red line) in which the AZ total fit has been imposed as a constraint: Solid curves indicate the main structural conformations (α-helices, β-sheets, and random coils). Fitting results are summarized in Table 2.

2.3. Principal Component Analysis of DBD, AZ, and the DBD:AZ Complex

Different combinations of scores for the first three principal components (PC1, PC2, and PC3) have been used to build two-dimensional plots; in the following, the components releasing the highest structural information will be shown. Figure 6A shows the PCA scores of PC1 vs. PC2 components (providing about the 90% of the total variance) for the Fermi Doublet region relative to Tyr residues (790–870 cm^{-1}) for the AZ and DBD isolated molecules and for the DBD:AZ complex. In the scatter plot, two distinct groupings along the PC1 axis can be identified (see the ellipses drawn as a guide). Indeed, the AZ scores (blue symbols) are located in the positive portion of the plot along PC1 with a low spread along both the axes (10 and 5 along PC1 and PC2, respectively), while the DBD and DBD:AZ scores (magenta and green symbols, respectively) occupy the negative range of PC1 values. Along PC1, a larger variability is detected for DBD with respect to AZ and DBD:AZ. Additionally, along PC2, for AZ and DBD, negative values of PC2 are obtained while positive values are detected for DBD. To correlate the position of the scores in the plot with the samples' spectral features, we have analyzed the loadings with the variables mostly contributing to the PCA scores. As shown in Figure 6B,C, high levels of variance are detected in correspondence with the peaks at 829 cm^{-1} and 851 cm^{-1} related to the Fermi Doublet of Tyr modes and with a weaker peak at 805 cm^{-1}, associated with Tyr [25,32,42]; the latter provides the largest variance for PC2 loadings (see Figure 6B) [42]. These results show that Tyr vibrational modes are responsible for sample differentiation in PCA, consistent with our previous study supporting the important role played by Tyr modes as structural markers [25,42].

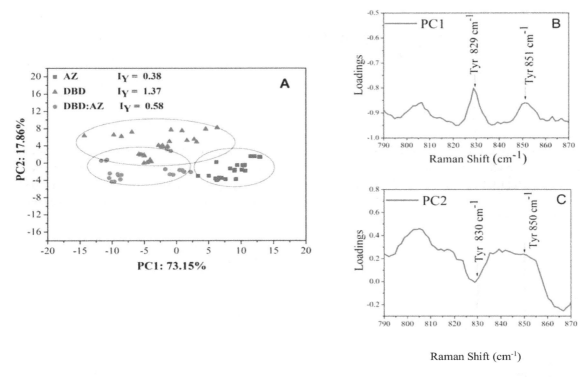

Figure 6. (**A**) Two-dimensional scores plot PC1 versus PC2 of the Raman spectra for AZ (blue squares), DBD (magenta triangles), and DBD:AZ complex (green circles) in the PBS performed on Fermi Doublet of Tyr region (790–870 cm^{-1}): The three groupings are indicated by ellipses. The Fermi Doublet ratio for Tyr residues are also reported. (**B**) PC1 (73% of total variance) and (**C**) PC2 (18% of total variance) one-dimensional loadings plot versus frequency. The Raman markers are indicated.

We then applied PCA to the Fermi Doublet region relative to Trp residues (1310–1380 cm^{-1}), with the PC1 and PC2 components providing about 91% of the total variance (see Figure 7A). AZ clusters at the upper side, DBD clusters at the middle, while DBD:AZ clusters at the lower region in correspondence to negative values of PC2 axis. Along PC1, DBD, and DBD:AZ, scores are mixed within an overlapped cloud, while AZ are well-clustered in a well-separated group. Concerning the loading plots, shown in Figure 8B,C, PC2 presents a broad band with a loading positive value of about 0.2. At 1360 cm^{-1}, a single evident peak emerges as ascribed to one of the Fermi Doublet of the Trp modes. This suggests that the separation among the three groups along PC2 depends on Trp vibrational modes. Since such a frequency changes according to the different Trp side-chain environment taken into consideration [28], a different spatial arrangement of this residue should be envisaged in the DBD isolated molecule and in DBD:AZ complex.

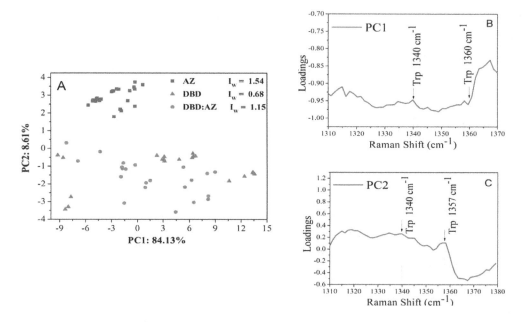

Figure 7. (**A**) Two-dimensional scores plot PC1 versus PC2 of the Raman spectra for AZ (blue squares), DBD (magenta triangles), and DBD:AZ complex (green circles) in the PBS performed on Fermi Doublet of Trp region (1310–1380 cm^{-1}): The Fermi Doublet ratio for Tyr residues are also reported. (**B**) PC1 (84% of total variance) and (**C**) PC2 (8% of total variance) one-dimensional loadings plot versus frequency. The Raman markers are indicated.

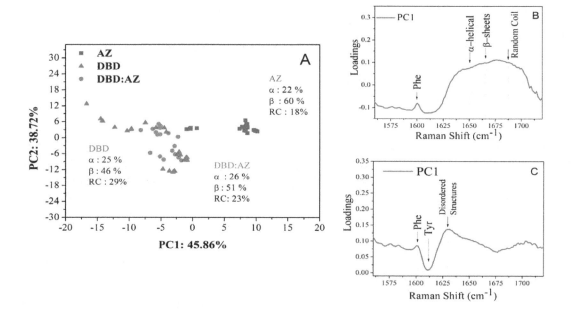

Figure 8. (**A**) Two-dimensional scores plot PC1 versus PC2 of the Raman spectra for AZ (blue squares), DBD (magenta triangles), and DBD:AZ complex (green circles) in the PBS performed on the Amide I band (1560–1720 cm^{-1}): The secondary structure percentages as obtained by the curve fitting analysis are also reported. (**B**) PC1 (46% of total variance) and (**C**) PC2 (39% of total variance) one-dimensional loadings plot versus frequency. The Raman markers are indicated.

Finally, the PCA was performed on the Amide I Raman band of AZ, DBD, and the DBD:AZ complex. From the scatter plot of PC1 versus PC2 (see Figure 8A), AZ data (blue squares) cluster at the positive side of the PC1 axis, with a low variance along both of the components, while DBD (magenta triangles) and DBD:AZ data (green circles) are characterized by negative values of PC1, with some overlap between them. A significantly larger variability is detected in DBD with respect to that of DBD:AZ.

The PC1 loading curve, accounting for 46% of the variance (Figure 8B), is characterized by a very broad band from 1625 to 1730 cm^{-1}, including all the frequencies related to the secondary structure of a protein. Such a band shows the lowest value for the 1650 cm^{-1} frequency and the highest for the 1680 cm^{-1} one, which are associated to the α-helices and disordered structures, respectively [19,24]. In PC2, accounting for a 39% of the variation in the spectra, the major source of variance comes from peak at 1627 cm^{-1}, consistent with the disordered structures component (Figure 8C) [19]. This indicates that PC1 discriminates the data based on different amounts of secondary structure of the sample, while PC2 reflects the amount of conformational disorder. Indeed, the AZ scores are very close to each other, indicating a very low variability in the secondary structure within different batches of samples, with this being consistent with the AZ ordered secondary structure [32]. Additionally, the superposition of DBD and DBD:AZ data along PC1 can be explained by assuming that the intrinsic disordered nature of DBD is able to populate an ensemble of different conformations. On the other hand, DBD:AZ distribution on the plot is narrower than that of DBD, reflecting a lower degree of disorder in the complex. These results confirm that PCA is a good reporter of the different structural differences among AZ, DBD, and DBD:AZ. Moreover, PCA is sensitive to changes in the conformational heterogeneity of DBD in the presence of AZ.

3. Materials and Methods

3.1. Sample Preparation

AZ (purity > 80%; MW = 14.6 kDa) was purchased from Sigma–Aldrich (St. Louis, MO, USA). The effective purity of the sample was checked by determining the ratio of spectral absorption at 630 nm and at 280 nm. AZ batches with a ratio value higher than 0.48 were used, with this indicating a good degree of purity [43]. AZ was dissolved in MilliQ water at a concentration of 200 μM. DBD (a.a 89–293; MW= 23 kDa) was purchased from GenScript (Piscataway, NJ, USA). DBD were dissolved in Phosphate Buffer Solution (PBS; 95.3% H_2O, 3.8% NaCl, 0.1% di KCl, 0.7% Na_2HPO_4, 0.1% KH_2PO_4; pH = 7.4), reaching a final concentration of 40 μM. The DBD:AZ complex in PBS solution were prepared by mixing equimolar amounts of the components.

3.2. Raman Spectroscopy

Raman measurements were carried out using a Super Labram confocal spectrometer (Horiba, France), equipped with several objectives, a diode-pumped solid-state laser (532 nm) and a spectrograph, with an 1800 g/mm grating allowing a resolution of 5 cm^{-1}. Raman spectra were collected by means of a liquid nitrogen-cooled charged coupled device (CCD) (back illuminated; pixel format: 1024 × 128 detector) and in the back-scattering geometry in which a notch filter was used to reject the elastic contribution. All the experiments were performed using a laser power of 10 mW (4.4 mW on the sample) and a 50× objective with a numerical aperture NA = 0.6 (laser spot diameter reaching the sample was about 1 μm). A large confocal diaphragm (400 μm) and a slit of 200 μm were used to obtain a good Raman signal.

Protein drops (10–15 μL) were deposited onto an optical glass, and spectra were acquired on partially dried samples. Indeed, it was demonstrated that there are no significant differences between the Raman spectra of protein in solution and the corresponding drop coating deposition, in which the protein remains substantially hydrated and the secondary structure is largely preserved [44].

Each Raman spectrum was acquired at room temperature by averaging 10 scans of 10 s integration time. For each sample, twenty-five Raman spectra were collected from different regions of the drops. Raman data processing and analysis were performed with OPUS software version 6.5 (Bruker Optics, Ettlingen, Germany). All the spectra were normalized with respect to the phenylalanine (Phe) ring breathing band at 1002 cm^{-1} due to its insensitivity to conformation or microenvironment [45], and the fluorescence background was removed by applying a rubber band baseline correction [46]. Finally, the spectra used for the structural analysis were obtained by averaging five measurements to improve the spectral signal/noise ratio.

3.3. Analysis of the Raman Spectra

The secondary structure content of isolated DBD and AZ was quantified through a deconvolution procedure of the Amide I Raman bands by using three pseudo-Voigt profiles. The model parameters were optimized with the Levenberg–Marquardt minimization algorithm (LMA), and the goodness of the fit was assessed by the reduced chi-square value. The AZ curves as extracted from the fit were used in the DBD:AZ complex analysis, under the hypothesis that the AZ secondary structure does not change upon the interaction [38]. The three pseudo-Voigt profiles were centered at 1650–1656, 1664–1670, and 1680 cm^{-1} and assigned to α-helix, β-strand, and random coil conformations, respectively, as validated on other IDPs [22,23,27]. In each fitting analysis, additional peaks had to be included in the band-fitting protocol to account for aromatic residue modes (1550, 1580, 1604, and 1615 cm^{-1}) and for disordered structure and/or vibronic coupling (1637 cm^{-1}) not baseline separated from Amide I features [22]. The errors relative to secondary structure percentages were evaluated by repeating the curve-fitting procedure on five different spectra and the accuracy associated with the determined secondary structure content was about 10% for each sample.

In order to improve the performance of deconvolution analysis, we performed a dimensionality reduction of the Raman spectra based on principal component analysis (PCA) [33]. The PCA transforms the original data set into a new data set with transformed variables (principal components) that are linear combinations of the original variables. The principal components were arranged in a swat that the variability of the original data set was contained in descending order in the first principal components. PCA was applied to the isolated DBD and AZ molecules and to the DBD:AZ complex (number of spectra n = 75) in three different spectral regions: (i) Fermi doublets relative to Tyr (830 and 850 cm^{-1}), satisfactorily described by a number of components N = 79; (ii) Fermi doublets relative to Trp (1340 and 1360 cm^{-1}) described by N = 79; and (iii) Amide I Raman band (1600–1700 cm^{-1}) described by N = 77. The number of components of the correlation matrix to be considered was defined as the number required to explain at least 80% of the total variance. STATISTICA 7.0 software (StatSoft Inc., Tulsa, OK, USA, 2004) was used for all the analyses.

4. Conclusions

The structural and conformational changes in the DBD region of the intrinsically disordered protein p53 upon interacting with the anticancer blue copper protein AZ were investigated by applying Raman spectroscopy. A careful inspection of the Raman spectra combined with a PCA analysis on the Fermi doublets of the Raman markers corresponding to the tyrosine and tryptophan residues allowed us to monitor the changes in their microenvironment as induced by the formation of a complex between DBD and AZ. Interestingly, we found a direct involvement of DBDTrp[146] in the complex formation, as suggested by other experimental investigations. Additionally, a deconvolution of the Amide I band, remarkably sensitive to the α-helix, β-sheets, and random coil structures, allowed us to quantify the main secondary structural motifs of the DBD and its changes as induced upon binding to AZ. We found that DBD undergoes a slight increase of the β-conformation, with a concomitant lowering of its disordered portions as well as of its conformational heterogeneity. These findings are in agreement with our previous computational results and suggest that the binding of AZ to some unstructured motifs of DBD can restrain their flexibility. Collectively, the observed modulation the DBD structure when bound to AZ may represent a ground for understanding the molecular mechanisms of the AZ anticancer activity and could provide some hints for designing other molecules for p53-targeted therapies. Finally, we would remark that our Raman-based approach can be applied to investigate the structural changes of other biomolecules undergoing specific complex formation in order also to elucidate the molecular mechanisms which regulate their biological functions.

Author Contributions: S.S. data curation, formal analysis, investigation, and writing—original draft preparation; S.C. and A.R.B. writing—review and editing; S.C. and A.R.B. conceptualization supervision.

References

1. Vousden, K.H.; Lane, D.P. p53 in health and disease. *Nat. Rev. Mol. Cell Biol.* **2007**, *8*, 275–283. [CrossRef] [PubMed]
2. Kruiswijk, F.; Labuschagne, C.F.; Vousden, K.H. p53 in survival, death and metabolic health: A lifeguard with a licence to kill. *Nat. Rev. Mol. Cell Biol.* **2015**, *16*, 393–405. [CrossRef] [PubMed]
3. Uversky, V.N. Unusual biophysics of intrinsically disordered proteins. *Biochim. Biophys. Acta (BBA)-Proteins Proteom.* **2013**, *1834*, 932–951. [CrossRef] [PubMed]
4. Tompa, P. Intrinsically unstructured proteins. *Trends Biochem. Sci.* **2002**, *27*, 527–533. [CrossRef]
5. Habchi, J.; Tompa, P.; Longhi, S.; Uversky, V.N. Introducing Protein Intrinsic Disorder. *Chem. Rev.* **2014**, *114*, 6561–6588. [CrossRef] [PubMed]
6. Minde, D.P.; Halff, E.F.; Tans, S. Designing disorder. *Intrinsically Disord. Proteins* **2013**, *1*, e26790. [CrossRef] [PubMed]
7. Cañadillas, J.M.P.; Tidow, H.; Freund, S.M.V.; Rutherford, T.J.; Ang, H.C.; Fersht, A.R. Solution structure of p53 core domain: Structural basis for its instability. *Proc. Natl. Acad. Sci. USA* **2006**, *103*, 2109–2114. [CrossRef]
8. Cho, Y.; Gorina, S.; Jeffrey, P.D.; Pavletich, N.P. Crystal structure of a p53 tumor suppressor-DNA complex: Understanding tumorigenic mutations. *Science* **1994**, *265*, 346–355. [CrossRef]
9. Pagano, B.; Jama, A.; Martinez, P.; Akanho, E.; Bui, T.T.T.; Drake, A.F.; Fraternali, F.; Nikolova, P.V. Structure and stability insights into tumour suppressor p53 evolutionary related proteins. *PLoS ONE* **2013**, *8*, e76014. [CrossRef]
10. Bell, S.; Klein, C.; Müller, L.; Hansen, S.; Buchner, J. P53 Contains Large Unstructured Regions in Its Native State. *J. Mol. Biol.* **2002**, *322*, 917–927. [CrossRef]
11. Berlow, R.B.; Dyson, H.J.; Wright, P.E. Functional advantages of dynamic protein disorder. *FEBS Lett.* **2015**, *589*, 2433–2440. [CrossRef]
12. Punj, V.; Das Gupta, T.K.; Chakrabarty, A.M. Bacterial cupredoxin azurin and its interactions with the tumor suppressor protein p53. *Biochem. Biophys. Res. Commun.* **2003**, *312*, 109–114. [CrossRef]
13. Yamada, T.; Goto, M.; Punj, V.; Zaborina, O.; Chen, M.L.; Kimbara, K.; Majumdar, D.; Cunningham, E.; Das Gupta, T.K.; Chakrabarty, A.M. Bacterial redox protein azurin, tumor suppressor protein p53, and regression of cancer. *Proc. Natl. Acad. Sci. USA* **2002**, *99*, 14098–14103. [CrossRef]
14. Goto, M.; Yamada, T.; Kimbara, K.; Horner, J.; Newcomb, M.; Gupta, T.K.; Chakrabarty, A.M. Induction of apoptosis in macrophages by Pseudomonas aeruginosa azurin: Tumour-suppressor protein p53 and reactive oxygen species, but not redox activity, as critical elements in cytotoxicity. *Mol. Microbiol.* **2003**, *47*, 549–559. [CrossRef]
15. Yamada, T.; Hiraoka, Y.; Ikehata, M.; Kimbara, K.; Avner, B.S.; Das Gupta, T.K.; Chakrabarty, A.M. Apoptosis or growth arrest: Modulation of tumor suppressor p53's specificity by bacterial redox protein azurin. *Proc. Natl. Acad. Sci. USA* **2004**, *101*, 4770–4775. [CrossRef]
16. Yamada, T.; Fialho, A.M.; Punj, V.; Bratescu, L.; Gupta, T.K.; Chakrabarty, A.M. Internalization of bacterial redox protein azurin in mammalian cells: Entry domain and specificity. *Cell. Microbiol.* **2005**, *7*, 1418–1431. [CrossRef]
17. Yamada, T.; Gupta, E.; Beattie, C.W. p28-Mediated Activation of p53 in G2–M Phase of the Cell Cycle Enhances the Efficacy of DNA Damaging and Antimitotic Chemotherapy. *Cancer Res.* **2016**, *76*, 2354–2365. [CrossRef]
18. Signorelli, S.; Santini, S.; Yamada, T.; Bizzarri, A.R.; Beattie, C.W.; Cannistraro, S. Binding of Amphipathic Cell Penetrating Peptide p28 to Wild Type and Mutated p53 as studied by Raman, Atomic Force and Surface Plasmon Resonance spectroscopies. *Biochim. Biophys Acta Gen. Subj.* **2017**, *1861*, 910–921. [CrossRef]
19. Domenici, F.; Frasconi, M.; Mazzei, F.; D'Orazi, G.; Bizzarri, A.R.; Cannistraro, S. Azurin modulates the association of Mdm2 with p53: SPR evidence from interaction of the full-length proteins. *J. Mol. Recognit.* **2011**, *24*, 707–714. [CrossRef]

20.   Funari, G.; Domenici, F.; Nardinocchi, L.; Puca, R.; D'Orazi, G.; Bizzarri, A.R.; Cannistraro, S. Interaction of p53 with Mdm2 and azurin as studied by atomic force spectroscopy. *J. Mol. Recognit.* **2010**, *23*, 343–351. [CrossRef]

21.   De Grandis, V.; Bizzarri, A.R.; Cannistraro, S. Docking study and free energy simulation of the complex between p53 DNA-binding domain and azurin. *J. Mol. Recognit.* **2007**, *20*, 215–226. [CrossRef]

22.   Maiti, N.C.; Apetri, M.M.; Zagorski, M.G.; Carey, P.R.; Anderson, V.E. Raman spectroscopic characterization of secondary structure in natively unfolded proteins: Alpha-synuclein. *J. Am. Chem. Soc.* **2004**, *126*, 2399–2408. [CrossRef]

23.   Signorelli, S.; Cannistraro, S.; Bizzarri, A.R. Structural Characterization of the Intrinsically Disordered Protein p53 Using Raman Spectroscopy. *Appl. Spectrosc.* **2016**, *71*, 823–832. [CrossRef]

24.   Yamada, T.; Signorelli, S.; Cannistraro, S.; Beattie, C.W.; Bizzarri, A.R. Chirality switching within an anionic cell-penetrating peptide inhibits translocation without affecting preferential entry. *Mol. Pharm.* **2015**, *12*, 140–149. [CrossRef]

25.   Siamwiza, M.N.; Lord, R.C.; Chen, M.C.; Takamatsu, T.; Harada, I.; Matsuura, H.; Shimanouchi, T. Interpretation of the doublet at 850 and 830 cm-1 in the Raman spectra of tyrosyl residues in proteins and certain model compounds. *Biochemistry* **1975**, *14*, 4870–4876. [CrossRef]

26.   Arp, Z.; Autrey, D.; Laane, J.; Overman, S.A.; Thomas, G.J. Tyrosine Raman signatures of the filamentous virus Ff are diagnostic of non-hydrogen-bonded phenoxyls: Demonstration by Raman and infrared spectroscopy of p-cresol vapor. *Biochemistry* **2001**, *40*, 2522–2529. [CrossRef]

27.   Tuma, R. Raman spectroscopy of proteins: From peptides to large assemblies. *J. Raman Spectrosc.* **2005**, *36*, 307–319. [CrossRef]

28.   Torreggiani, A.; Fini, G. Raman spectroscopic studies of ligand-protein interactions: The binding of biotin analogues by avidin. *J. Raman Spectrosc.* **1998**, *29*, 229–236. [CrossRef]

29.   Krimm, S.; Bandekar, J. Vibrational spectroscopy and conformation of peptides, polypeptides, and proteins. *Adv. Protein Chem.* **1986**, *38*, 181–364. [CrossRef]

30.   Altose, M.D.; Zheng, Y.; Dong, J.; Palfey, B.A.; Carey, P.R. Comparing protein-ligand interactions in solution and single crystals by Raman spectroscopy. *Proc. Natl. Acad. Sci. USA* **2001**, *98*, 3006–3011. [CrossRef]

31.   Carey, P.R. *Biochemical Applications of Raman and Resonance Raman Spectroscopies*; Academic Press: Cambridge, UK, 1982; ISBN 9780121596507.

32.   Harada, I.; Miura, T.; Takeuchi, H. Origin of the doublet at 1360 and 1340 cm−1 in the Raman spectra of tryptophan and related compounds. *Spectrochim. Acta. Part A Mol. Spectrosc.* **1986**, *42*, 307–312. [CrossRef]

33.   David, C.C.; Jacobs, D.J. Principal component analysis: A method for determining the essential dynamics of proteins. *Methods Mol. Biol.* **2014**, *1084*, 193–226. [CrossRef]

34.   Thomas, G.J. Raman spectroscopy of protein and nucleic acid assemblies. *Annu. Rev. Biophys. Biomol. Struct.* **1999**, *28*, 1–27. [CrossRef]

35.   Nar, H.; Messerschmidt, A.; Huber, R.; Van De Kamp, M.; Canters, G.W. Crystal structure of Pseudomonas aeruginosa apo-azurin at 1.85 A resolution. *FEBS Lett.* **1992**, *306*, 119–124. [CrossRef]

36.   Natan, E.; Baloglu, C.; Pagel, K.; Freund, S.M.V.; Morgner, N.; Robinson, C.V.; Fersht, A.R.; Joerger, A.C. Interaction of the p53 DNA-binding domain with its n-terminal extension modulates the stability of the p53 tetramer. *J. Mol. Biol.* **2011**, *409*, 358–368. [CrossRef]

37.   Wen, Z.Q. Raman spectroscopy of protein pharmaceuticals. *J. Pharm. Sci.* **2007**, *96*, 2861–2878. [CrossRef]

38.   Apiyo, D.; Wittung-Stafshede, P. Unique complex between bacterial azurin and tumor-suppressor protein p53. *Biochem. Biophys. Res. Commun.* **2005**, *332*, 965–968. [CrossRef]

39.   Uversky, V.N.; Oldfield, C.J.; Midic, U.; Xie, H.; Xue, B.; Vucetic, S.; Iakoucheva, L.M.; Obradovic, Z.; Dunker, A.K. Unfoldomics of human diseases: Linking protein intrinsic disorder with diseases. *BMC Genom.* **2009**, *10*, S7. [CrossRef]

40.   Yamada, T.; Christov, K.; Shilkaitis, A.; Bratescu, L.; Green, A.; Santini, S.; Bizzarri, A.R.; Cannistraro, S.; Gupta, T.K.; Beattie, C.W. p28, A first in class peptide inhibitor of cop1 binding to p53. *Br. J. Cancer* **2013**, *108*, 2495–2504. [CrossRef]

41.   Minde, D.P.; Dunker, A.K.; Lilley, K.S. Time, space, and disorder in the expanding proteome universe. *Proteomics* **2017**, *17*, 1600399. [CrossRef]

42.   Kengne-Momo, R.P.; Daniel, P.; Lagarde, F.; Jeyachandran, Y.L.; Pilard, J.F.; Durand-Thouand, M.J.;

Thouand, G. Protein Interactions Investigated by the Raman Spectroscopy for Biosensor Applications. *Int. J. Spectrosc.* **2012**, *2012*, 1–7. [CrossRef]

43. Domenici, F.; Bizzarri, A.R.; Cannistraro, S. Surface-enhanced Raman scattering detection of wild-type and mutant p53 proteins at very low concentration in human serum. *Anal. Biochem.* **2012**, *421*, 9–15. [CrossRef] [PubMed]

44. Ortiz, C.; Zhang, D.; Xie, Y.; Ribbe, A.E.; Ben-Amotz, D. Validation of the drop coating deposition Raman method for protein analysis. *Anal. Biochem.* **2006**, *353*, 157–166. [CrossRef] [PubMed]

45. Krafft, C.; Hinrichs, W.; Orth, P.; Saenger, W.; Welfle, H. Interaction of Tet repressor with operator DNA and with tetracycline studied by infrared and Raman spectroscopy. *Biophys. J.* **1998**, *74*, 63–71. [CrossRef]

46. Yang, H.; Yang, S.; Kong, J.; Dong, A.; Yu, S. Obtaining information about protein secondary structures in aqueous solution using Fourier transform IR spectroscopy. *Nat. Protoc.* **2015**, *10*, 382–396. [CrossRef] [PubMed]

6

In Silico Study of Rett Syndrome Treatment-Related Genes, *MECP2*, *CDKL5* and *FOXG1* by Evolutionary Classification and Disordered Region Assessment

Muhamad Fahmi [1], **Gen Yasui** [1], **Kaito Seki** [1], **Syouichi Katayama** [2], **Takako Kaneko-Kawano** [2], **Tetsuya Inazu** [2], **Yukihiko Kubota** [3] and **Masahiro Ito** [1,3,*]

[1]   Advanced Life Sciences Program, Graduate School of Life Sciences, Ritsumeikan University, Kusatsu, Shiga 525-8577, Japan; gr0343rp@ed.ritsumei.ac.jp (M.F.); sj0048hh@ed.ritsumei.ac.jp (G.Y.); sj0036kf@ed.ritsumei.ac.jp (K.S.)

[2]   Department of Pharmacy, College of Pharmaceutical Sciences, Ritsumeikan University, Kusatsu, Shiga 525-8577, Japan; s-kata@fc.ritsumei.ac.jp (S.K.); takanek@fc.ritsumei.ac.jp (T.K.-K.); tinazu@fc.ritsumei.ac.jp (T.I.)

[3]   Department of Bioinformatics, College of Life Sciences, Ritsumeikan University, Kusatsu, Shiga 525-8577, Japan; yukubota@fc.ritsumei.ac.jp

*   Correspondence: maito@sk.ritsumei.ac.jp

Abstract: Rett syndrome (RTT), a neurodevelopmental disorder, is mainly caused by mutations in methyl CpG-binding protein 2 (*MECP2*), which has multiple functions such as binding to methylated DNA or interacting with a transcriptional co-repressor complex. It has been established that alterations in cyclin-dependent kinase-like 5 (*CDKL5*) or forkhead box protein G1 (*FOXG1*) correspond to distinct neurodevelopmental disorders, given that a series of studies have indicated that RTT is also caused by alterations in either one of these genes. We investigated the evolution and molecular features of MeCP2, CDKL5, and FOXG1 and their binding partners using phylogenetic profiling to gain a better understanding of their similarities. We also predicted the structural order–disorder propensity and assessed the evolutionary rates per site of MeCP2, CDKL5, and FOXG1 to investigate the relationships between disordered structure and other related properties with RTT. Here, we provide insight to the structural characteristics, evolution and interaction landscapes of those three proteins. We also uncovered the disordered structure properties and evolution of those proteins which may provide valuable information for the development of therapeutic strategies of RTT.

Keywords: Rett syndrome; intrinsically disordered region; phylogenetic profile analysis; post-transcriptional modification; methyl-CpG-binding protein 2; cyclin-dependent kinase-like 5; forkhead box protein G1

---

1. Introduction

Rett syndrome (RTT; OMIM entry #312750) is a rare disease that was first described by Andreas Rett in 1966 [1]. It is characterized by severe impairment such as deceleration of head growth, loss of speech, seizures, ataxia, movement disorder, and breathing disturbance [2]. Alterations in methyl CpG-binding protein (*MECP*)2, an X-linked gene involved in the regulation of RNA splicing and chromatin remodeling, were confirmed in approximately 95% of individuals diagnosed with RTT [3], while the others were confirmed in either cyclin-dependent kinase-like (*CDKL*)5 or forkhead box protein (*FOXG*)1 alterations as atypical cases of RTT [4,5]. The mutations in *MECP2* are generally paternally derived. Thus, this syndrome mainly affects girls, and the age of onset varies from 6 to

18 months [2,6]. Additionally, Rett syndrome can also affect males with severe phenotype and early lethality following the inactivation of the sole X-linked copy of *MECP2* [7]. In a rare case, it can also exist as somatic mosaicism or co-occur with Klinefelter syndrome in males [8,9]. Even though the causative genes have been determined, the infrequent clinical phenotypes yield to the difficulty in diagnosis. Further, diagnosis may be challenging as many of the clinical features overlap with those of other neurological and neurodevelopmental disorders, and mutation in *MECP2, FOXG1*, and *CDKL5* can also cause neurodevelopmental disorders distinct from RTT [10]. As a result, subsequent studies have suggested that alterations in either *CDKL5* or *FOXG1* should be classified as a distinct disorder from RTT as the majority of cases showed some differences in clinical features [11–13] Moreover, recent studies have suggested that RTT is a monogenic disorder caused by mutations that alter the functionality of the methyl-CpG-binding domain (MBD) and the NCoR/SMRT interaction domain (NID) in *MECP2* [14–16]. This may simplify the complication of developing a treatment strategy. But, elucidation on the overlapped symptoms between those three proteins comprehensively on the molecular basis also seems necessary as the study about it remains scarce and it may provide meaningful insight, particularly for RTT.

The MeCP2 structure has been determined using various experimental methods, while the structure of FOXG1 has only been investigated by predictions [17,18]. In the case of CDKL5, the structure of the amino-terminal kinase domain has already been identified, but that of the long carboxy-terminal tail has not been clarified [19]. These proteins have been suggested to contain polypeptide segments that are unable to fold spontaneously into three-dimensional structures; the so-called intrinsically disordered regions (IDRs) exist as dynamic ensembles that rapidly interconvert from molten globule (collapsed) to coiled or pre-molten globule (extended) as a result of the relatively flat energy landscapes [20,21]. The different entities of IDRs and ordered regions (displaying tertiary structures in native conditions) are dictated by the amino acid sequence; the former generally lack bulky hydrophobic residues [22]. Proteins are composed of either fully structured or fully disordered regions (with the latter referred to as intrinsically disordered proteins (IDPs) or a combination of the two, which is the case for most eukaryotic proteins [23]. Although protein function has traditionally been elucidated based on a well-defined structure, it is now widely acknowledged that IDRs contribute to diverse functions, which can be classified into six types: entropic chain activity, display site, chaperone, molecular effector, molecular assembler, and molecular scavenger [23–26]. Excluding entropic chain activity, IDRs adopt specific tertiary conformations—at least locally—in order to perform those functions by binding to other proteins, nucleic acids, membranes, and small molecules or responding to changes in their environment [20,27]. Hence, IDR structure varies over time—i.e., it exhibits spatiotemporal heterogeneity. Moreover, long IDRs contain more modification sites than fully ordered regions, and their flexibility provides more opportunities for displaying these sites [28,29]. These features explain how proteins with IDRs or IDPs interact with and are tightly regulated by various factors to ensure that appropriate levels of proteins are available at the right time to minimize the possibility of inappropriate protein–protein interactions [26]. Thus, misfolding and altered availability of proteins with IDRs or IDPs are more likely to be associated with disease states. Given a similarity in those properties, we proposed that a study concerning the link between MeCP2, CDKL5, and FOXG1 disordered structure properties with RTT or RTT-like syndrome collectively is necessary.

Restoring *Mecp2* gene function in an animal model abolished the symptoms of RTT. Growth factor stimulation (e.g., insulin-like growth factor 1) and the activation of neurotransmitter pathways (e.g., β2-adrenergic receptor pathway) can also partially rescue phenotypes of *Mecp2* knockout mice (RTT model mice), suggesting that the disorder is treatable [15,30,31]. In addition to gene therapy, reactivation of an inactivated X chromosome is known to be a new therapeutic method [32,33]. The therapeutic strategies of RTT are under development, and elucidation on this enigmatic disorder needs various points of view to make advances in understanding. Even though RTT has been determined as a monogenic disorder, the complex biological system compels us to necessarily broaden our perspective; moreover, MeCP2 contains an extensive amount of disordered regions which may facilitate binding with multiple partners. Considering several points above, we investigated the evolution and molecular features of MeCP2, CDKL5, and FOXG1 and their binding partners using phylogenetic profiling to gain a better understanding of their similarities. Additionally, we predicted the structural order–disorder propensity and assessed the evolutionary rates per site of MeCP2, CDKL5, and FOXG1 to investigate the relationships between disordered structure and other related properties with RTT.

2. Results

2.1. Structural Order–Disorder Properties of RTT and RTT-like Causing Proteins during Chordate Evolution

We retrieved 97, 113, and 108 chordates sequences of MeCP2, CDKL5, and FOXG1, respectively, and constructed a heat map of the structural order–disorder propensity for each protein of these genes according to aligned sequences and taxonomic position in the phylogenetic tree (Supplementary Table S1 and Figure 1). This analysis was conducted in order to investigate the evolutionary patterns of structural properties. The results showed that all proteins harbored both ordered and disordered regions; by comparing their distribution to domain and non-domain regions, we found that the catalytic domain and non-domain regions of CDKL5 were ordered and disordered, respectively (Figure 1B). While most regions of MeCP2 were predicted to be disordered, some ordered structures were observed in the MBD (Figure 1A). Furthermore, FOXG1 showed a varied distribution of ordered–disordered regions corresponding to domain and non-domain regions, with the former predicted to be fully ordered (Figure 1C). Although insertions and deletions were frequently detected in disordered regions, particularly in MeCP2 and FOXG1 (Figure 1A,C), the structural order–disorder of all proteins showed to be stable in chordates, excluding a few conformational transitions of FOXG1 and CDKL5 in mammals and fishes, respectively. This indicated that the disordered regions of MeCP2, CDKL5, and FOXG1 tend to be functional either as an entropic chain, transient binding site, or permanent binding site in chordates. Additionally, insertions and deletions were frequently detected in disordered regions. This is caused by their flexibility, which makes sequence alignment difficult; a tendency of linear motifs to lie among the flexible disordered regions; and the permutation of functional modules with respect to others during evolution that is possible in disordered regions, such as SUMO modification sites in *Drosophila melanogaster* and human p53 that are located before and after the oligomerization domain, respectively [26,34].

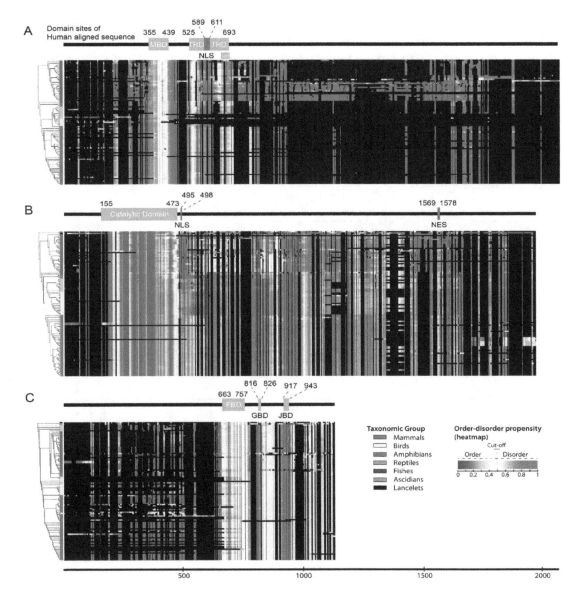

Figure 1. The order–disorder propensity of RTT and RTT-like causing proteins in chordates. Heat maps of the order–disorder propensity were generated according to the taxonomic positions in the phylogenetic tree (rows) and multiple sequence alignment (columns). The heat maps show a color gradient of blue (ordered) to red (disordered), with white as the boundary between the two and black as gaps. Colored boxes between the trees and heat maps indicate the taxonomic group, and bars above the heat maps indicate domain position in the multiple sequence alignment, with light blue and black areas indicating the domain and absence of a domain, respectively. (**A**–**C**) Heat maps for MeCP2 (**A**), CDKL5 (**B**), and FOXG1 (**C**) are shown. MBD, TRD, NID, FBD, GBD, JBD, NLS, and NES indicate methyl-CpG-binding domain, transcriptional repression domain, NCoR/SMRT interaction domain, forkhead binding domain, Groucho-binding domain, JARID1B binding domain, nuclear localization signal, and nuclear export signal, respectively.

2.2. Rate of Evolution per Site in RTT and RTT-like Causing Proteins

We calculated the evolutionary rates of MeCP2, CDKL5, and FOXG1 in chordates to investigate their relationships with structural features and the distribution of missense point mutations that have previously been suggested to contribute to RTT or RTT-like syndrome. We used the human sequence as a reference and determined standardized evolutionary rate scores (Z scores), with values greater than or less than zero reflecting evolution at a faster and slower than average rate, respectively (Figure 2 and Supplementary Table S2). Evolutionary rates per site showed similar patterns in all proteins, with low rates of evolution more commonly observed in domains and ordered regions; some exceptional cases such as the transcriptional repression domain (TRD) of MeCP2 showed a partial higher rate of amino acid substitution. On the other hand, non-domain regions that were also usually disordered—excluding the ordered region surrounding a domain in FOXG1—typically exhibited a higher evolutionary rate, although some regions with low rates of evolution were nonetheless detected (Figure 2). This was corroborated by the distribution of evolutionary rates for predicted structural order–disorder residues in the three proteins, with disordered residues showing a wide and overlapping distribution that reflected their conservation. The evolutionary rates of ordered and disordered regions are significantly distinct in those three proteins ($p < 2.2e-16$ for CDKL5 and FOXG1 and $p < 6.409e-08$ for MeCP2, Mann–Whitney U-test; Figure S1).

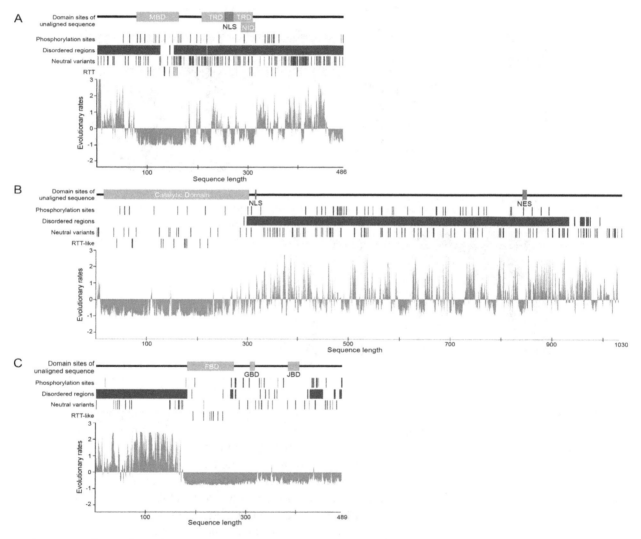

Figure 2. Rate of evolution per site in human RTT-related proteins. (**A–C**) Rates of amino acid substitution in MeCP2 (**A**), CDKL5 (**B**), and FOXG1 (**C**) are shown as blue areas. The bars above charts indicate the position of the domain in the human sequence, with light blue areas indicating the domain and black lines indicating no domain. Conserved phosphorylation sites, disordered region, single nucleotide polymorphisms in the general population, and pathogenic missense point mutation are plotted in green, purple, blue, and red lines, respectively. The x and y axes represent the sequence length and Z score of the evolutionary rates, respectively.

We identified structurally conserved disordered regions, with slowly and rapidly evolving residues reflecting constrained disorder and flexible disorder, respectively [26]. The flexible disorder has a constrained disordered structure despite having rapid evolution of residues; the amino acid substitutions of this property are constrained to residues that confer structural flexibility as the change from structurally disordered to ordered can affect protein function. This type of IDR typically functions as an entropic spring, flexible linker, or spacer without becoming structured and is frequently located outside the domain region [26,35–37]. In contrast, constrained disorder is associated with protein–protein interaction interfaces that adopt a structured conformation or undergo folding upon binding and are thus constrained in terms of sequence, while still requiring flexibility. This module can be present as short linear motifs (SLiMs) or intrinsically disordered domains (IDDs) [26,38]. These regions commonly have secondary structures that may be important for binding and, hence, slowing their evolutionary rates [36,39]. IDDs were observed in the MBD—which was predicted to be partly disordered—and in the TRD and NID of MeCP2; it is in accordance with previous reports that structured regions are found only in the MBD, while other regions are extensively disordered [17,18,40]. Most domains with conserved disordered regions are involved in DNA, RNA, and protein binding, which has been demonstrated by those domains of MeCP2 [41]. SLiMs are frequently located outside the domain and may display modification site. In this study, we predicted the constrained disorder regions and conserved phosphorylation sites located outside the domain to be associated with SLiMs, such as the region that spans after the catalytic domain to the C-terminus of human CDKL5.

2.3. Post-Translational Modifications (PTMs)

Phosphorylation is important for modulating the balance of proteins between the bound and unbound states, and previous studies reported that kinases target disordered proteins as many as twice, on average, the number of times they target structured proteins [42,43]. In this study, we predicted PTM (phosphorylation) sites in chordate sequences of MeCP2, CDKL5, and FOXG1 and predicted the conserved human phosphorylation sites to chordates in order to investigate the dynamics of their phosphorylation-related function. We found numerous conserved phosphorylation sites including 60/82 in CDKL5, 30/45 in MeCP2, and all 23 sites in FOXG1 in human (Figure 2 green lines and Supplementary Table S3). Most predicted human phosphorylation sites in MeCP2, CDKL5, and FOXG1 are conserved across chordates and are located in disordered regions; one exception is FOXG1, in which almost half of the phosphorylation sites are located in predicted ordered regions; structural disorder makes such sites accessible for phosphorylation. As PTMs affect the stability, turnover, interaction potential, and localization of proteins within the cell, proteins with disordered regions are more likely to be multifunctional [26]; accordingly, it has shown that MeCP2, CDKL5, and FOXG1 play multiple roles in the molecular basis.

2.4. Disease-Associated Missense Mutation Distribution in the Sequence of RTT and RTT-like Causing Proteins

Plotting missense mutations associated with diseases may yield crucial information on structure–function relationships and the features of the protein. We investigated missense mutations in human MeCP2, CDKL5, and FOXG1 that were previously associated with pathogenic RTT from RettBASE and examined the features of the associated sequences. There were 7, 12, and 18 individual amino acid sites in FOXG1, CDKL5, and MeCP2, respectively, that harbored pathogenic missense mutations associated or previously suggested to be associated with pathogenic RTT (Figure 2 and Supplementary Table S4). When the frequencies were combined with those of cases observed for each mutation, MeCP2 had a higher number of cases (1225) than CDKL5 (30) and FOXG1 (8) (Supplementary Tables S4 and S9). Pathogenic RTT or RTT-like-associated missense mutations were more frequently detected in domain regions for all proteins, and in ordered and slowly evolving regions for MeCP2 and CDKL5 (Supplementary Table S9). On the other hand, many mutation sites in MeCP2 were located close to (or in the case of Ser346Arg and Ser134Cys, overlapped with) phosphorylation sites (Figure 2), although the frequency of cases harboring these mutation sites was low (only one for each).

2.5. Phylogenetic Profiling of RTT and RTT-like Causing Proteins and Their Interaction Partners

We retrieved 240 human proteins interacting with MeCP2, CDKL5, and FOXG1 from BioGRID and UniProt databases (Supplementary Table S5) [44,45]. To illuminate the interconnection of MeCP2, CDKL5, and FOXG1 binding partners as well as their evolutionary relationship, we conducted phylogenetic profiling and cluster analysis of 326 eukaryotes using the retrieved sequences and the sequences of the three proteins, MeCP2, CDKL5, and FOXG1, as queries (Figure 3, Supplementary Table S6). The results showed that the dataset was divided into four clusters, which were defined as Classes 1 to 4. There were 58 conserved proteins in chordates of Class 1, 92 in metazoans of Class 2, 17 in multicellular of Class 3, and 73 in eukaryotes of Class 4. MeCP2 and CDKL5 belonged to Class 1, whereas FOXG1 belonged to Class 2 (Figure 3). FOXG1 and MECP2 showed to have many binding partners that act as a transcription factor or gene expression regulator. In contrast, CDKL5 tend to bind to a fewer number of proteins having functions in regulating cell adhesion, ciliogenesis, and cell proliferation; however, this protein has been shown to interact with MeCP2. As RTT has been determined to occur from the altered functionality of MBD and NID of MECP2, we focused on the widely known binding partners of these domains, such as SIN3 transcription regulator family member A (SIN3A), histone deacetylase (HDAC)1, and nuclear receptor corepressor (NCOR) which play roles as co-repressor complexes. Even though FOXG1 does not directly bind to MeCP2, we found that the binding partners of MeCP2 co-repressor complex are also associated with FOXG1 binding partners that also act as co-repressor complexes such as special AT-rich sequence-binding protein (SATB)2, lysine-specific histone demethylase (KDM)1A, SWI/SNF-related matrix-associated actin-dependent regulator of chromatin subfamily (SMARC)A member 5, A-kinase anchor protein (AKAP)8, of which are ancient proteins within Classes 3 and 4.

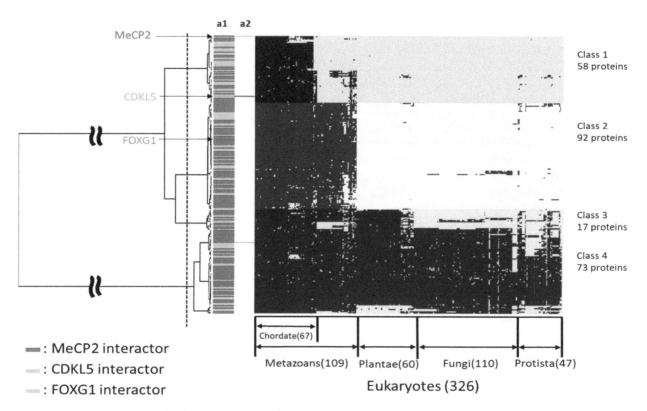

Figure 3. Phylogenetic profiling of MeCP2, CDKL5, and FOXG1 proteins and their interaction partners. The horizontal axis shows 326 eukaryotes for which whole genome sequences are available, and the vertical axis shows 240 human proteins related to RTT. Bar in a1 and a2 shows MeCP2-interactor (red), CDKL5-interactor (green), FOXG1-interactor (blue), respectively. The human orthologous proteins in each species are shown in black. The phylogenetic tree was divided into four clusters (Class 1–4); those conserved across chordates, metazoan, multicellular, and eukaryotes are shown.

2.6. Subcellular Localization and Gene Ontology (GO) Analysis

We predicted the subcellular localization of each protein and GO categories in each class for the evolutionary classification (Figure 4, Supplementary Table S7). Specific GO categories included epigenetic regulation of gene expression, transcriptional regulation, and organogenesis or organ morphogenesis (Figure 4). We confirmed the evolutionary trends of proteins with specific GO categories and their subcellular localization and found that 129 and 48 proteins in Classes 1–4 were expressed in the nucleus only or the nucleus and cytoplasm, respectively. Proteins in Classes 1–4 were represented in the epigenetic regulation of gene expression category, whereas transcriptional regulation was observed only in Classes 1 and 2, and organogenesis and organ morphogenesis were mainly observed in Class 2 (Figure 4).

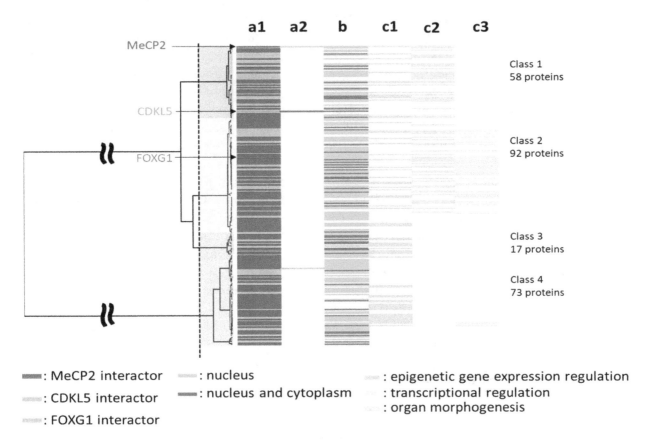

Figure 4. Subcellular localization and specific GO categories of human RTT-related proteins: Phylogenetic trees show interactors, subcellular localization, and specific GO categories for each protein. The vertical axis shows 240 RTT-related proteins, and each bar shows MeCP2-interactor (red), CDKL5-interactor (green), and FOXG1-interactor (blue) (a1 and a2); cellular localization (b); epigenetic regulation of gene expression (c1); transcriptional regulation (c2); and organogenesis (c3).

2.7. Tissue and Organ Localization

Tissue and organ expression data for 237 proteins were extracted from The Human Protein Atlas as transcripts per million (TPM) values [46]. In addition, four proteins were not expressed in the cerebral cortex. Tissues and organs with specific expression were identified using 195 RTT-related human proteins as queries (Figure S2, Supplementary Table S8). There were nine proteins that were specifically expressed in the cerebral cortex including apolipoprotein E, CDKL5, SATB2, spalt-like transcription factor (SALL)1, zinc finger protein (ZNF)483, FOXG1, (sex-determining region Y)-box (SOX)2, homeodomain-interacting protein kinase (HIPK)2, and histone cluster 2 H3 family member A.

3. Discussion

RTT is a progressive postnatal neurodevelopmental disorder; three individual genes, *MECP2*, *CDKL5*, and *FOXG1*, have previously been thought to be the cause of its variants with the altered *MECP2* as the major contributor. Later, it was suggested that RTT is a monogenic disorder caused by either null mutations or mutations that alter the MBD or NID functions of *MECP2* [15,16,47]. MBD and NID facilitate the binding of MeCP2 to modified cytosine in chromatin and recruitment of the NCOR-SMRT complex, respectively; their combination is vital for MeCP2's role as a repressor [48,49]. The altered forms in the other two genes which were previously characterized as variants of RTT were designed as distinct disorders with several overlapping symptoms to RTT. The three proteins have similar extensive amount of disordered regions and play important roles in the brain. The disordered structure itself is a unique property in protein that may contribute to the interaction with a diverse binding partner and the versatility of a protein. While the three proteins may show similar symptoms in the altered form, the investigation on their similarity in the molecular basis remains scarce, particularly on the disordered structure properties and their binding partners. Focusing on RTT, we investigated the evolution of their disordered structures and their binding partners through prediction and phylogenetic profiling, respectively. This approach is important to give an insight into the similarity of biological systems of those proteins structurally and evolutionarily, which may provide useful information for the development of a RTT therapy strategy. RTT itself has attracted considerable attention as its causative protein displays features related to epigenetics and have been shown to have partially or fully disordered structures.

All three proteins have been experimentally determined to play roles and are abundant in the brain, especially the MeCP2_e1 and hCDKL5_1 isoforms [50,51]. It is confirmed by the emergence of neurological impairments in the altered availability or forms of either protein. Through evolutionary analysis and IDRs properties, we provide an additional point of view for that feature. Phylogenetic profiling analysis of MeCP2, CDKL5, and FOXG1 and their interacting proteins showed that 240 molecules formed four clusters—i.e., chordates, metazoans, multicellular, and eukaryotes. Among the three, only *FOXG1* was a member of Class 2, which comprises genes acquired during metazoan evolution, whereas the acquisition of *MECP2* and *CDKL5* was correlated with chordate evolution. The acquisition of *CDKL5* and *MECP2*, and *FOXG1* may contribute to the development of the chordate brain and metazoan nervous system during evolution, respectively. Additionally, order–disorder structure predictions revealed that all three proteins had order–disorder structures that were relatively conserved across chordates. Human MeCP2, CDKL5, and FOXG1 phosphorylation sites were also shown to be relatively conserved to chordates. IDRs properties provide proteins with more interaction areas and PTMs sites, spatiotemporal heterogeneity of structure, and ability to associate and dissociate easily with binding partners. Hence, proteins with long IDRs are likely to have a capacity to bind to many different partners. Accordingly, all three proteins were shown to have multiple binding partners, and FOXG1 and MeCP2 displayed the highest number of partners, some of which were evolutionarily acquired before the metazoan evolved. By cooperating with various proteins partners, particularly the co-repressor complex, FOXG1 or MeCP2 can modulate the expression and suppression of different genes [15,52]. The co-repressor complex itself denotes a conserved mechanism that manifests in diverse forms and may have several functional entities depending on the context in which they are recruited [53]. This indicates the necessity to regulate either FOXG1 or MeCP2 concentration precisely; otherwise, altered availability is likely to be deleterious. Several studies have shown that either overexpression or under-expression of MeCP2 and FOXG1 corresponds to neurological deficits; this phenomenon may not independent from their co-repressor complex that has been showed to play roles in neurogenesis and neuron maturation for FOXG1, and MeCP2, respectively [7,15,52]. On the other hand, CDKL5 binds to a fewer number of proteins that have functions in regulating cell adhesion, ciliogenesis, and cell proliferation. We hypothetically suggest that the amount of CDKL5 binding partners is underestimated since this protein was predicted to have relatively long disordered regions with many constrained disorder features and phosphorylation sites; it also has fewer insertions and deletions than either MeCP2 or FOXG1 along the evolution.

FOXG1 is a transcriptional factor playing an essential role in ventral telencephalon development; it serves as a hallmark of the telencephalon in vertebrates [52,54]. Among the 237 Class 1 or 2 genes, 233 were detected in the cerebral cortex, with nine expressed at a high level (Figure S2). Seven genes were acquired during metazoan evolution, of which four and three encode MeCP2- and FOXG1-interacting molecules, respectively. Since FOXG1 was also acquired during metazoan evolution, acquisition of FOXG1, SATB2, and SALL1 may have played essential roles in development of the neocortex. FOXG1 is transiently expressed in neuronal progenitor cells and regulates their migration to the cortical plate [55]. During this process, FOXG1 expression is upregulated, which contributes to cortical plate development [56]. Similarly, the FOXG1-interacting chromatin remodeling factor SATB2 was found to be expressed in the cortical plate and regulates neocortical development [54,55]. Therefore, it is conceivable that transcriptional co-operation between FOXG1 and SATB2 mediates the laminarization of the neocortex. In support of this possibility, patients with the SATB2 mutation exhibit an RTT-like phenotype [57,58]. There is no direct interplay reported for MeCP2 and FOXG1. The causative regions in the altered form of these proteins that result in the development of RTT or RTT-like disorder exhibited similar functions in regulating the other genes' expression, but likely via a distinct pathway. We suggest that FOXG1 is not a potential target for developing treatment for RTT. However, induced pluripotent stem cell (iPSC)-derived neurons generated from FOXG1+/− patients and patients with MECP2 and CDKL5 mutations reportedly exhibited a similar increase in synaptic cell adhesion protein orphan glutamate receptor δ-1 subunit (GluD1) expression; this result indicates the need for further study to reveal the mechanism of each protein and might be implicated in the clinical symptom overlap among FOXG1-, CDKL5- and MECP2-related syndromes [52,59,60].

CDKL5 belongs to the same molecular pathway of MeCP2. MeCP2 was acquired during chordate evolution; a prerequisite for this step was the acquisition of MeCP2-interacting molecules such as ZNF483, SOX2, HIPK2, and HIST2H2A. The MeCP2 kinase HIPK2 was shown to be required for the induction of apoptotic cell death in neuronal and other cell types via phosphorylation of the MeCP2 N-terminus [61]. Given that CDKL5, another MeCP2 kinase was also acquired during chordate evolution; it is possible that HIPK2 and CDKL5 cooperate to activate MeCP2 during neocortical development. Since apoptotic cell death increased in *Cdkl5* knockout mouse brain, CDKL5 probably has a suppressive function in the apoptosis process in contrast to HIPK2 [62]. Therefore, functional division of their kinases through phosphorylation of MeCP2 is an important issue. Indeed, the CDKL5-interacting domain was shown to be associated with the C-terminus of MeCP2 [63]. Hence, CDKL5 may phosphorylate the carboxy terminus. Thus, both HIPK2 and CDKL5 may activate MeCP2 by phosphorylating different regions of the protein. It has been suggested that MeCP2 also suppresses CDKL5 transcription and that CDKL5 overexpression may also contribute to the typical RTT symptoms [64]. Hence, aiming the catalytic domain of CDKL5 as the key target for developing alternative strategies to treat classical RTT may be essential since its sole impairment resulted in some symptoms that overlapped with those of classical RTT. Additionally, the CDKL5 disordered region, which spans after the catalytic domain to the C-terminus, is suggested to have many SLiMs. The linear motifs theoretically help to determine the various fates of a protein including subcellular localization, stability, and degradation; these motifs are also able to promote recruitment of binding factors and facilitating post-translational modifications [26,38]. Since these motifs typically regulate low-affinity interactions, they can bind to molecules with different structures of similar affinity and facilitate transient-binding, which are favorable properties for drug targets. Accordingly, this region appears to be a potential target for classical RTT treatment. However, this should also consider the expression levels of CDKL5 which are highly modulated spatiotemporally [64,65].

IDRs show unique properties within protein which challenges the traditional viewpoint of the protein structure paradigm. They have differences in residue composition, intramolecular contacts, and functions to ordered regions which cause different evolutionary rates. Generally, they evolve more rapidly than ordered regions, owing to the different accepted point mutations. However, some disordered regions can be highly constrained as they may play crucial roles and have multiple functions; assessing the evolutionary rate of IDRs may thus reveal crucial protein-specific amino acids in the biological system [66]. In this study, we found a unique relationship between evolutionary rates of disordered regions and symptoms of a disease caused by FOXG1. The N-terminus residues of FOXG1 are highly variable and constrained to be disordered, while the residues from FBD to the C-terminus are constrained and contain an ordered structure. It has been reported that mutations in the N-terminal are more likely to be associated with severe phenotypes, and mutations in the C-terminal are associated with milder phenotypes [52]. We reported and predicted a phosphorylation site located in Ser 19 to be conserved in chordates even though it is located among flexible disordered regions; casein kinase 1 (CK1) modifies this site and promotes the nuclear import of FOXG1, which corresponds to neurogenesis in the forebrain [67]. This explains that a flexible disordered region can retain its functional module from phosphorylation, despite harboring numerous insertions and deletions, and that severe phenotypes may result from the altered function of Ser 19 of FOXG1.

Among 236 male testis expressing RTT-related genes, 47 genes expressed at a high level. Because paternal-derived de novo mutation has been shown to affect X-linked MeCP2-related female Rett syndrome [6,68], paternally expressing mutation in these genes may affect the sperm-derived genetic and/or epigenetic inheritance that influence the cause of Rett syndrome in a daughter. Further studies are required to analyze these possibilities.

It is important to remember that the features of structural order–disorder and phosphorylation sites in this study have been inferred using linear sequence predictors and that the sequences and mutation points were retrieved from databases whose data have been collected from studies with various methods. It should be considered that we use canonical isoforms instead of predominant brain isoforms, this option may be able to be applied computationally but should be of concern experimentally. This study provides suggestive or hypothetical conclusions, thus further experimental study is important to verify the findings of this study. Ultimately, the results can still be used and considered as a basis for further identification.

4. Materials and Methods

4.1. Sequence Retrieval, Alignment, and Phylogenetic Analysis of MeCP2, CDKL5, and FOXG1 Proteins

Orthologous sequences of human RTT and RTT-like causing proteins (MeCP2, CDKL5, and FOXG1) in chordates were retrieved from the Kyoto Encyclopedia of Genes and Genomes (KEGG) sequence similarity database (https://www.kegg.jp/kegg/ssdb/) with a Smith–Waterman similarity score threshold of 100 and the bidirectional best hits (best–best hits) option [69]. We primarily used the canonical isoforms MeCP2_e2 and hCDKL_5 instead of those the predominant isoforms in the human brain, MeCP2_e1, and hCDKL5_1. MeCP2_e2 is the most characterized isoform relative to MeCP2_e1, and RettBASE has chosen to name the variants MeCP2_e2 due to historical reason. Variants specific to MeCP2_e1 are still reported in RettBASE with the prefix MeCP2_e1 in the database, but we decided to exclude them in our analysis as we only found one variant that meets our criteria and it cannot be included within the MeCP2_e2 sequence as they differ in the N-terminal region; however, we still reported that variant in our Supplementary Data. CDKL5 has a similar case as MeCP2, but the differences of sequences between hCDKL_5 and hCDKL5_1 are located in the C-terminal region (905–1030 a.a) which does not shift the reported Rett-like variants in the catalytic domain. We selected this option as we primarily collected the RTT and RTT-like variants from RettBASE. The used isoforms do not differ greatly to those predominant brain isoforms. The highest similarity score for each species was used for each of those proteins to minimize redundancy. Datasets were created for each protein and then aligned using MAFFT v.7 (https://mafft.cbrc.jp/alignment/software/) with the iterative refinement method (FFT-NS-i), with a maximum of 1000 iterations [70]. Phylogenetic trees were constructed with the maximum likelihood method using RAxML-HPC2 BlackBox with the RAxML automatic bootstrapping option using Jones, Taylor, and Thornton amino acid substitutions with the + F method and gamma shape parameter (JTT + F + G) model for MeCP2 and CDKL5, and the JTT + G model for FOXG1, which were selected as the best fit models under the Bayesian information criterion (BIC) by ModelTest-NG [71,72]. The outgroup for each tree was selected based on the NCBI Taxonomy Common Tree for the common ancestor within the dataset [73]. Reconstruction of phylogenetic trees and calculation of models were performed in CIPRES Science Gateway (http://www.phylo.org/) [74].

4.2. Structural Order–Disorder Prediction and Secondary Structure Predictions

The structural order–disorder propensity of each protein was predicted using IUPred2A (https://iupred2a.elte.hu/) [75] using the option for long disordered regions. This prediction had values ranging from 0 (strong propensity for an ordered structure) to 1 (strong propensity for a disordered structure), with 0.5 as the cut-off between the propensity for order and disorder. The results for each site of each protein were mapped onto its sequence alignment and taxon position in the phylogenetic tree using iTOL (https://itol.embl.de/) [76].

4.3. Rate of Evolution per Site

We calculated the rate of evolution per site of human CDKL5, FOXG1, and MeCP2 relative to their orthologs using Rate4site (https://m.tau.ac.il/~{}itaymay/cp/rate4site.html) [77]. The aligned sequences of each protein dataset were calculated using the empirical Bayesian principle with the JTT model and 16 discrete categories of the prior gamma distribution. Gaps were treated as missing data, and outputs were standardized as Z scores. The results of the rate of evolution of each residue were then integrated with the structural order–disorder prediction result, and the distribution of the rate of evolution in the structural order and disorder of each protein was evaluated with the Mann–Whitney U-test using R software.

4.4. PTM Prediction

We predicted phosphorylation sites using NetPhos 3.1 (http://www.cbs.dtu.dk/services/NetPhos/) [78] to infer PTM sites conserved between human CDKL5, FOXG1, and MeCP2 sequences and their orthologs. The predictions had values ranging from 0 (strong propensity for obtaining a negative result) to 1 (strong propensity for obtaining a positive result); we used 0.75 as a cut-off to divide the negative and positive results. The prediction results for each sequence were plotted following multiple sequence alignment of each protein dataset. Predicted PTM sites in each dataset were considered as conserved through evolution if they had a positive value according to the 50% majority rule of the amount of sequence in the alignment.

4.5. Point Mutations in MeCP2, CDKL5, and FOXG1

Point mutations in CDKL5, FOXG1, and MeCP2 were identified from RettBASE (http://mecp2.chw.edu.au/) [79]. The amount of mutations variants in general in RettBASE are 929, 298, and 44 for MeCP2, CDKL5, and FOXG1, respectively. We only selected missense mutations that were associated with pathogenic RTT. Additionally, non-pathogenic polymorphisms in the general population for comparison were extracted from the Exome Aggregation Consortium database (http://exac.broadinstitute.org) [80].

4.6. Phylogenetic Profiling and Cluster Analyses of Human MeCP2, CDKL5, and FOXG1 and Their Interacting Proteins

Sequences of human MeCP2, CDKL5, and FOXG1 and their interaction partners identified with BioGRID (https://thebiogrid.org/; release 2019_03) were obtained from the UniProtKB/Swiss-Prot database (https://www.uniprot.org/help/uniprotkb; release 2019_04) and used as the dataset [45,46]. We generated phylogenetic profiles of 326 eukaryotes in the KEGG database (https://www.genome.jp/kegg/) using the dataset as a query [81]. Phylogenetic profiling is a method for detecting the presence or absence of orthologous proteins in a target organism [82]. The presence or absence of proteins homologous to the query in each species was determined using KEGG Ortholog Cluster (https://www.genome.jp/tools/oc/; release 2019_04), this tool uses Smith–Waterman similarity scores of ≥150 and symmetric similarity measures to classify the ortholog genes [83]. We suggest that it is a reliable tool to get ortholog data. Profiles were determined based on the Manhattan distance and then clustered using Ward's method [84].

4.7. Protein Expression in Human Tissues

Expression levels of human RTT-related proteins in each tissue were extracted from the Human Protein Atlas (https://www.proteinatlas.org/; release 2019_4) [45] and classified into 37 tissues. The protein expression level was determined using the TPM value, which was corrected for protein expression by gene length. Comparisons of protein expression levels were not shown as a ratio so that proteins with high expression did not skew the results (Equations (1)–(3)). The mean and standard deviation were derived from Equations (1) and (2), and the range was obtained from Equation (3). The range in Equation (3) was taken as the tissue for each of the specifically expressed proteins—i.e., the value was "1" when included in the range of Equation (3) and "0" when it was not included in the expression level of each protein expressed as a percentage. The procedure yielded human protein-specific expression profiles in the context of RTT.

$$\mu = \frac{1}{n} \sum_{i=0}^{n} x_i \tag{1}$$

$$s = \sqrt{\frac{1}{n} \sum_{i=0}^{n} (x_i - \mu)^2} \tag{2}$$

$$\mu + 1.65 \times s < x \tag{3}$$

Here, μ, s, n, and x are the mean, standard deviation, number of samples, and one sample, respectively. The value of 1.65 in Equation (3) is the standard confidence factor for extracting data outside the 90% confidence interval.

4.8. GO Analysis

Specific GO categories in the target protein group were obtained using the Panther tool [85]. Categories with an appearance frequency of $p < 0.05$ were defined as protein group-specific. In this study, we obtained GO categories specific for human proteins related to RTT that were classified based on defined functions.

5. Conclusions

In the last two decades, effort on elucidating RTT has shown a promising trend towards developing a reliable treatment for this disorder. Given a similarity in IDR properties and several overlapping symptoms, we investigated the evolution of MeCP2, CDKL5, and FOXG1 disordered structures and their binding partners through prediction and phylogenetic profiling, respectively. Here, we provided insight to the structural characteristics, evolution and interaction landscapes of those three proteins related to RTT. We suggested that the disordered structures of MECP2, CDKL5, and FOXG1 contribute to the versatility in brain development and may play a crucial role in brain evolution in chordates. We hypothetically suggested that CDKL5 could be a potential target for RTT treatment, particularly by targeting its disordered structure that spans after the catalytic domain to the C-terminus, which shows abundant linear motifs that can bind to molecules with different structures of similar affinity. Finally, this study may provide valuable guidance for experimental research, particularly on the relationship between RTT and disordered regions.

Author Contributions: Conceptualization, M.F., Y.K. and M.I.; methodology, M.F., G.Y., Y.K. and M.I..; software, M.F. and G.Y.; validation, M.F., G.Y. and K.S.; formal analysis, M.F. and G.Y.; investigation, M.F., G.Y., K.S., S.K., T.K.-K., T.I., Y.K. and M.I.; resources, S.K, T.K.-K. and T.I.; data curation, M.F., Y.K. and M.I.; writing—original draft preparation, M.F..; writing—review and editing, S.K., T.K.-K., T.I., Y.K. and M.I.; visualization, M.F., G.Y. and K.S.; supervision, M.I.; project administration, M.I.; funding acquisition, T.I. and M.I.

Acknowledgments: We would like to thank Takahiro Nakamura and Tadasu Shin-I for support and helpful comments.

Abbreviations

a.a	Amino acids
AKAP8	A-kinase anchor protein 8
APOE	Apolipoprotein E
BioGRID	Biological General Repository for Interaction Datasets
CDKL5	Cyclin-dependent kinase-like 5
CIPRES	Cyberinfrastructure for Phylogenetic Research
CK1	Casein kinase 1
DOI	Digital object identifier
FBD	Forkhead box domain
FOXG1	Forkhead box protein G1
GBD	Groucho-binding domain
GluD1	Glutamate dehydrogenase 1
GO	Gene Ontology
HDAC1	Histone deacetylase 1
HIPK2	Homeodomain-interacting protein kinase 2
IDDs	Intrinsically disordered domains
IDPs	Intrinsically disordered proteins
IDRs	Intrinsically disordered regions
iPSC	Induced pluripotent stem cell
iTOL	Interactive Tree of Life
IUPred	Prediction of Intrinsically Unstructured Proteins
JBD	JARID1B-binding domain
JARID1B	Histone Demethylase Jumonji AT-rich Interactive Domain
JTT	The Jones, Taylor, and Thornton
KDM1A	Lysine-specific histone demethylase 1A
KEGG	Kyoto Encyclopedia of Genes and Genomes
MAFFT	Modified Multiple Alignment Fast Fourier Transform
MBD	Methyl-CpG-binding domain
MeCP2	Methyl-CpG-binding protein 2
NCOR	Nuclear receptor corepressor
NCoR/SMRT	Nuclear receptor co-repressor/silencing mediator of retinoic acid and thyroid hormone receptor
NES	Nuclear export signal
NLS	Nuclear localization signal
NID	NCoR/SMRT interaction domain
OMIM	Online Mendelian Inheritance in Man
PTM	Post-translational modification
RAxML-HPC2	Randomized Axelerated Maximum Likelihood for High-Performance Computing 2
RTT	Rett syndrome
RettBASE	Rett syndrome Variation Database
SALL1	Spalt-like transcription factor 1
SATB2	Special AT-rich sequence-binding protein 2
SIN3A	SIN3 transcription regulator family member A
SLiMs	Short linear motifs
SMARCA5	SWI/SNF-related matrix-associated actin-dependent regulator of chromatin subfamily A member 5
SOX2	SRY-box transcription factor 2

SSDB	Sequence Similarity DataBase
TRD	Transcriptional repression domain
TPM	Transcripts per million
ZNF483	zinc finger protein (ZNF)483

References

1. Rett, A. On a unusual brain atrophy syndrome in hyperammonemia in childhood. *Wien. Med. Wochenschr.* **1966**, *116*, 723–726.
2. Hanefeld, F. The clinical pattern of the Rett syndrome. *Brain Dev.* **1985**, *7*, 320–325. [CrossRef]
3. Laurvick, C.L.; De Klerk, N.; Bower, C.; Christodoulou, J.; Ravine, D.; Ellaway, C.; Williamson, S.; Leonard, H. Rett syndrome in Australia: A review of the epidemiology. *J. Pediatr.* **2006**, *148*, 347–352. [CrossRef]
4. Ariani, F.; Hayek, G.; Rondinella, D.; Artuso, R.; Mencarelli, M.A.; Spanhol-Rosseto, A.; Pollazzon, M.; Buoni, S.; Spiga, O.; Ricciardi, S.; et al. FOXG1 is responsible for the congenital variant of Rett syndrome. *Am. J. Hum. Genet.* **2008**, *83*, 89–93. [CrossRef] [PubMed]
5. Weaving, L.S.; Christodoulou, J.; Williamson, S.L.; Friend, K.L.; McKenzie, O.L.; Archer, H.; Evans, J.; Clarke, A.; Pelka, G.J.; Tam, P.P.; et al. Mutations of CDKL5 cause a severe neurodevelopmental disorder with infantile spasms and mental retardation. *Am. J. Hum. Genet.* **2004**, *75*, 1079–1093. [CrossRef] [PubMed]
6. Trappe, R.; Laccone, F.; Cobilanschi, J.; Meins, M.; Huppke, P.; Hanefeld, F.; Engel, W. MECP2. mutations in sporadic cases of Rett syndrome are almost exclusively of paternal origin. *Am. J. Hum. Genet.* **2001**, *68*, 1093–1101. [CrossRef] [PubMed]
7. Van Esch, H.; Bauters, M.; Ignatius, J.; Jansen, M.; Raynaud, M.; Hollanders, K.; Lugtenberg, D.; Bienvenu, T.; Jensen, L.R.; Gecz, J.; et al. Duplication of the MECP2 region is a frequent cause of severe mental retardation and progressive neurological symptoms in males. *Am. J. Hum. Genet.* **2005**, *77*, 442–453. [CrossRef]
8. Clayton-Smith, J.; Watson, P.; Ramsden, S.; Black, G.C.M. Somatic mutation in MECP2 as a non-fatal neurodevelopmental disorder in males. *Lancet* **2000**, *356*, 830–832. [CrossRef]
9. Ben Zeev, B.; Yaron, Y.; Schanen, N.C.; Wolf, H.; Brandt, N.; Ginot, N.; Shomrat, R.; Orr-Urtreger, A. Rett syndrome: Clinical manifestations in males with MECP2 mutations. *J. Child. Neurol.* **2002**, *17*, 20–24. [CrossRef]
10. Neul, J.L. The relationship of Rett syndrome and MECP2 disorders to autism. *Dialogues Clin. Neurosci.* **2012**, *14*, 253–262.
11. Fehr, S.; Wilson, M.; Downs, J.; Williams, S.; Murgia, A.; Sartori, S.; Vecchi, M.; Ho, G.; Polli, R.; Psoni, S.; et al. The CDKL5 disorder is an independent clinical entity associated with early-onset encephalopathy. *Eur. J. Hum. Genet.* **2013**, *21*, 266–273. [CrossRef] [PubMed]
12. Hector, R.D.; Kalscheuer, V.M.; Hennig, F.; Leonard, H.; Downs, J.; Clarke, A.; Benke, T.A.; Armstrong, J.; Pineda, M.; Bailey, M.E.S.; et al. CDKL5 variants: Improving our understanding of a rare neurologic disorder. *Neurol. Genet.* **2017**, *3*, e200. [CrossRef] [PubMed]
13. Kortum, F.; Das, S.; Flindt, M.; Morris-Rosendahl, D.J.; Stefanova, I.; Goldstein, A.; Horn, D.; Klopocki, E.; Kluger, G.; Martin, P.; et al. The core FOXG1 syndrome phenotype consists of postnatal microcephaly, severe mental retardation, absent language, dyskinesia, and corpus callosum hypogenesis. *J. Med. Genet.* **2011**, *48*, 396–406. [CrossRef] [PubMed]
14. Lyst, M.J.; Ekiert, R.; Ebert, D.H.; Merusi, C.; Nowak, J.; Selfridge, J.; Guy, J.; Kastan, N.R.; Robinson, N.D.; de Lima Alves, F.; et al. Rett syndrome mutations abolish the interaction of MeCP2 with the NCoR/SMRT co-repressor. *Nat. Neurosci.* **2013**, *16*, 898–902. [CrossRef] [PubMed]
15. Lyst, M.J.; Bird, A. Rett syndrome: A complex disorder with simple roots. *Nat. Rev. Genet.* **2015**, *16*, 261–275. [CrossRef]
16. Tillotson, R.; Selfridge, J.; Koerner, M.V.; Gadalla, K.K.E.; Guy, J.; De Sousa, D.; Hector, R.D.; Cobb, S.R.; Bird, A. Radically truncated MeCP2 rescues Rett syndrome-like neurological defects. *Nature* **2017**, *550*, 398–401. [CrossRef]
17. Ghosh, R.P.; Nikitina, T.; Horowitz-Scherer, R.A.; Gierasch, L.M.; Uversky, V.N.; Hite, K.; Hansen, J.C.; Woodcock, C.L. Unique physical properties and interactions of the domains of methylated DNA binding protein 2. *Biochemistry* **2010**, *49*, 4395–4410. [CrossRef]
18. Toth-Petroczy, A.; Palmedo, P.; Ingraham, J.; Hopf, T.A.; Berger, B.; Sander, C.; Marks, D.S. Structured States of Disordered Proteins from Genomic Sequences. *Cell* **2016**, *167*, 158–170. [CrossRef]

19. Canning, P.; Park, K.; Goncalves, J.; Li, C.; Howard, C.J.; Sharpe, T.D.; Holt, L.J.; Pelletier, L.; Bullock, A.N.; Leroux, M.R. CDKL Family Kinases Have Evolved Distinct Structural Features and Ciliary Function. *Cell Rep.* **2018**, *22*, 885–894. [CrossRef]

20. Dunker, A.K.; Lawson, J.D.; Brown, C.J.; Williams, R.M.; Romero, P.; Oh, J.S.; Oldfield, C.J.; Campen, A.M.; Ratliff, C.M.; Hipps, K.W.; et al. Intrinsically disordered protein. *J. Mol. Graph. Model.* **2001**, *19*, 26–59. [CrossRef]

21. Uversky, V.N. Protein folding revisited. A polypeptide chain at the folding-misfolding-nonfolding cross-roads: Which way to go? *Cell Mol. Life Sci.* **2003**, *60*, 1852–1871. [CrossRef] [PubMed]

22. Dyson, H.J.; Wright, P.E. Equilibrium NMR studies of unfolded and partially folded proteins. *Nat. Struct. Biol.* **1998**, *5*, 499–503. [CrossRef] [PubMed]

23. Dunker, A.K.; Babu, M.M.; Barbar, E.; Blackledge, M.; Bondos, S.E.; Dosztanyi, Z.; Dyson, H.J.; Forman-Kay, J.; Fuxreiter, M.; Gsponer, J.; et al. What's in a name? Why these proteins are intrinsically disordered: Why these proteins are intrinsically disordered. *Intrinsically Disord. Proteins* **2013**, *1*, e24157. [CrossRef] [PubMed]

24. Tompa, P. Intrinsically unstructured proteins. *Trends Biochem. Sci.* **2002**, *27*, 527–533. [CrossRef]

25. Tompa, P. The interplay between structure and function in intrinsically unstructured proteins. *FEBS Lett.* **2005**, *579*, 3346–3354. [CrossRef] [PubMed]

26. Van der Lee, R.; Buljan, M.; Lang, B.; Weatheritt, R.J.; Daughdrill, G.W.; Dunker, A.K.; Fuxreiter, M.; Gough, J.; Gsponer, J.; Jones, D.T.; et al. Classification of intrinsically disordered regions and proteins. *Chem. Rev.* **2014**, *114*, 6589–6631. [CrossRef]

27. Uversky, V.N.; Oldfield, C.J.; Dunker, A.K. Intrinsically disordered proteins in human diseases: Introducing the D2 concept. *Annu. Rev. Biophys.* **2008**, *37*, 215–246. [CrossRef]

28. Diella, F.; Haslam, N.; Chica, C.; Budd, A.; Michael, S.; Brown, N.P.; Trave, G.; Gibson, T.J. Understanding eukaryotic linear motifs and their role in cell signaling and regulation. *Front. Biosci.* **2008**, *13*, 6580–6603. [CrossRef]

29. Galea, C.A.; Wang, Y.; Sivakolundu, S.G.; Kriwacki, R.W. Regulation of cell division by intrinsically unstructured proteins: Intrinsic flexibility, modularity, and signaling conduits. *Biochemistry* **2008**, *47*, 7598–7609. [CrossRef]

30. Mellios, N.; Woodson, J.; Garcia, R.I.; Crawford, B.; Sharma, J.; Sheridan, S.D.; Haggarty, S.J.; Sur, M. beta2-Adrenergic receptor agonist ameliorates phenotypes and corrects microRNA-mediated IGF1 deficits in a mouse model of Rett syndrome. *Proc. Natl. Acad. Sci. USA* **2014**, *111*, 9947–9952. [CrossRef]

31. Tropea, D.; Giacometti, E.; Wilson, N.R.; Beard, C.; McCurry, C.; Fu, D.D.; Flannery, R.; Jaenisch, R.; Sur, M. Partial reversal of Rett Syndrome-like symptoms in MeCP2 mutant mice. *Proc. Natl. Acad. Sci. USA* **2009**, *106*, 2029–2034. [CrossRef] [PubMed]

32. Carrette, L.L.G.; Wang, C.Y.; Wei, C.; Press, W.; Ma, W.; Kelleher, R.J., 3rd; Lee, J.T. A mixed modality approach towards Xi reactivation for Rett syndrome and other X-linked disorders. *Proc. Natl. Acad. Sci. USA* **2018**, *115*, E668–E675. [CrossRef] [PubMed]

33. Shah, R.R.; Bird, A.P. MeCP2 mutations: Progress towards understanding and treating Rett syndrome. *Genome Med.* **2017**, *9*, 17. [CrossRef] [PubMed]

34. Mauri, F.; McNamee, L.M.; Lunardi, A.; Chiacchiera, F.; Del Sal, G.; Brodsky, M.H.; Collavin, L. Modification of Drosophila p53 by SUMO modulates its transactivation and pro-apoptotic functions. *J. Biol. Chem.* **2008**, *283*, 20848–20856. [CrossRef]

35. Dyson, H.J.; Wright, P.E. Intrinsically unstructured proteins and their functions. *Nat. Rev. Mol. Cell. Biol.* **2005**, *6*, 197–208. [CrossRef]

36. Fahmi, M.; Ito, M. Evolutionary Approach of Intrinsically Disordered CIP/KIP Proteins. *Sci. Rep.* **2019**, *9*, 1575. [CrossRef]

37. Gsponer, J.; Babu, M.M. The rules of disorder or why disorder rules. *Prog. Biophys. Mol. Biol.* **2009**, *99*, 94–103. [CrossRef]

38. Van Roey, K.; Uyar, B.; Weatheritt, R.J.; Dinkel, H.; Seiler, M.; Budd, A.; Gibson, T.J.; Davey, N.E. Short linear motifs: Ubiquitous and functionally diverse protein interaction modules directing cell regulation. *Chem. Rev.* **2014**, *114*, 6733–6778. [CrossRef]

39. Ahrens, J.; Rahaman, J.; Siltberg-Liberles, J. Large-Scale Analyses of Site-Specific Evolutionary Rates across Eukaryote Proteomes Reveal Confounding Interactions between Intrinsic Disorder, Secondary Structure, and Functional Domains. *Genes* **2018**, *11*, 553. [CrossRef]

40. Wakefield, R.I.; Smith, B.O.; Nan, X.; Free, A.; Soteriou, A.; Uhrin, D.; Bird, A.P.; Barlow, P.N. The solution structure of the domain from MeCP2 that binds to methylated DNA. *J. Mol. Biol.* **1999**, *291*, 1055–1065. [CrossRef]

41. Chen, J.W.; Romero, P.; Uversky, V.N.; Dunker, A.K. Conservation of intrinsic disorder in protein domains and families: II. functions of conserved disorder. *J. Proteome Res.* **2006**, *5*, 888–898. [CrossRef] [PubMed]

42. Grimmler, M.; Wang, Y.; Mund, T.; Cilensek, Z.; Keidel, E.M.; Waddell, M.B.; Jakel, H.; Kullmann, M.; Kriwacki, R.W.; Hengst, L. Cdk-inhibitory activity and stability of p27Kip1 are directly regulated by oncogenic tyrosine kinases. *Cell* **2007**, *128*, 269–280. [CrossRef] [PubMed]

43. Gsponer, J.; Futschik, M.E.; Teichmann, S.A.; Babu, M.M. Tight regulation of unstructured proteins: From transcript synthesis to protein degradation. *Science* **2008**, *322*, 1365–1368. [CrossRef] [PubMed]

44. UniProt, C. The Universal Protein Resource (UniProt) in 2010. *Nucleic Acids Res.* **2010**, *38*, D142–D148. [CrossRef]

45. Chatr-Aryamontri, A.; Oughtred, R.; Boucher, L.; Rust, J.; Chang, C.; Kolas, N.K.; O'Donnell, L.; Oster, S.; Theesfeld, C.; Sellam, A.; et al. The BioGRID interaction database: 2017 update. *Nucleic Acids Res.* **2017**, *45*, D369–D379. [CrossRef]

46. Uhlen, M.; Fagerberg, L.; Hallstrom, B.M.; Lindskog, C.; Oksvold, P.; Mardinoglu, A.; Sivertsson, A.; Kampf, C.; Sjostedt, E.; Asplund, A.; et al. Proteomics. Tissue-based map of the human proteome. *Science* **2015**, *347*, 1260419. [CrossRef]

47. Guy, J.; Alexander-Howden, B.; FitzPatrick, L.; DeSousa, D.; Koerner, M.V.; Selfridge, J.; Bird, A. A mutation-led search for novel functional domains in MeCP2. *Hum. Mol. Genet.* **2018**, *27*, 2531–2545. [CrossRef]

48. Ballestar, E.; Yusufzai, T.M.; Wolffe, A.P. Effects of Rett syndrome mutations of the methyl-CpG binding domain of the transcriptional repressor MeCP2 on selectivity for association with methylated DNA. *Biochemistry* **2000**, *39*, 7100–7106. [CrossRef]

49. Yusufzai, T.M.; Wolffe, A.P. Functional consequences of Rett syndrome mutations on human MeCP2. *Nucleic Acids Res.* **2000**, *28*, 4172–4179. [CrossRef]

50. Mnatzakanian, G.N.; Lohi, H.; Munteanu, I.; Alfred, S.E.; Yamada, T.; MacLeod, P.J.; Jones, J.R.; Scherer, S.W.; Schanen, N.C.; Friez, M.J.; et al. A previously unidentified MECP2 open reading frame defines a new protein isoform relevant to Rett syndrome. *Nat. Genet.* **2004**, *36*, 339. [CrossRef]

51. Williamson, S.L.; Giudici, L.; Kilstrup-Nielsen, C.; Gold, W.; Pelka, G.J.; Tam, P.P.; Grimm, A.; Prodi, D.; Landsberger, N.; Christodoulou, J. A novel transcript of cyclin-dependent kinase-like 5 (CDKL5) has an alternative C-terminus and is the predominant transcript in brain. *Hum. Genet.* **2012**, *131*, 187–200. [CrossRef] [PubMed]

52. Wong, L.C.; Singh, S.; Wang, H.P.; Hsu, C.J.; Hu, S.C.; Lee, W.T. FOXG1-Related Syndrome: From Clinical to Molecular Genetics and Pathogenic Mechanisms. *Int. J. Mol. Sci.* **2019**, *20*, 4176. [CrossRef] [PubMed]

53. Payankaulam, S.; Li, L.M.; Arnosti, D.N. Transcriptional repression: Conserved and evolved features. *Curr. Biol.* **2010**, *17*, R764–R771. [CrossRef] [PubMed]

54. Toresson, H.; Martinez-Barbera, J.P.; Bardsley, A.; Caubit, X.; Krauss, S. Conservation of BF-1 expression in amphioxus and zebrafish suggests evolutionary ancestry of anterior cell types that contribute to the vertebrate telencephalon. *Dev. Genes Evol.* **1998**, *208*, 431–439. [CrossRef]

55. Miyoshi, G.; Fishell, G. Dynamic FoxG1 expression coordinates the integration of multipolar pyramidal neuron precursors into the cortical plate. *Neuron* **2012**, *74*, 1045–1058. [CrossRef]

56. Kumamoto, T.; Toma, K.; Gunadi; McKenna, W.L.; Kasukawa, T.; Katzman, S.; Chen, B.; Hanashima, C. Foxg1 coordinates the switch from nonradially to radially migrating glutamatergic subtypes in the neocortex through spatiotemporal repression. *Cell Rep.* **2013**, *3*, 931–945. [CrossRef]

57. Docker, D.; Schubach, M.; Menzel, M.; Munz, M.; Spaich, C.; Biskup, S.; Bartholdi, D. Further delineation of the SATB2 phenotype. *Eur. J. Hum. Genet.* **2014**, *22*, 1034–1039. [CrossRef]

58. Lee, J.S.; Yoo, Y.; Lim, B.C.; Kim, K.J.; Choi, M.; Chae, J.H. SATB2-associated syndrome presenting with Rett-like phenotypes. *Clin. Genet.* **2016**, *89*, 728–732. [CrossRef]

59. Livide, G.; Patriarchi, T.; Amenduni, M.; Amabile, S.; Yasui, D.; Calcagno, E.; Lo Rizzo, C.; De Falco, G.; Ulivieri, C.; Ariani, F.; et al. GluD1 is a common altered player in neuronal differentiation from both MECP2-mutated and CDKL5-mutated iPS cells. *Eur. J. Hum. Genet.* **2015**, *23*, 195–201. [CrossRef]

60. Patriarchi, T.; Amabile, S.; Frullanti, E.; Landucci, E.; Rizzo, C.L.; Ariani, F.; Costa, M.; Olimpico, F.; Hell, J.W.; Vaccarino, F.M.; et al. Imbalance of excitatory/inhibitory synaptic protein expression in iPSC-derived neurons from FOXG1(+/−) patients and in foxg1(+/−) mice. *Eur. J. Hum. Genet.* **2016**, *24*, 871–880. [CrossRef]

61. Bracaglia, G.; Conca, B.; Bergo, A.; Rusconi, L.; Zhou, Z.; Greenberg, M.E.; Landsberger, N.; Soddu, S.; Kilstrup-Nielsen, C. Methyl-CpG-binding protein 2 is phosphorylated by homeodomain-interacting protein kinase 2 and contributes to apoptosis. *EMBO Rep.* **2009**, *10*, 1327–1333. [CrossRef] [PubMed]

62. Fuchs, C.; Trazzi, S.; Torricella, R.; Viggiano, R.; De Franceschi, M.; Amendola, E.; Gross, C.; Calza, L.; Bartesaghi, R.; Ciani, E. Loss of CDKL5 impairs survival and dendritic growth of newborn neurons by altering AKT/GSK-3beta signaling. *Neurobiol. Dis.* **2014**, *70*, 53–68. [CrossRef] [PubMed]

63. Mari, F.; Azimonti, S.; Bertani, I.; Bolognese, F.; Colombo, E.; Caselli, R.; Scala, E.; Longo, I.; Grosso, S.; Pescucci, C.; et al. CDKL5 belongs to the same molecular pathway of MeCP2 and it is responsible for the early-onset seizure variant of Rett syndrome. *Hum. Mol. Genet.* **2005**, *14*, 1935–1946. [CrossRef] [PubMed]

64. Carouge, D.; Host, L.; Aunis, D.; Zwiller, J.; Anglard, P. CDKL5 is a brain MeCP2 target gene regulated by DNA methylation. *Neurobiol. Dis.* **2010**, *38*, 414–424. [CrossRef] [PubMed]

65. Rusconi, L.; Salvatoni, L.; Giudici, L.; Bertani, I.; Kilstrup-Nielsen, C.; Broccoli, V.; Landsberger, N. CDKL5 expression is modulated during neuronal development and its subcellular distribution is tightly regulated by the C-terminal tail. *J. Biol. Chem.* **2008**, *283*, 30101–30111. [CrossRef] [PubMed]

66. Brown, C.J.; Johnson, A.K.; Dunker, A.K.; Daughdrill, G.W. Evolution and disorder. *Curr. Opin. Struct. Biol.* **2011**, *21*, 441–446. [CrossRef]

67. Regad, T.; Roth, M.; Bredenkamp, N.; Illing, N.; Papalopulu, N. The neural progenitor-specifying activity of FoxG1 is antagonistically regulated by CKI and FGF. *Nat. Cell Biol.* **2007**, *9*, 531–540. [CrossRef]

68. Zhang, Q.; Yang, X.; Wang, J.; Li, J.; Wu, Q.; Wen, Y.; Zhao, Y.; Zhang, X.; Yao, H.; Wu, X.; et al. Genomic mosaicism in the pathogenesis and inheritance of a Rett syndrome cohort. *Genet. Med.* **2019**, *21*, 1330–1338. [CrossRef]

69. Sato, Y.; Nakaya, A.; Shiraishi, K.; Kawashima, S.; Goto, S.; Kanehisa, M. Ssdb: Sequence similarity database in kegg. *Genome Inf.* **2001**, *12*, 230–231. [CrossRef]

70. Katoh, K.; Standley, D.M. MAFFT multiple sequence alignment software version 7: Improvements in performance and usability. *Mol. Biol. Evol.* **2013**, *30*, 772–780. [CrossRef]

71. Stamatakis, A. RAxML-VI-HPC: Maximum likelihood-based phylogenetic analyses with thousands of taxa and mixed models. *Bioinformatics* **2006**, *22*, 2688–2690. [CrossRef] [PubMed]

72. Darriba, D.; Posada, D.; Kozlov, A.M.; Stamatakis, A.; Morel, B.; Flouri, T. ModelTest-NG: A new and scalable tool for the selection of DNA and protein evolutionary models. *Mol. Biol. Evol.* **2019**. [CrossRef] [PubMed]

73. Federhen, S. The NCBI Taxonomy database. *Nucleic Acids Res.* **2012**, *40*, D136–D143. [CrossRef] [PubMed]

74. Miller, M.A.; Pfeiffer, W.; Schwartz, T. Creating the CIPRES Science Gateway for inference of large phylogenetic trees. In Proceedings of the 2010 Gateway Computing Environments Workshop (GCE), New Orleans, LA, USA, 14 November 2010; pp. 1–8.

75. Meszaros, B.; Erdos, G.; Dosztanyi, Z. IUPred2A: Context-dependent prediction of protein disorder as a function of redox state and protein binding. *Nucleic Acids Res.* **2018**, *46*, W329–W337. [CrossRef]

76. Letunic, I.; Bork, P. Interactive Tree of Life (iTOL): An online tool for phylogenetic tree display and annotation. *Bioinformatics* **2007**, *23*, 127–128. [CrossRef]

77. Pupko, T.; Bell, R.E.; Mayrose, I.; Glaser, F.; Ben-Tal, N. Rate4Site: An algorithmic tool for the identification of functional regions in proteins by surface mapping of evolutionary determinants within their homologues. *Bioinformatics* **2002**, *18* (Suppl. 1), S71–S77. [CrossRef]

78. Blom, N.; Gammeltoft, S.; Brunak, S. Sequence and structure-based prediction of eukaryotic protein phosphorylation sites. *J. Mol. Biol.* **1999**, *294*, 1351–1362. [CrossRef]

79. Krishnaraj, R.; Ho, G.; Christodoulou, J. RettBASE: Rett syndrome database update. *Hum. Mutat.* **2017**, *38*, 922–931. [CrossRef]

80. Lek, M.; Karczewski, K.J.; Minikel, E.V.; Samocha, K.E.; Banks, E.; Fennell, T.; O'Donnell-Luria, A.H.; Ware, J.S.; Hill, A.J.; Cummings, B.B.; et al. Analysis of protein-coding genetic variation in 60,706 humans. *Nature* **2016**, *536*, 285–291. [CrossRef]

81. Kanehisa, M.; Furumichi, M.; Tanabe, M.; Sato, Y.; Morishima, K. KEGG: New perspectives on genomes, pathways, diseases and drugs. *Nucleic Acids Res.* **2017**, *45*, D353–D361. [CrossRef]

82. Pellegrini, M.; Marcotte, E.M.; Thompson, M.J.; Eisenberg, D.; Yeates, T.O. Assigning protein functions by

comparative genome analysis: Protein phylogenetic profiles. *Proc. Natl. Acad. Sci. USA* **1999**, *96*, 4285–4288. [CrossRef] [PubMed]

83.   Nakaya, A.; Katayama, T.; Itoh, M.; Hiranuka, K.; Kawashima, S.; Moriya, Y.; Okuda, S.; Tanaka, M.; Tokimatsu, T.; Yamanishi, Y.; et al. KEGG OC: A large-scale automatic construction of taxonomy-based ortholog clusters. *Nucleic Acids Res.* **2013**, *41*, D353–D357. [CrossRef] [PubMed]

84.   Ward, J.H. Hierarchical Grouping to Optimize an Objective Function. *J. Am. Stat. Assoc.* **1963**, *58*, 236–244. [CrossRef]

85.   Mi, H.; Muruganujan, A.; Ebert, D.; Huang, X.; Thomas, P.D. PANTHER version 14: More genomes, a new PANTHER GO-slim and improvements in enrichment analysis tools. *Nucleic Acids Res.* **2019**, *47*, D419–D426. [CrossRef] [PubMed]

7

Modulation of Disordered Proteins with a Focus on Neurodegenerative Diseases and Other Pathologies

Anne H. S. Martinelli [1,†], Fernanda C. Lopes [2,3,†], Elisa B. O. John [2,3], Célia R. Carlini [3,4,5,*] and Rodrigo Ligabue-Braun [6,*]

[1]   Department of Molecular Biology and Biotechnology & Department of Biophysics, Biosciences Institute-IB, (UFRGS), Porto Alegre CEP 91501-970, RS, Brazil; ahsmartinelli@gmail.com
[2]   Center for Biotechnology, Universidade Federal do Rio Grande do Sul (UFRGS), Porto Alegre CEP 91501-970, RS, Brazil; fernandacortezlopes@gmail.com (F.C.L.); elisabeajohn@gmail.com (E.B.O.J.)
[3]   Graduate Program in Cell and Molecular Biology, Universidade Federal do Rio Grande do Sul (UFRGS), Porto Alegre CEP 91501-970, RS, Brazil
[4]   Graduate Program in Medicine and Health Sciences, Pontifícia Universidade Católica do Rio Grande do Sul (PUCRS), Porto Alegre CEP 91410-000, RS, Brazil
[5]   Brain Institute-InsCer, Laboratory of Neurotoxins, Pontifícia Universidade Católica do Rio Grande do Sul (PUCRS), Porto Alegre CEP 90610-000, RS, Brazil
[6]   Department of Pharmaceutical Sciences, Universidade Federal de Ciências da Saúde de Porto Alegre (UFCSPA), Porto Alegre CEP 90050-170, RS, Brazil
*   Correspondence: celia.carlini@pucrs.br or celia.carlini@pq.cnpq.br (C.R.C.); rodrigolb@ufcspa.edu.br (R.L.-B.)
†   These authors contributed equally to this work.

Abstract: Intrinsically disordered proteins (IDPs) do not have rigid 3D structures, showing changes in their folding depending on the environment or ligands. Intrinsically disordered proteins are widely spread in eukaryotic genomes, and these proteins participate in many cell regulatory metabolism processes. Some IDPs, when aberrantly folded, can be the cause of some diseases such as Alzheimer's, Parkinson's, and prionic, among others. In these diseases, there are modifications in parts of the protein or in its entirety. A common conformational variation of these IDPs is misfolding and aggregation, forming, for instance, neurotoxic amyloid plaques. In this review, we discuss some IDPs that are involved in neurodegenerative diseases (such as beta amyloid, alpha synuclein, tau, and the "IDP-like" PrP), cancer (p53, c-Myc), and diabetes (amylin), focusing on the structural changes of these IDPs that are linked to such pathologies. We also present the IDP modulation mechanisms that can be explored in new strategies for drug design. Lastly, we show some candidate drugs that can be used in the future for the treatment of diseases caused by misfolded IDPs, considering that cancer therapy has more advanced research in comparison to other diseases, while also discussing recent and future developments in this area of research. Therefore, we aim to provide support to the study of IDPs and their modulation mechanisms as promising approaches to combat such severe diseases.

Keywords: intrinsically disordered proteins (IDPs); neurodegenerative diseases; aggregation; drugs; drug discovery

1. Introduction

The protein structure–function paradigm was established in the 20th century. The key point of this paradigm is that an ordered (rigid) and unique 3D structure of a protein is an obligatory prerequisite for protein function [1,2]. Nevertheless, recent studies have provided broad and convincing evidences that some proteins do not adopt only one structure, but still are fully functional [3].

The different possible protein conformations are structured (folded), molten globular, pre-molten globular, and unstructured (unfolded) [4].

Since the beginning of the 2000s, a new class of unstructured proteins started to be studied more due to the improvement of techniques to elucidate protein structure. Crystal-structure analysis using X-ray diffraction cannot provide information on unstructured states, with only the absence of electron density in some regions being observed. However, the nuclear magnetic resonance (NMR) technique allowed for the better characterization of these disordered proteins, confirming the flexibility of protein segments that are missing in crystallography experiments [5]. They are defined mainly as intrinsically disordered proteins (IDPs), in spite of some authors defining these proteins as natively denatured [6], natively unfolded [7], intrinsically unstructured [8], and natively disordered proteins [9], among other definitions. We will use the IDP definition to refer to these proteins. The disorder could be also present in some regions of proteins; these regions are named intrinsically disordered regions (IDRs). Intrinsically disordered proteins /IDRs have no single, well-defined equilibrium structure and exist as heterogeneous ensembles of conformers [10].

There are significant differences between the amino acid sequences of IDPs/IDRs in comparison with structured globular proteins and/or domains. These differences are related to amino acid composition, sequence complexity, hydrophobicity, aromaticity, charge, flexibility, type and rate of amino acid substitutions over evolutionary time [11]. Some features of IDPs are the low content of hydrophobic residues and the high load of charged residues [12]. Intrinsically disordered proteins/IDPRs present large hydrodynamic volumes, low content of ordered secondary structure, and high structural heterogeneity. These proteins are very flexible. However, some of them show transitions from the disordered to the ordered state in the presence of natural ligands [10]. The ability of IDPs to return to the highly flexible conformations after performing their biological function, and their predisposition to acquire different conformations according to the environment, are unique properties of IDPs [13] (Figure 1).

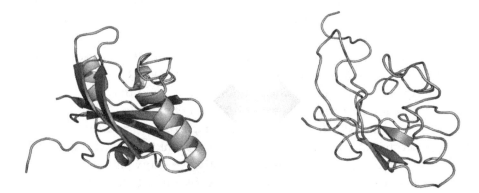

Figure 1. Example of intrinsically disordered proteins (IDP) conformational plasticity. Shown are ordered and disordered extremes in the conformational ensemble described for the photoactive yellow protein from *Halorhodospira halophile* (PDB IDs 3PHY and 2KX6).

It was demonstrated that these IDPs are highly prevalent in many genomes, including humans', and are important in several cellular processes, such as regulation of transcription and translation, cell cycle control, and signaling [3]. It is important to highlight that they are much more common in eukaryotes, in comparison to Eubacteria and Archaea, reflecting the greater importance of disorder-associated signaling and regulation for eukaryotic cells [13]. Intrinsically disordered proteins are present in major disease pathways, such as cancer, amyloidosis, diabetes, cardiovascular, and neurodegenerative diseases. Changes in the environment and/or mutation(s) of IDPs would be expected to affect their normal function, leading to misidentification and missignaling. Consequently, it can result in misfolding and aggregation, which are known to be associated with the pathogenesis of numerous diseases. Some IDPs, such as α-synuclein, tau protein, p53, and BRCA1 are important in neurodegenerative diseases and cancer, being attractive targets for drugs modulating protein–protein interactions. Based on these IDPs and other examples, novel strategies for drug discovery have been developed [11,13]. The ability to modulate the interactions of these proteins offers tremendous opportunities of investigation in chemical biology and molecular therapeutics. Several recent small molecules, such as potential drugs, have been shown to act by blocking protein–protein interactions based on intrinsic disorder of one of the partners [14].

In this review, we will focus on IDPs involved in some neurodegenerative diseases, such as α-synuclein, amyloid β-peptide, and tau protein, while also commenting on cancer associated IDPs, such as p53 and c-Myc, and diabetes-related amylin. In addition, we will summarize the strategies to modulate IDPs action in some diseases and the promising drugs in this field, which are currently more developed for non-neurodegenerative disorders, prompting the need of focusing strategies on IDP-centered drug development for them.

2. Intrinsically Disordered Proteins in Some Diseases

Inside the cell, protein folding is promoted by chaperone machinery that allows the protein to adopt a folded, biologically active form [15]. However, IDPs remain partially or totally unfolded and could cause many neurodegenerative disorders due to some changes in their folding [10]. Neurodegenerative diseases are disorders characterized by progressive loss of neurons associated with deposition of proteins showing altered physicochemical properties in the brain and in peripheral organs. These proteins show misbehavior and disarrangement, affecting negatively their processing, functioning, and/or folding [16,17]. In some of these disorders, there is a conversion of the functional state of specific proteins into an aggregate state that can accumulate as fibrils, causing loss of native function, and consequent gain of a toxic function. The toxicity of these fibrils is caused by disrupting intracellular transport, overwhelming protein degradation pathways, and/or disturbing vital cell functions [16,18]. Misfolding and aggregation of IDPs/IDPRs are especially common in neurodegeneration [16,19,20].

If these misfolded proteins accumulate as deposits of aggregates, they can originate many neurodegenerative diseases such as Alzheimer's, Parkinson's, Huntington's, and prionic diseases, among others [21]. Proteins that accumulate as amyloid fibrils are called amyloidogenic proteins. In order to facilitate the understanding, they can be divided in two groups: 1) proteins that present a well-defined structure with only part of the molecule being disordered, as in the case of prion protein; 2) IDPs like amyloid-β (Aβ), tau and α-synuclein, that show changes in the entire protein [22]. In addition to neurodegenerative diseases, IDPs are also involved in diabetes and different types of cancer. Here, we briefly summarize and cover the general characteristics of some IDPs that can accumulate as fibril aggregates rich in β-structure, and their association with some neurodegenerative diseases, as well as features of cancer- and diabetes-related IDPs.

2.1. α-Synuclein and Parkinson's Disease

Synucleinopathies refer to a group of neurodegenerative diseases, namely Parkinson's disease (PD), dementia with Lewy bodies, and multiple system atrophy, characterized histologically by the presence of inclusions (Lewy bodies and Lewy neurites) composed of aggregated α-synuclein in the central nervous system (CNS) [23,24]. Aggregates containing α-synuclein can be found also in microglia and astrocytes, and in neurons of the peripheral nervous system associated with rarer autonomic diseases. This protein, encoded by the *SNCA* gene located on chromosome 4 region q21, is predominantly expressed in the brain, where it concentrates in nerve terminals. Three isoforms of synuclein, α, β, and γ, are known, but only the α isoform is found in Lewy bodies and neurites. α-synuclein is a single chain with 140 amino acids, and displays 61% identity compared to β-synuclein (134 amino acids), and its sequence contains seven imperfect repeats of eleven amino acids, each with a -KTKEGV- conserved core, separated by nine amino acid residues [25]. Weinreb and co-workers [7] reported, in 1996, the intrinsically disordered nature of α-synuclein in solution, and the protein was found to maintain its disordered state in physiological cell conditions [26]. Upon reversible binding to negatively charged phospholipids, α-synuclein oligomerizes and undergoes structural changes to assume a highly dynamic α-helical conformation while still maintaining partially disordered stretches [27,28].

The physiological role of α-synuclein is still elusive. Mice lacking all three synucleins developed only mild neurodegenerative pathology [29,30]. Lipid-bound α-synuclein accumulates in the plasma membrane of synaptic terminals and synaptic vesicles suggesting a role in neurotransmitter release [31]. The protein has been shown to possess some chaperone activity, interacting with components of the SNARE complex [32] and promoting dilatation of the exocytotic fusion pore [33]. In synucleinopathies, misfolding of lipid-bound α-synuclein occurs leading to β-sheet rich amyloid fibrils, found as the main component of Lewy bodies and Lewy neurites [20,34]. The core of fibrillated protein comprises about 70 amino acids of its repeat region, organized in parallel, in-register β-sheets in a Greek key topology [35].

In contrast to its normal, physiological form, pathological aggregated α-synuclein is extensively phosphorylated at S129 and S87. Other posttranslational modifications present in pathological α-synuclein include nitration, oxidation (for which oxidized by-products of dopamine might contribute) and truncation. Not all of these modifications contribute to accelerate the fibrillation process, since nitration and oxidation decrease fibril formation and stabilize oligomers and protofibrils of α-synuclein. On the other hand, truncated α-synuclein, typically at its C-terminal, shows increased propensity towards fibrillation. In some of the familial forms of PD, point mutations in α-synuclein (E46K, A30P, A53T) alter its propensity to fibrillate [36]. However, about 90% of Parkinson's disease cases are idiopathic [23]. The identification of which α-synuclein species are indeed toxic is yet incomplete and is an intense field of debate. There is a growing perception that soluble oligomeric forms of α-synuclein are the most relevant in terms of toxicity, suggesting that Lewy inclusions might represent a protective response, and that interventions to favor the fibrillation process could be of therapeutic value.

Aggregation of α-synuclein apparently starts in the synapses and the aggregates propagate to nearby neurons through a prion-like mechanism [37,38]. Initial brain structures accumulating intracellular α-synuclein inclusions are the olfactory bulb, glossopharyngeal, and vagal nerves; then the Lewy pathology spreads to other regions of the brain reaching the amygdala and substantia nigra, where it causes death of dopaminergic neurons consequently leading to the motor symptoms characteristic of PD. In the more advanced cases, Lewy bodies and neurites are found in the neocortex, accounting for the cognitive impairment associated to the disease [23,39,40].

2.2. Amyloid β-Peptide, Tau Protein, and Alzheimer's Disease

Alzheimer's disease (AD) is one of the most prevalent neurodegenerative diseases that affects the learning and memory processes beyond the reduction of the brain area, degenerationand death of neurons [41,42]. Diagnosis of AD requires the identification of senile plaques composed by fibril β-amyloid peptides and tangles of tau protein aggregates [41,43]. Amyloid-β (Aβ) peptide is a well-known IDP with several oligomeric forms [44]. Amyloid-β aggregates are formed mainly by peptides containing 39 to 43 amino acids yielded by proteolytic cleavage of amyloid precursor protein (APP) [45–47]. Its aggregated form is significantly linked to Alzheimer's disease, and the generation of Aβ and plaque pathology is linked to the presence of mutations or transport defects related to this protein [48,49].

The amyloid precursor protein (APP) is a transmembrane glycoprotein (type I) that is suggested to be involved in the development of the neurosystem, acting as a cell adhesion molecule [50]. The gene that encodes APP is located in human chromosome 21 [51,52] and this gene yields different isoforms by alternative splicing. Nevertheless, the function of APP is still not understood [43]. The APP proteolytic processing occurs via α, β, and γ-secretase [53]. This process can happen via two pathways: the non-amyloidogenic and the amyloidogenic route (producing toxic $A\beta_{1-40/42}$) [54]. Aβ peptides occur in two major lengths, $A\beta_{1-40}$ and $A\beta_{1-42}$ amino acids, both present in senile plaques [11,46].

Some studies showed that $A\beta_{1-42}$ accumulate as an early event in neuronal dysfunction, acting as seeding in the formation of amyloid plaques [55,56].

The two alloforms, $A\beta_{1-40}$ and $A\beta_{1-42}$, have identical sequences with the exception of two residues in the C-terminus of $A\beta_{1-42}$, causing major differences in conformational behavior, with $A\beta_{1-42}$ being much more folded than $A\beta_{1-40}$ [57]. The amyloid plaques could also be associated to other molecules and metal ions, playing an important role in their assembly and toxicity [58,59]. If some mutations occur in the substrate (APP) or in the γ-secretase regulator proteins (prenisilin-1 and prenisilin-2) it may cause an alteration of APP processing, increasing the levels of $A\beta_{1-42}$ or $A\beta_{1-43}$ peptides formed [60,61]. These mutations are known to be involved in development of early onset AD [62–65].

In order to support the idea that $A\beta$ peptides possess an important role in AD, Simmons and co-workers [66] demonstrated that aggregation of $A\beta$ increased the neurotoxic effect in rat embryonic neuronal cells. Kirkitadze and co-workers [67] studied the $A\beta_{1-40}$ and $A\beta_{1-42}$ oligomerization and assembly into fibrils, showing that the early features of fibril assembly were the increase of intermediates containing α-helix and then their decrease by the assembly of fibrils. Yan and Wang [68] showed that $A\beta_{1-42}$ possesses more tendencies to aggregate in comparison with $A\beta_{1-40}$, and that their C-terminal domain is more rigid.

A structural model for amyloid $A\beta_{1-40}$ using solid state NMR (ssNMR) spectroscopy was proposed. It was found that the first 10 residues are disordered, a β-strand conformation forming β-sheet structure was found between residues 12–24 and 30–40 [69]. After that, other studies were performed with different forms of preparation of the fibrils, with the binding of Cu^{2+}, with mutant forms of the peptide, among others [70–73].

Recently, the peptide $A\beta_{1-42}$ was studied also using ssNMR and in one of the studies it displayed triple parallel β-sheet segments, which is formed by three β-sheets encompassing residues 12–18 (β1), 24–33 (β2), and 36–40 (β3) [74]. Another NMR study of $A\beta_{1-42}$, demonstrated that the fibril core is formed by a dimeric form of the peptide, containing four β-strands in an S-shaped amyloid fold [75]. Wälti and coworkers [76] found similar results: the fibril in dimeric form, forming a double-horseshoe. The different results of these groups were probably due to the differences in the preparation of the fibrils, such as pH, peptide concentration, agitation and ionic strength, as well as the source of the peptide (recombinant or synthetic) [42]. For a detailed review about the structural features of the two peptides, see Reference [42]. Besides the importance of $A\beta_{1-40}$ and $A\beta_{1-42}$, some studies demonstrated that the presence of minor isoforms of $A\beta$ peptides could be involved in aggregation and/or or neurotoxicity [49,77,78], although their effect in AD is not fully understood.

Tau is a microtubule-associated protein initially identified as a protein involved in microtubule (MT) assembly and stabilization [79] and in the axonal transport of proteins [80]. Nowadays, the list of physiological functions of tau has expanded to include diverse roles such as protection against DNA damage and cell signaling [81]. Recent data revealed that tau physiologically interacts with various proteins and subcellular structures, and upon release from neurons, it may even act on other cells, widening the spectrum of its repercussions in health and in diseased states [82].

The single gene encoding the tau protein is present in one copy in the human genome, located in chromosome 17q21 [83,84]. Alternative splicing of this gene can yield six different isoforms of tau with polypeptide chains varying from 352 to 441 amino acids [85,86], all containing either three or four tandem repeats of 31 or 32 amino acid residues, the so-called microtubule binding repeats [81,82]. Tau is composed of 25 to 30% of charged amino acids and contains many proline residues, rendering it full intrinsically disordered. Tau undergoes many types of posttranslational modifications such as phosphorylation, glycosylation, methylation, acetylation, ubiquitinilation, SUMOylation (interaction with Small Ubiquitin-like Modifiers), nitration, among others, which are thought to finely regulate the involvement of the protein in its various biological functions. As a result of "abnormal" phosphorylation, glycosylation, oxidation, truncation or other posttranslational modification [6,82], tau becomes prone to aggregation and forms intracellular deposits, a feature of several neurodegenerative diseases collectively known as "tauopathies". The abnormal tau adopts many transient local foldings among which β-structures of hydrophobic regions, characteristic of neurofibrillary tangles, and paired helical filaments of its microtubule binding domains [87,88] (for a review, see Reference 81]). Tau aggregates can mediate the spreading of the neuropathology to neighboring cells through its paired helical filaments, emerging as a possible target for taoupathy therapies [89]. Oxidation status of tau cysteine residues plays an important role in aggregation. While the formation of intermolecular disulfide bridges aggregates the protein, intramolecular cystine bonds prevent aggregation [90]. Truncation and/or proteolysis of tau yielding lower molecular mass forms of the protein, either in the intracellular or extracellular compartments, were also reported to lead to conformational changes that culminate in toxic, aggregated fibrillar tau [91,92].

In the most common tauopathy, Alzheimer's disease, and in some forms of frontotemporal dementia, the sites of neurodegeneration correlate with deposits of an aberrant hyperphosphorylated tau. All six isoforms of hyperphosphorylated tau are found in tauopathies, resulting in loss of the protein's ability to bind to microtubules and causing disturbance of axonal transport [82]. Phosphorylation of tau may occur in more than 85 putative sites, and distinct kinases and phosphatases are involved in controlling the protein's phosphate content. On the other hand, the glycosylation and/or acetylation status of tau determines its phosphorylation pattern [93].

As a consequence of its disordered nature, tau interacts with a diverse array of partners inside the cell, among which are proteins, small molecules, nucleic acids, and metal ions, with many of these interactions modifying tau's structural properties and biological functions. At least 33 distinct protein partners bind to tau's different domains or motifs, as reviewed in Reference [81]. The multifunctionality of tau resulting from the combination of the wide range of its binding partners and a plethora of posttranslational modifications guarantees its place among true moonlighting proteins [94]. One of such interactions is with the β-amyloid peptide, in a manner that the neurotoxicity of both partners is thought to be reinforced [95,96].

2.3. Prion Protein in Prion Diseases

The term prion was introduced to describe a small proteinaceous agent that was causing neurodegenerative disease in humans and other animals [97]. It was identified as an abnormal form of the prion protein [98,99]. The prion protein (PrP), encoded by the *Prnp* gene, is a glycoprotein, natively found in cells and that could be involved in the maintenance of myelin in neurons among other functions [100–102]. Structural studies of PrP using NMR demonstrated that the N-terminal portion of the recombinant murine PrP is unstructured and flexible, and that the C-terminal portion is globular, containing 3 α-helices and a short anti-parallel β-sheet [103]. A similar structure was found for murine PrP [103], hamster PrP [104], human PrP [105] and bovine Prp [106]. Prions are not considered IDPs per se due to their mixed structural features. Some authors argue in favor of prion-specific classification [107], while others consider them to be IDP-like or IDR-containing proteins [11,108].

In prion diseases, PrP changes are predominantly from an α-helical conformation (PrPC) into a β-sheet-rich structure acquiring a PrPSc form that is misfolded, aggregated and that causes transmissible and fatal neurodegenerative diseases [18,109]. Three kinds of prion diseases have been reported: sporadic, infectious, and hereditary forms, including human disorders like Creutzfeldt–Jakob disease (CJD), Gerstmann–Sträussler–Scheinker disease (GSS), familial atypical dementia, Kuru, and veterinary disorders such as scrapie in sheep, goats, mouse, etc. [110,111]. The basic neurocytological characteristics of these diseases are a progressive vacuolation of neurons and gray matter changing to a spongiform aspect with extensive neuronal loss [112].

Interspecies transmission of prions has been postulated [113], although some interspecific barrier for transmission of PrPSc prions has been established [114]. One factor involved in this barrier could be the difference between the donor and host amino acids sequence [115–117]. A recent study brings new insights on prion replication during species transition [118].

The structural modifications involved in prion propagation and infectivity is the transition of α-helices of PrPc into aggregated β-sheet of PrPSC [109,119]. The presence of PrPSC abnormal form seems to stimulate and serves as template for transition of PrPC into the infectious conformation [120]. Makarava and colleagues [121] reported that prion disease could be induced in wild-type animals by injection of recombinant PrPC fibrils. In order to understand how this transition occurs, Stahl and co-workers performed a study using mass spectrometry and Edman sequencing. They demonstrated that the primary structures of PrPSc were the same as the one predicted for the *PrPC* gene, suggesting that the difference between them is not in RNA modification nor splicing events. In the same study, no covalent modifications were identified in this transition [122]. In another study, using Fourier-transform infrared (FTIR) spectroscopy and circular dichroism (CD), it was shown that PrPC contains around 42% of α-helices in its structure and only 3% of β-sheet content. On the other hand, the modified isoform PrPSc contain a higher content of β-sheet (43%) and a lower content of α-helices (30%) [109].

The PrPSc aggregates present resistance to proteolytic degradation at the C-terminal region, differently from the PrPc normal form [123]. Saverioni and co-workers [124] demonstrated that human PrPSc isolates showed strain-specific differences in their resistance to proteolytic digestion, something that could be linked to aggregate stability. Such aggregates can have heterogeneous sizes [124,125].

When PrPc obtains a β-sheet-rich conformation and misfolded form, it has a tendency to accumulate as amyloid fibers, a useful characteristic for detection and diagnostic of diseases [126,127]. In spite of that, the formation of amyloid plaques is not an obligatory event in prion infectivity [128]. Thinking in a therapeutic target for prion diseases, one approach would be blocking the conversion of PrPC into PrPSc [129].

2.4. p53, c-Myc, and Cancer

Several human diseases, such as cancer, diabetes, and autoimmune disorders, have been found to be associated with deregulation of transcription factors [130]. Carcinogenesis is a multi-step process, resulting in uncontrolled cell growth. Mutations in DNA that lead to cancer disturb these orderly processes by disrupting their regulation. This disruption results in uncontrolled cell division leading to cancer development [131]. Deregulation of multiple transcription factors has been reported in cancer progression. Extensively studied transcription factors that have shown a major role in progression of different types of cancer are p53 and c-Myc, two intrinsically disordered proteins [132,133].

Fifty percent of all human cancer present mutations in *TP53*, and on many other cancers, the function of the p53 protein is compromised. Thus, p53 is a very important target in cancer therapy [134]. Mutations in p53 are found in several types of cancer such as colon, lung, esophagus, breast, liver, brain, reticuloendothelial, and hemopoietic tissues [135]. Additionally, many p53 mutants, instead of losing functions, acquire oncogenic properties, enabling them to promote invasion, metastasis, proliferation, and cell survival [136].

p53 is a key transcription factor involved in the regulation of cell proliferation, apoptosis, DNA repair, angiogenesis, and senescence. It acts as an important defense protein against cancer onset and evolution and is negatively regulated by interaction with the oncoprotein MDM2 (murine double minute 2). In human cancers, the *TP53* gene is frequently mutated or deleted, or the wild-type p53 function is inhibited by high levels of MDM2, leading to the downregulation of tumor suppressive p53 pathways [137–139]. When DNA damage occurs, p53 is activated to promote the elimination or repair of the damaged cells. p53 is phosphorylated by DNA damage response (DDR) kinase, leading to cell cycle arrest, senescence, or apoptosis. In addition, p53 stimulates DNA repair by activating genes encoding components of the DNA repair machinery [140].

Human p53 is a homotetramer of 393 amino acids composed of an intrinsically disordered N-terminal transactivation domain (TAD), followed by a conserved proline-rich domain, a central and structured DNA-binding domain, and an intrinsically disordered C-terminal encoding its nuclear localization signals and oligomerization domain required for transcriptional activity [138,141–147]. Natively unfolded regions account for about 40% of the full-length protein and the disordered regions are extensively used to mediate and modulate interactions with other proteins. Disorder is crucial for p53 function, since its numerous posttranslational modifications are majorly found within the disordered regions [11,146,148]. The full TAD of p53 consists of the N-terminal containing 73 residues and with a net charge of −17, due to its richness in acidic amino acid residues, such as aspartic acid and glutamic acid [143]. The C-terminus, on the other hand, is rich in basic amino acids (mainly lysines) and binds DNA non-specifically [146].

Transactivation domain is a promiscuous binding site for several interacting proteins, including negative regulators as MDM2 and MDM4 [146,149–151]. Transactivation domain is an IDR that undergoes coupled folding and binding when interacting with partner proteins like the E3 ligase, RPA70 (the 70 kDa subunit of replication protein A) and MDM2. p53 forms an amphipathic helix when it binds to the MDM2 in a hydrophobic cleft in its N-terminal domain [137,138,149,152–155]. The p53–MDM2 interaction blocks the binding of p53 to several transcription factors. In addition, MDM2 tags p53 for ubiquitination and consequent degradation by the proteasome and the p53–MDM2 complex tends to be exported from the nucleus, preventing p53 to act as a "cellular gatekeeper" [138,144,156].

The proto-oncogene c-MYC encodes a transcription factor that is implicated in various cellular processes such as cell growth, proliferation, loss of differentiation and apoptosis [157]. Elevated or deregulated expression of c-MYC has been detected in various human cancers and is frequently associated with aggressive and poorly differentiated tumors. Some of these cancers include breast, colon, cervical, small-cell lung carcinomas, osteosarcomas, glioblastomas, melanoma, and myeloid leukemia [158–160]. c-Myc is a very important protein for understanding and developing therapeutics against cancers and cancer stem cells [161].

c-Myc is an IDP and becomes transcriptionally functional when it forms an heterodimer with its obligate partner Max to assume a coiled-coil structure that recognizes the E-box (enhancer-box)-sequence 5'-CACGTG-3'. The c-Myc N-terminus, its TAD, can activate transcription in mammalian cells when fused to a heterologous DNA-binding domain. The C-terminus of this protein contains a basic-helix-loop-helix-leucine zipper (b-HLH-LZ) domain, and it promotes its interaction with Max, that has the same (b-HLH-LZ) domain, and the sequence-specific DNA binding mentioned above [162–165]. Nuclear magnetic resonance studies of c-Myc disordered region have attributed to it the protein functional plasticity and multiprotein complex formation capacity [166]. Computational and experimental investigations show that c-Myc extensively employs its disorder regions to perform diverse interactions with other partners [161].

It is important to highlight that Max protein is critical for c-Myc's transcriptional activities, both gene activation and repression [162,167]. Considering c-Myc as a target for cancer therapy, one approach to c-Myc inhibition has been to disrupt the formation of this dimeric complex [132]. However, the disruption of c-Myc-Max dimerization is not easy, since both proteins are IDPs and protein–protein interaction involving large flat surface areas are difficult to target with small molecules, such as drugs [168,169].

2.5. Amylin and Diabetes

Diabetes (Type II) is a multifactorial disease characterized by dysfunction of insulin action (insulin resistance) and failure of insulin secretion by pancreatic β-cells [11,170]. One hallmark feature of this disease is the accumulation of amyloid fibrils into pancreatic islets (islets of Langerhans). These amyloid deposits are majority composed by islet amyloid polypeptides (IAPP), also called amylin. Islet amyloid polypeptides are IDPs composed of 37 amino acid residues, co-secreted with insulin by the same pancreatic cells, and its gene is located on chromosome 12 in humans [171–173].

The process of aggregation of IAPP seems to be initiated by interaction of one IAPP monomer to another, progressively leading to the formation of aggregates [174,175]. Analysis of human IAPP using circular dichroism spectroscopy demonstrated that the fibril formation was accompanied by a conformational change of random coil to β-sheet/α-helical structure [176]. These transient conformations were further confirmed by other studies [177–179].

Cytotoxicity of IAPP accumulated as amyloid deposits could be associated with loss of pancreatic β-cells functions and cells apoptosis [180,181]. Recent reviews of computational studies provided mechanistic insights of IAPP structure as monomers and oligomers and their interaction with lipid bilayers in order to understand the IAPP cytotoxicity mediated by membranes [175,182].

3. Strategies for IDP Modulation

Intrinsically disordered proteins can rapidly populate different conformations in solution, usually not assuming a well-defined three-dimensional structure in their native state, as a result of their signature low-sequence complexity and low proportion of bulky hydrophobic amino acids (instead, charged and hydrophilic residues are common) that lead to a flexible, dynamically disordered behavior and larger interaction surface areas than analogous folded regions in globular proteins [183–186]. Considering that IDPs participate in numerous key processes in cell metabolism, it is expected that their activity would be regulated by multiple mechanisms at transcriptional, post-transcriptional, and translational levels, which makes active IDPs accessible in shorter periods compared to structured proteins [187,188]. Some aspects of IDP modulation (mainly for IDPs involved in cell signaling) are covered in this section, arbitrarily grouped in mechanisms that engage in direct structural changes on IDPs in order to achieve stabilization (coupled folding and binding, post-translational modifications), and mechanisms that control IDPs abundance in the cell (mRNA decay, IDP proteasomal degradation, nanny model for stabilization).

3.1. Regulation of IDP Activity through Structural Changes

Intrinsically disordered proteins acting in intracellular pathways contain conserved motifs for interaction with nucleic acids and other proteins, and frequently form low-affinity complexes advantageous to processes like signal transduction [184]. The recognition elements (being often called "SLiMs"—short linear motifs ranging from 3 to 10 amino acids [189–191]) determine a pivotal feature associated with disordered proteins, that is their binding promiscuity, which is carried out through "one-to-many" and "many-to-one" mechanisms [192]. Interestingly, many IDPs are able to adopt ordered structures when interacting with certain targets, characterizing the coupled folding and binding phenomenon. Also, structural polymorphisms can emerge from the IDP conformational landscape in cases where the same disordered protein can assume different defined structures as it binds to different targets. Some of the recognition elements are also targeted for post-translational modifications by regulatory enzymes, enabling disordered-to-ordered transitions in IDPs. However, induced folding in IDPs is not mandatory for activity, as many regions that remain disordered upon partner binding are important to function, constituting "fuzzy" complexes [193,194].

3.1.1. Coupled Folding and Binding

The folding induction by partner interaction is possibly one of the most reported characteristics of IDPs, despite not being a phenomenon absolutely widespread across this class of proteins (considering the fuzzy complexes), nor a completely understood process. There are numerous examples of disordered-to-ordered transitions in proteins implicated in the regulation of gene expression, like transcription factors that assume folded motifs when interacting with DNA. One particular example is the leucine zipper protein GCN4, which presents a basic region that is unstructured in the absence of DNA but becomes a stable helical structure when interacting to its cognate AP-1 site [195]. The transition begins with transient nascent helical forms, observed in the unbound state, that interact with DNA and lead to dramatic structural changes, explained by a reduction of the entropic cost of DNA binding due to restriction of the conformational space accessible to the basic region [8,196]. Thus, the induced folding is usually explained by loss of conformational entropy (from the unbound IDP state) upon target binding and compensatory favorable contributions from reduction in exposed hydrophobic surfaces and enhanced electrostatic interactions [196]. Exhaustive kinetic studies are necessary for the investigation on what order the events of binding and folding occur, and segregate them in schemes of induced fit (IF, when the IDP binds to a partner and then folds) or conformational selection (CS, when the partner only binds to IDPs in a certain conformation) [197]. However, trying to define the coupled folding and binding in two categories may offer a rather simplistic explanation for the IDPs behavior, since the mechanisms can overlap depending upon the system conditions [197,198].

3.1.2. Post-Translational Modifications

Because of their accessibility to modifying enzymes, IDPs are frequent targets for post-translational modifications (PTMs), which expand their functional versatility [185,199]. The occurrence of PTMs causes structural changes on IDPs by affecting their energy landscapes, due to modifications of the physicochemical properties of the primary sequence [185]. Post-translational modifications engage in the addition of chemical functional groups (usually small radicals such as phosphoryl, alkyl, acyl or glycosyl) or involve the direct modification of residues through reactions of oxidation, deimidation, and deamidation. Intrinsically disordered proteins suffering PTMs may have their electrostatic, steric, and hydrophobic properties modified, possibly inducing transformations on the structure due to enhancement/inhibition of contacts of motifs within the IDP chain or with binding partners [185,200–202]. Phosphorylation is one of the most prevalent PTM and constitutes a major regulatory mechanism in various cellular processes involving signal transduction. Replacing a neutral hydroxyl group with a tetrahedral phosphoryl results in new possibilities for intra- and intermolecular electrostatic interactions (e.g., salt bridges and hydrogen bonds). Many other types of PTMs work in a similar fashion (but modulating different chemical properties), providing new forms of interaction that can ultimately cause alterations in IDPs activity, including disordered to ordered transitions.

3.2. Regulation of IDPs Abundance

Intrinsically disordered proteins levels are carefully monitored in the cell, and changes in their abundance are associated with disease, mainly due to defective signal transduction (linked to the occurrence of some cancers [11,203]) and non-specific interactions that generate fibrillar aggregates (present in many neurodegenerative disorders [199,204]). Tight regulation can be achieved in different levels, controlling the half-lives of mRNAs encoding IDPs and the abundance of IDPs themselves. Obviously, there are outliers for these global trends and certain IDPs are present in cells in large amounts or for long periods of time [205]—usually not the IDPs involved in dynamic processes such as cell signaling. Examples include the fibrous muscle protein titin [206,207] and the curious case of tardigrade-specific IDPs, that are constitutively expressed and upregulated in some tardigrade species and are essential for desiccation tolerance [208].

3.2.1. IDP-Encoding mRNAs

A robust study monitoring gene expression in *Saccharomyces cerevisiae* (with similar trends detected for human genes [205]) demonstrated that mRNAs encoding sequences categorized as "highly unstructured" have lower half-lives than mRNAs encoding more structured proteins, having a comparable number of transcription factors regulating them [206]. One of the reasons for the increased mRNA decay is hypothesized to be the short poly(A) tails observed in IDP-encoding mRNAs, that foment RNA degradation pathways. Moreover, other factors related to transcript instability, like the binding of RNA-binding PUF proteins (that usually facilitate deadenylation and subsequent RNA clearance [209,210]) were found to be increased in mRNA coding for disordered proteins [206].

3.2.2. Proteasomal Degradation

Intrinsically disordered proteins undergo proteasomal degradation through two different (but not mutually exclusive) pathways, ubiquitin-dependent (UD) and ubiquitin-independent (UI) [211]. The UD pathway relies on the addition of ubiquitin to the substrate to be degraded, in a process regulated by a series of enzymes and is mediated collectively by the 26S proteasome [212]. It was found that IDPs present a high content of predicted ubiquitination sites [185,205], and, in an analysis of ubiquitinated proteins, there seems to be a correlation between confirmed degradation sites and regions of disorder [213]. Alternatively, the UI pathway is mainly orchestrated by the core of the 20S proteasome, being a default process for degradation of free disordered proteins. Some evidences support the hypothesis that the flexible and extended structures of IDPs (as well as disordered terminal segments in folded proteins) facilitates the interaction with the proteasome, considering that bulky particles have reduced cleavage rates [214,215].

3.2.3. Stabilization through "Nanny" Proteins

As IDPs are usually prone to degradation, there is a need for protein stabilization in some contexts. The ubiquitous enzyme NQO1 functions as a "gatekeeper" of the 20S proteasomes, binding and regulating the degradation of some IDPs, in a mechanism consuming NADH [211,216]. An analogous mechanism characterizes the "nanny" model for IDP protection from cleavage, where there is sequestration of ID segments by interactions with other proteins, resulting in evasion from 20S proteasomal digestion. The binding of the nanny is transient, beginning at the initial stage of IDPs' life cycle, when they are newly synthesized (assuming they are more sensible to digestion in this stage) [154]. Despite the fact that nanny proteins also bind to nascent polypeptide chains, they are not considered chaperones because they do not induce a fixed three-dimensional organization on targets, only assisting on IDP conservation without permanently affecting their disordered structure.

3.3. Modulation of IDPs by Chaperones and Co-Chaperones

Aggregates produced in neurodegenerative diseases have been shown to respond to changes in levels of molecular chaperones, suggesting the possibility of therapeutic intervention and a role for chaperones in disease pathogenesis [217]. The heat shock protein Hsp90 promotes neurodegenerative disorders indirectly [218]. Tau protein accumulation is regulated by a (Hsp90) chaperone system. This chaperone is able to bind Tau, causing a conformational change that allows tau's phosphorylation by glycogen synthase kinase (GSK3β), leading to tau aggregation [219]. The inhibition of this chaperone results in the reduction of tau phosphorylation levels, due to reduction of GSK3β levels [220]. Another approach was performed, the use of a co-chaperone of Hsp90, ATPase homolog 1 (Aha1), this protein is an activator of Hsp90. This approach promoted the increase of the production of aggregated tau in vitro and in mouse model of neurodegenerative disease. Moreover, inhibition of Aha1 reduced tau accumulation in cultured cells. Thus, Aha1 is an interesting target to the treatment of Alzheimer's disease [221].

The Hsp70 is a protein stabilizer, has a cellular protection against neurodegeneration of the central nervous system [218]. Members of the Hsp70 family, such as Hsp70 and Hsc70, bind to misfolded proteins and somehow send them to the lysosome–autophagy pathway or ubiquitin–proteasome system for degradation [222,223]. Folding and degradation of proteins are linked through co-chaperones, such as C-terminus of HSP70-interacting protein (CHIP) and HSJ1 (DNAJB2) [224] which regulate the decisions determining whether misfolded proteins are refolded or degraded. The CHIP is associated with α-synuclein inclusions and act as a co-chaperone, altering its aggregation and enhancing the degradation of the misfolded α-synuclein [225].

Another important chaperone is cyclophilin 40 (CyP40) that is a cis/trans peptidyl-prolyl isomerase (PPIase) and is involved in regulation and orientation of proline residues [226,227]. Tau protein is rich in proline residues and its residues, usually found in β-turns, are involved in tau aggregation propensity [228]. Based on this information, Baker and coworkers [229] demonstrated that CyP40 possess the ability to dissolve amyloids fibrils in vitro. Nuclear magnetic resonance experiments showed that CyP40 acts specifically on proline rich residues performing the disaggregation of tau fibrils and oligomers. This cyclophilin could also interact with others aggregated proteins containing proline, like α-synuclein [229].

4. Known Drugs Acting on IDPs

Despite IDPs abundance in eukaryotes, currently there are no FDA-approved drugs specifically targeting these proteins, only experimental and speculative ones (i.e. drugs that have been evaluated by the United States Food and Drug Administration agency and had their marketing sanctioned). Some experimental drug examples are prevalent in the literature, such as those targeting p53-MDM2, c-Myc-Max, and EWS-Fli1 complexes, while some others are less discussed [230–233]. In this section we provide an overview of the pharmacological modulation of IDPs, neurodegenerative and otherwise.

Intrinsically disordered proteins are normally considered aggregation-avoidant, due to their high proportion of charged residues (as opposed to patches of folding-inducing, hydrophobic residues). Such "non-folding" plasticity is proposed to be advantageous for proteins with multiple partners [234,235]. However, some of the IDPs, as described earlier, are found in conformational diseases and amyloid formation (e.g., in Alzheimer's disease, Parkinson's disease). The suppression of fibril formation, thus, is of therapeutic interest. Drug candidates in this front include molecular tweezers (Table 1), which are ligands designed to bind lysine and arginine specifically, perturbing aggregation [236–238]. These positively-charged residues are prone to interact with negatively-charged regions in the fibril-forming monomers [237]. The SEN1576 compound, a 5-aryloxypyrimidine inhibitor of synaptotoxic Aβ aggregation (Table 1) was shown to be safe and orally bioavailable with good brain penetration [239].

Fragments of amyloid fibrils also served as templates for non-natural amino acid inhibitors of amyloid fibril formation (D-TLKIVW) [240], while the ELN484228 (Table 1) compound was shown to be protective in cell models for vesicular dysfunction via α-Synuclein [235]. Alterations of the neuroleptic agent chlorpromazine allowed for enhanced 20S proteasome activation, inducing degradation of IDPs, such as tau and α-synuclein, but not of structured proteins [241]. These chlorpromazine-derived molecules, despite showing noteworthy potential (even as tools to study the proteasome 20S gate regulation), may interfere with other, physiological, non-pathologic disordered proteins to a still unstudied extent. A naphthoquinone-tryptophan hybrid (NQTrp) (Table 1) was shown to be effective in model systems for tau aggregation [242].

Regarding tumor-associated IDPs, the most commonly mutated gene in human cancers, the tumor suppressor protein p53, is key to cell cycle signaling. It is regulated by binding to various partners, including MDM2 and Taz2 [243–245]. Despite being highly disordered, p53 is not itself the target for the currently screened drug candidates aiming the p53-MDM2 complex. Instead, these ligands aim to occupy the p53-binding site in MDM2. The inhibitors nutlins (Table 1) (cis-imidazoline analogs currently in phase-I clinical trials), were shown to be potent against multiple cancerous cell lines, including breast cancer, colorectal cancer, lung cancer, osteosarcoma, prostate cancer, and renal cancer [246–249].

The c-Myc-Max complex involves the IDP transcription factor c-Myc that is activated by binding to Max, being expressed constitutively in various cancer cells [250]. Inhibitor candidates target the disordered c-Myc in this case, including peptidomimetic inhibitors [249,250], the small molecules 10058-F4, 10074-G5 (Table 1), and some others [250–257]. The oncogenic fusion protein EWS-Fli1 is an IDP exclusively present in Ewing's sarcoma [257]. As with c-Myc-Max inhibitors, the small molecule inhibitor YK-4–279 (Table 1) targets the disordered EWS-Fli1 protein directly [257,258].

The AF9-AF4 dimer is found in acute leukemias and is composed by two disordered fusion proteins. The AF9 protein is responsible for turning hematopoietic cells oncogenic [259,260]. The AF4-derived peptide of amino acid sequence PFWT was shown to inhibit AF9 when used in combination with established chemotherapeutic agents [261,262], while some non-peptidic inhibitor candidates have been identified by high-throughput screening [263]. The protein-tyrosine phosphatase 1B (PTP1B), a reticular non-transmembrane enzyme, has been validated as therapeutic target for diabetes, obesity, and breast cancer, due to its role as negative regulator of insulin and leptin signaling [264]. Protein-tyrosine phosphatase 1B has an elongated disordered carboxy-terminus, to which trodusquemine (Table 1) (MSI-1436, a natural product) binds [265]. This aminosterol acts allosterically, stabilizing an inactive form of the enzyme by binding to a non-catalytic disordered site [265].

Table 1. Known drugs acting on IDPs (selected examples).

Compound Name *	Targets	Compound Structure
CLR01 (Molecular tweezers)	Lysine and arginine residues in amyloid proteins	
ELN484228	α-Synuclein	
SEN1576	Amyloid β	
NQTrp	PHF6 (Tau protein)	
Nutlin-3	p53-MDM2 complex	
10058-F4	c-Myc-Max complex	
10074-G5	c-Myc-Max complex	
YK-4–279	EWS-Fli1	
Trodusquemine	PTP1B	

* Unless otherwise specified, these names are directly taken from their lead-compound coding and have no meaning on their own. EWS: Ewing Sarcoma; MDM2: oncoprotein (murine double minute 2); Myc: Myelocytomatosis-transcription factor homolog; NQTrp: naphthoquinone-tryptophan hybrid; PHF6: Tau-derived peptide (amino acid sequence VQIVYK); PTP1B: protein-tyrosine phosphatase 1B.

Inhibitors of α-synuclein aggregation are considered a promising approach. From in vitro studies, a few lead molecules were identified, such as EGCG (epigallocatechin gallate) [266]. Iron is known to induce the aggregation of α-synuclein. Deferiprone is an iron chelator used in thalassemic patients. Two clinical trials have shown a decrease in iron content in the substantia nigra of some PD patients, while a trend for improved motor scores was seen for all enrolled patients [267]. Although copper can induce aggregation of α-synuclein in vitro, levels of copper in the substantia nigra of PD patients are up to 50% lower than that of age-matched controls. The Cu^{2+} complex of diacetylbis-(4-methylthiosemicarbazone), called Cu^{2+}(atsm), showed neuroprotective action in different animal models of PD [268], prompting for a phase I trial. From in vivo studies, promising results were obtained for KYP-2047, an inhibitor of prolyl oligopeptidase, an enzyme shown to interact with α-synuclein [269–271]; the diphenyl-pyrazole compound anle138b, was shown to cross the blood-brain barrier of mice and to reduce aggregation of α-synuclein [269]. The NPT200-11 compound prevented the formation of oligomers of α-synuclein and improved neuropathological symptoms in transgenic mice [272] and it has already been subjected to a phase I clinical trial. Another promising compound, NPT088, is a fusion protein of a general amyloid interaction motif derived from a bacteriophage [273] and a fragment of human immunoglobin, developed by Proclara Biosciences, entered a phase I clinical trial for AD [274]. Phase II clinical trial of NPT088 is expected to include PD patients. Intrabodies, a single chain variable fragment of immunoglobulin expressed intracellularly, have been developed to target oligomeric and fibrillary α-synuclein, conferred neuroprotection, apparently by shifting the dynamics of the aggregation process [275–277]. Addition of a proteasome-addressing sequence to intrabodies targeted pathological forms of α-synuclein to degradation, NbSyn87PEST, directed towards the C-terminal region, and VH14PEST, directed against the NAC hydrophobic interaction domain, effectively degraded α-synuclein in cultured cells [278]. In rats overexpressing wild-type α-synuclein, these proteasome-targeted intrabodies (or nanobodies) decreased the levels of pathological aggregates, increased striatal dopamine levels and improved motor function [279]. Research in this promising field moves to find ways to deliver these compounds in adequate levels in specific areas of the brain, probably by using viral vectors.

Immunotherapies against α-synuclein, based on the evidence of an extracellular pathological protein during spreading of PD to different brain structures, show promising results in animal models [280]. Besides opsonization of the pathological protein for clearance, it is likely that antibodies could block further oligomerization of α-synuclein. Both passive (humanized monoclonal antibodies) and active (vaccine) immunization are being pursued. Pharmaceutical companies have joined the efforts and early clinical trials have been concluded or are under way. A brief description of the more advanced planned immunotherapies follows. A phase I trial was conducted by Roche for PRX002, a monoclonal antibody against the C-terminus of α-synuclein. It was well tolerated and reduced by 96% the levels of serum α-synuclein [281,282]. Affitope PD01A, a synthetic α-synuclein-mimicking peptide developed by Affiris for active immunization, had the first pilot study in 21 PD patients concluded in May 2018. It elicited a specific antibody response and showed good safety and tolerability profiles in a long-term (4 years) outpatient setting. Results of Affitope PD03A phase I clinical trial indicated no severe off-target effects, and a dose-dependent production of antibodies that cross-reacted with the intended α-synuclein epitope. Results from animal studies demonstrated that the antibodies raised against these antigens crossed the blood-brain barrier, decreasing the levels of aggregated α-synuclein, thereby improving motor function [283,284].

Regarding tauopathies, various therapeutic approaches have been tested, aiming to inhibit aggregation of tau, either directly or by preventing its interaction with some partners, and removal of toxic conformers and fibrillated tau [80]. However, the enormous effort put on finding ways to revert or delay the neurodegeneration symptoms associated to fibrillar tau, or to prevent the onset of tauopathies, has been so far unsuccessful, partly due to the intrinsically disordered nature of tau, which hampers drug design based on structural approaches. Small molecules that inhibit tau aggregation in vitro are considered promising leads to anti-tauopathy drugs [285] and the number of new tau inhibitory molecules grows steadily [286]. Nevertheless, there are unanswered questions regarding their effectiveness in vivo and the potential non-specific effects on normal tau physiology that could impact heavily on the CNS. The most studied small inhibitory molecules belong to distinct chemical groups, such as phenotiazines, cyanines, rhodanines, and arylmethines [287,288]. Peptides derived from neuroprotective proteins like NAP (amino acid sequence NAPVSIPQ) and D-SAL (all D-amino acid sequence SALLRSIPA) [289,290], enantiomeric peptides [291], and RNA/DNA aptamers [292] are also attractive components of future anti-tauopathy therapies. Some natural molecules present in cellular medium, such vitamin B_{12} [293] and 8-nitro-$_C$GMP [294], are known to inhibit tau aggregation through oxidation of its cysteine residues. Drugs that bind tau, inducing formation of intramolecular disulfide bonds, such as methylene blue [295] or cinnamon derivatives [296], are potential frames for developing specific tau aggregation inhibitors. Taking into account that accumulation of hyperphosphorylated tau is a hallmark of AD and other neurodegenerative disorders, inhibitors of kinases, particularly of glycogen sintase kinase 3β and of Fyn, a member the Src-family of non-receptor tyrosine kinases, have drawn much attention for their anti-tauopathy potential [80]. Another strategy focuses on dual inhibitors that would interfere on tau aggregability and simultaneously block its interaction with protein partners, particularly kinases [297,298]. Other attempts to develop an anti-tauopathy drug have focused on inhibiting tau interaction with proteases like beta-secretases [299], caspases [300], and calpain [301], and chaperones such as Hsp90 [302], among others.

Tau-targeted immunotherapy began in 2013 [303], and since then a dozen of different types of immunological strategies were subject of clinical trials, including two active immunizations (vaccines) and humanized monoclonal antibodies directed towards distinct tau epitopes aiming passive immunization (reviewed in References [304,305]). These clinical trials are still at early phases and only limited data on the outcomes have been disclosed so far [306,307]. To achieve a successful immunotherapy to treat tauopathies, antibodies should be capable of neutralizing at least one of the many diseased isoforms of tau, either intracellularly or in the extracellular space, and interrupt the processes that lead to tau fibrillation and the neuron-to-neuron spreading. Ideally, the antibodies do not bind to normal tau and are able to cross the blood–brain barrier. In the case of a tau vaccine, the senescence of the immune system of the elderly has to be considered [304]. AADvac1 was conceived as the first vaccine against AD, using as immunogen a tau peptide previously identified to be essential for its pathological aggregation. Active immunization with this peptide elicits antibodies against a stretch of tau's primary sequence (amino acid residues 294–305) and to conformational epitopes as well, targeting mainly extracellular tau, reducing its oligomerization. Tested in different animal models of AD, AADvac1 raised a protective humoral immune response with antibodies that discriminated between normal and pathological tau, reduced the level of neurofibrillary pathology in rat brains and lowered the content of disease-specific hyperphosphorylated tau [308]. Phase I clinical trial of AADvac1, conducted in 2013–2015 in patients aged 50–85 years with mild-to-moderate AD immunized weekly for 12 weeks, revealed a favorable safety profile and 29 out of 30 patients given AADvac1 developed an IgG response [309]. After 72 weeks, and booster doses of AADvac1, patients who had developed higher IgG titers showed lower hippocampal atrophy and cognitive decline rates and only mild adverse side effects [310]. A second active immunotherapy against tau has the compound ACI-35 as the immunogen, a peptide containing tau's phospho-epitope pS396/pS404, in a liposome-based formulation able to elicit antibodies against abnormal hyperphosphorylated tau in P301L tau-mice [311].

Attempts of passive immunotherapy utilize humanized monoclonal antibodies, mostly of IgG1 or IgG4 isotypes, which are directed towards stretches of tau's primary sequence known to be involved in the oligomerization of the protein, or to extracellular seeding-capable forms of truncated tau [304]. Phase I clinical trial of ABBV-8E12, one of such humanized monoclonal antibodies [312], revealed a satisfactory safety profile in 30 patients with the progressive supranuclear palsy tauopathy, receiving single doses (2.5 to 50 mg/kg) of the antibody, with no signs of immunogenicity against it [313]. Phase I trials of other two anti-tau antibodies, C2N-8E12 and BMS-986168, were also conducted [307].

Something that is noteworthy, and that demonstrates the close interplay between amyloid β peptide and tau in causing neurodegenerative diseases, therapeutic interventions aimed at one pathology can ameliorate symptoms of the other. This is the case of immunization of triple transgenic AD-like mice with a full-length DNA of amyloid β_{1-42} peptide, which showed a 40% reduction in the brain content of the amyloid β_{1-42} concomitant with a 25–50% decrease of total tau and different phosphorylated tau isoforms [314]. Conversely, passive immunization with antibodies against tau's fragments 6–18 and 184–195 protected triple transgenic AD-like mice by reducing amyloid precursor protein in the CA1 region of hypothalamus and in amyloid plaques [315,316].

5. Status and Challenges in Drug Development for IDPs

The limited number of drugs targeting IDPs currently available (see previous section) may look disappointing, considering the physiological relevance of these proteins. One should be careful, though, to not exaggerate the current lack of IDP-specific drugs as being a reflection of disorder as a limitation for drug development.

The rational drug design strategy has been used successfully since the 1980s [317–319]. It depends on knowledge of the three-dimensional structure of the target protein, based on which ligands (usually inhibitors) are planned with aid of computational tools [320]. By definition, IDPs do not have a single, major conformation, occurring in dynamic conformational ensembles [10], and there is difficulty in using such traditional techniques to design IDP ligands [321]. Hence, most cases of IDP drug development were carried out by experimental screening and not by rational design [322]. Still, detection of IDP "hits" (potential initial drug candidates) through high throughput screening of compounds has been challenging [321] and computational methods achieved some success in predicting good candidates [323].

For IDPs with recognizable/determined metastable structures in their conformational ensembles, such structures could be used for rational drug design. However, IDPs are expected to be promiscuous, acting as hubs for multiple cellular processes [324]. This scenario, described as "protein clouds" [325] has its complexity further increased, with IDP ligands being described as "ligand clouds around protein clouds" [326]. Such roles in protein–protein interactions (PPIs) make IDPs especially interesting as drug targets, but the development of molecules targeting PPIs has been in itself, challenging [327–330].

A large difference is expected between the entropic loss and the enthalpic gain upon binding of a small ligand to an IDP, but some of them were shown to be capable of forming adaptable, specific interfaces for small molecule binding [231,235,321]. Intrinsically disordered proteins are difficult targets, since their interactions with small molecules are weaker and more transient, and the entropic loss is greater, in comparison to structured proteins [331]. The fragment-based drug design approach allows fragments to sample large amounts of chemical space, reducing the number of compounds for screening, with different fragments that bind at different regions of IDPs being able to be linked together via an appropriate linker [331]. Such fragments usually require hydrogen bonds to achieve detectable binding, generating an enthalpic gain that compensates for the entropic loss upon binding of the small molecules, lowering the free energy of the protein upon binding [331].

Still, the over-representation of IDPs in disorders, as summarized by the D^2 concept (for "disorder in disorders") [11] and the D^3 concept (for "disorder in degenerative disorders") [16], points to these proteins as promising therapeutic targets. Attempts to detect potentially druggable cavities in IDPs have identified at least 14 targets that could be subjected to rational drug design [233]. A more general estimate is that 9% of detected cavities may be druggable in IDPs, in comparison to 5% in ordered proteins [233]. These observations are especially interesting, considering that current drugs target around 500 proteins, less than 10% of the estimated potential target list, with very strict classes (such as enzymes and G-protein coupled receptors) accounting for more than 70% of them [230,331,332].

There have been major advances in the detection and prediction of IDP features in protein sequences [293], something that will surely help in the identification of these special drug targets. Major breakthroughs are also being achieved by the combination of experimental methods (especially

NMR and fluorescence techniques) with computational modeling and molecular dynamics simulation of IDPs [294,322]. The latter is one of the few methods that allow for the description of IDPs in their conformational ensemble, instead of a single (or just a few) conformations [132,322,324,333–337].

6. Conclusion and Perspectives

The pharmacological strategies developed so far (and reviewed here) that target IDPs can be separated as binding directly to IDPs and hampering their aggregation by keeping them in the interaction incompetent conformation; interacting with the IDP and promoting the stabilization of non-toxic/ non-amyloidogenic oligometric species; and interacting with the amyloidogenic protein and greatly accelerating its aggregation to minimize the period of toxic oligomer formation [338]. Still, as we described here, there are many pathways acting on IDP control, and these are still unexplored targets for pharmaceutical interference. As the binding mechanisms of IDPs are being better described from a physical chemical standpoint [339,340], it is becoming clear that for candidate molecules to act on IDPs they must deviate from traditional prediction rules for drug-likeness [319,320]. One standout feature of IDP ligands is that they are larger and more three-dimensional than traditional drugs [341]. Adding another layer of complexity to this scenario, some proteins are shown to be conditionally unfolded [342], being disordered only under specific conditions.

Despite being abundant in eukaryotes, in which IDPs have evolutionarily conserved interaction partners [343], the occurrence of disorder in proteins from other organisms is being described. It includes the description of IDPs in Trypanosomatid parasites [344] and in some paramyxoviruses, including measles, Nipah and Hendra viruses [345]. These proteins constitute prospective targets for drug design endeavors, as was also observed for multiple disordered targets in prostate cancer [346]. Furthermore, recent evaluations indicate that IDP-targeted drug development may not be irreconcilable with structure-based drug design [347].

The development of drugs specifically tailored for IDPs is still in its infancy. As with the whole pipeline for drug discovery, there has been continuous progress in this area, and as we proposed in this work, there are many untapped pathways and unexplored targets regarding these proteins. As the biophysical techniques advance to catch up with the diversity of disordered behaviors in proteins, one can expect major developments in this front. Taking the limited but solid cases of success in IDP-specific drug design, we may face a future in which target disorder may be taken as the rule and not the exception.

Author Contributions: Conceptualization, C.R.C. and R.L.-B.; Writing—Original Draft Preparation, A.H.S.M., F.C.L., E.O.J., C.R.C., and R.L.-B.; Writing—Review & Editing, C.R.C. and R.L.-B.; Funding Acquisition, C.R.C.

Abbreviations

AD	Alzheimer's Disease
APP	Amyloid Precursor Protein
CJD	Creutzfeldt–Jakob Disease
CNS	Central Nervous System
CS	Conformational Selection
GSS	Gerstmann–Sträussler–Scheinker Disease
IAPP	Islet Amyloid Polypeptide (Amylin)
IDP	Intrinsically Disordered Protein
IDR	Intrinsically Disordered Region
IF	Induced Fit
MT	Microtubules
NMR	Nuclear Magnetic Resonance
PDB	RCSB Protein Databank

PrP   Prion Protein
PrPSc   Prion Protein, Alternate Conformation
PTM   Post-Translational Modification
SLiMs   Short Linear Motifs
ssNMR   Solid-State Nuclear Magnetic Resonance
TAD   Transactivation Domain (of p53)
UD   Ubiquitin-Dependent
UI   Ubiquitin-Independent

References

1.   Dunker, A.K.; Lawson, J.D.; Brown, C.J.; Williams, R.M.; Romero, P.; Oh, J.S.; Oldfield, C.J.; Campen, A.M.; Ratliff, C.M.; Hipps, K.W.; et al. Intrinsically disordered protein. *J. Mol. Graph. Model.* **2001**, *19*, 26–59. [CrossRef]

2.   Uversky, V.N. Natively unfolded proteins: A point where biology waits for physics. *Protein Sci.* **2002**, *11*, 739–756. [CrossRef] [PubMed]

3.   Wallin, S. Intrinsically disordered proteins: Structural and functional dynamics. *Res. Rep. Biol.* **2017**, *8*, 7–16. [CrossRef]

4.   Uversky, V.N. Protein folding revisited. A polypeptide chain at the folding—Misfolding—Nonfolding cross-roads: Which way to go? *Cell. Mol. Life Sci.* **2003**, *60*, 1852–1871. [CrossRef] [PubMed]

5.   Dyson, H.J.; Wright, P.E. Intrinsically unstructured proteins and their functions. *Nat. Rev. Mol. Cell Biol.* **2005**, *6*, 197–208. [CrossRef] [PubMed]

6.   Schweers, O.; Schönbrunn-Hanebeck, E.; Marx, A.; Mandelkow, E. Structural studies of tau protein and Alzheimer paired helical filaments show no evidence for beta-structure. *J. Biol. Chem.* **1994**, *269*, 24290–24297. [PubMed]

7.   Weinreb, P.H.; Zhen, W.; Poon, A.W.; Conway, K.A.; Lansbury, P.T. NACP, a protein implicated in Alzheimer's disease and learning, is natively unfolded. *Biochemistry* **1996**, *35*, 13709–13715. [CrossRef] [PubMed]

8.   Wright, P.E.; Dyson, H.J. Intrinsically unstructured proteins: Re-assessing the protein structure-function paradigm. *J. Mol. Biol.* **1999**, *293*, 321–331. [CrossRef] [PubMed]

9.   Daughdrill, G.W.; Pielak, G.J.; Uversky, V.N.; Cortese, M.S.; Dunker, A.K. Natively disordered proteins. In *Protein Fold Handbook*; Buchner, J., Kiefhaber, T., Eds.; Wiley VCH: Weinheim, Germany, 2005; pp. 275–357.

10. Uversky, V.N. A decade and a half of protein intrinsic disorder: Biology still waits for physics. *Protein Sci.* **2013**, *22*, 693–724. [CrossRef] [PubMed]

11. Uversky, V.N.; Oldfield, C.J.; Dunker, A.K. Intrinsically Disordered Proteins in Human Diseases: Introducing the D^2 Concept. *Annu. Rev. Biophys.* **2008**, *37*, 215–246. [CrossRef] [PubMed]

12. Uversky, V.N.; Gillespie, J.R.; Fink, A.L. Why are "natively unfolded" proteins unstructured under physiologic conditions? *Proteins Struct. Funct. Bioinform.* **2000**, *41*, 415–427. [CrossRef]

13. Uversky, V.N.; Dunker, A.K. Understanding protein non-folding. *Biochim. Biophys. Acta* **2010**, *1804*, 1231–1264. [CrossRef] [PubMed]

14. Dunker, A.K.; Oldfield, C.J.; Meng, J.; Romero, P.; Yang, J.Y.; Chen, J.W.; Vacic, V.; Obradovic, Z.; Uversky, V.N. The unfoldomics decade: An update on intrinsically disordered proteins. *BMC Genom.* **2008**, *26*, S1. [CrossRef] [PubMed]

15. Hartl, F.U. Molecular chaperones in cellular protein folding. *Nature* **1996**, *381*, 571–580. [CrossRef] [PubMed]

16. Uversky, V.N. The triple power of D^3: Protein intrinsic disorder in degenerative diseases. *Front. Biosci.* **2014**, *19*, 181–258. [CrossRef]

17. Kovacs, G.G. Concepts and classification of neurodegenerative diseases. *Handb. Clin. Neurol.* **2017**, *145*, 301–307. [PubMed]

18. Kransnoslobodtsev, A.V.; Shlyakhtenko, L.S.; Ukraintsev, E.; Zaikova, T.O.; Keana, J.F.W.; Lyubchenko, Y.L. Nanomedicine and protein misfolding diseases. *Nanomedicine* **2005**, *1*, 300–305. [CrossRef] [PubMed]

19. Uversky, V.N. Targeting intrinsically disordered proteins in neurodegenerative and protein dysfunction diseases: Another illustration of the D^2 concept. *Expert Rev. Proteom.* **2010**, *7*, 543–564. [CrossRef] [PubMed]

20. Breydo, L.; Uversky, V.N. Role of metal ions in aggregation of intrinsically disordered proteins in neurodegenerative diseases. *Metallomics* **2011**, *3*, 1163–1180. [CrossRef] [PubMed]

21. Eftekharzadeh, B.; Hyman, B.T.; Wegmann, S. Structural studies on the mechanism of protein aggregation in age related neurodegenerative diseases. *Mech. Ageing Dev.* **2016**, *156*, 1–13. [CrossRef] [PubMed]

22. Eisele, Y.S.; Monteiro, C.; Fearns, C.; Encalada, S.E.; Wiseman, R.L.; Powers, E.T.; Kelly, J.W. Targeting protein aggregation for the treatment of degenerative diseases. *Nat. Rev. Drug Discov.* **2015**, *14*, 759–780. [CrossRef] [PubMed]

23. Goedert, M.; Jakes, R.; Spillantini, M.G. The Synucleinopathies: Twenty Years On. *J. Parkinsons Dis.* **2017**, *7*, S51–S69. [CrossRef] [PubMed]

24. Spillantini, M.; Crowther, R.; Jakes, R.; Hasegawa, M.; Goedert, M. alpha-synuclein in filamentous inclusions of Lewy bodies from Parkinson's disease and dementia with Lewy bodies. *Proc. Natl. Acad. Sci. USA* **1998**, *95*, 6469–6473. [CrossRef] [PubMed]

25. Jakes, R.; Spillantini, M.G.; Goedert, M. Identification of two distinct synucleins from human brain. *FEBS Lett.* **1994**, *345*, 27–32. [CrossRef]

26. Theillet, F.X.; Binolfi, A.; Bekei, B.; Martorana, A.; Rose, H.M.; Stuiver, M.; Verzini, S.; Lorenz, D.; van Rossum, M.; Goldfarb, D.; et al. Structural disorder of monomeric α-synuclein persists in mammalian cells. *Nature* **2016**, *530*, 45–50. [CrossRef] [PubMed]

27. Ferreon, A.C.; Gambin, Y.; Lemke, E.A.; Deniz, A.A. Interplay of alpha-synuclein binding and conformational switching probed by single-molecule fluorescence. *Proc. Natl. Acad. Sci. USA* **2009**, *106*, 5645–5650. [CrossRef] [PubMed]

28. Choi, T.S.; Han, J.Y.; Heo, C.E.; Lee, S.W.; Kim, H.I. Electrostatic and hydrophobic interactions of lipid-associated α-synuclein: The role of a water-limited interfaces in amyloid fibrillation. *Biochim. Biophys. Acta Biomembr.* **2018**, *1860*, 1854–1862. [CrossRef] [PubMed]

29. Spillantini, M.G.; Goedert, M. Neurodegeneration and the ordered assembly of α-synuclein. *Cell Tissue Res.* **2018**, *373*, 137–148. [CrossRef] [PubMed]

30. Greten-Harrison, B.; Polydoro, M.; Morimoto-Tomita, M.; Diao, L.; Williams, A.M.; Nie, E.H.; Makani, S.; Tian, N.; Castillo, P.E.; Buchman, V.L.; et al. αβγ-Synuclein triple knockout mice reveal age-dependent neuronal dysfunction. *Proc. Natl. Acad. Sci. USA* **2010**, *107*, 19573–19578. [CrossRef]

31. Fortin, D.L.; Troyer, M.D.; Nakamura, K.; Kubo, S.; Anthony, M.D.; Edwards, R.H. Lipid rafts mediate the synaptic localization of alpha-synuclein. *J. Neurosci.* **2004**, *24*, 6715–6723. [CrossRef] [PubMed]

32. Burré, J.; Sharma, M.; Südhof, T.C. α-Synuclein assembles into higher-order multimers upon membrane binding to promote SNARE complex formation. *Proc. Natl. Acad. Sci. USA* **2014**, *111*, E4274–E4283. [CrossRef] [PubMed]

33. Logan, T.; Bendor, J.; Toupin, C.; Thorn, K.; Edwards, R.H. α-Synuclein promotes dilation of the exocytotic fusion pore. *Nat. Neurosci.* **2017**, *20*, 681–689. [CrossRef] [PubMed]

34. Gai, W.P.; Yuan, H.X.; Li, X.Q.; Power, J.T.; Blumbergs, P.C.; Jensen, P.H. In situ and in vitro study of colocalization and segregation of alpha-synuclein, ubiquitin, and lipids in Lewy bodies. *Exp. Neurol.* **2000**, *166*, 324–333. [CrossRef] [PubMed]

35. Tuttle, M.D.; Comellas, G.; Nieuwkoop, A.J.; Covell, D.J.; Berthold, D.A.; Kloepper, K.D.; Courtney, J.M.; Kim, J.K.; Barclay, A.M.; Kendall, A.; et al. Solid-state NMR structure of a pathogenic fibril of full-length human α-synuclein. *Nat. Struct. Mol. Biol.* **2016**, *23*, 409–415. [CrossRef] [PubMed]

36. Choi, W.; Zibaee, S.; Jakes, R.; Serpell, L.C.; Davletov, B.; Crowther, R.A.; Goedert, M. Mutation E46K increases phospholipid binding and assembly into filaments of human alpha-synuclein. *FEBS Lett.* **2004**, *576*, 363–368. [CrossRef] [PubMed]

37. Tofaris, G.K.; Goedert, M.; Spillantini, M.G. The Transcellular Propagation and Intracellular Trafficking of α-Synuclein. *Cold Spring Harb. Perspect. Med.* **2017**, *7*, a024380. [CrossRef] [PubMed]

38. Osterberg, V.R.; Spinelli, K.J.; Weston, L.J.; Luk, K.C.; Woltjer, R.L.; Unni, V.K. Progressive aggregation of alpha-synuclein and selective degeneration of Lewy inclusion-bearing neurons in a mouse model of parkinsonism. *Cell Rep.* **2015**, *10*, 1252–1260. [CrossRef] [PubMed]

39. Fusco, G.; Chen, S.W.; Williamson, P.T.F.; Cascella, R.; Perni, M.; Jarvis, J.A.; Cecchi, C.; Vendruscolo, M.; Chiti, F.; Cremades, N.; et al. Structural basis of membrane disruption and cellular toxicity by α-synuclein oligomers. *Science* **2017**, *358*, 1440–1443. [CrossRef] [PubMed]

40. Varela, J.A.; Rodrigues, M.; De, S.; Flagmeier, P.; Gandhi, S.; Dobson, C.M.; Klenerman, D.; Lee, S.F. Optical Structural Analysis of Individual α-Synuclein Oligomers. *Angew. Chem. Int. Ed. Engl.* **2018**, *57*, 4886–4890. [CrossRef] [PubMed]

41. Mattson, M.P. Pathways towards and away from Alzheimer's disease. *Nature* **2004**, *430*, 631–639. [CrossRef] [PubMed]

42. Aleksis, R.; Oleskovs, F.; Jaudzems, K.; Pahnke, J.; Biverstål, H. Structural studies of amyloid-β peptides: Unlocking the mechanism of aggregation and the associated toxicity. *Biochimie* **2017**, *140*, 176–192. [CrossRef] [PubMed]

43. Selkoe, D.J. Alzheimer's disease: Genes, proteins, and therapy. *Physiol. Rev.* **2001**, *81*, 741–766. [CrossRef] [PubMed]

44. Kumari, A.; Rajput, R.; Shrivastava, N.; Somvanshi, P.; Grover, A. Synergistic approaches unraveling regulation and aggregation of intrinsically disordered β-amyloids implicated in Alzheimer's disease. *Int. J. Biochem. Cell Biol.* **2018**, *99*, 19–27. [CrossRef] [PubMed]

45. Kang, J.; Lemaire, H.G.; Unterbeck, A.; Salbaum, J.M.; Masters, C.L.; Grzeschik, K.H.; Multhaup, G.; Beyreuther, K.; Müller-Hill, B. The precursor of Alzheimer's disease amyloid A4 protein resembles a cell-surface receptor. *Nature* **1987**, *325*, 733–736. [CrossRef] [PubMed]

46. Iwatsubo, T.; Odaka, A.; Suzuki, N.; Mizusawa, H.; Nukina, N.; Ihara, Y. Visualization of Aβ42(43) and Aβ40 in senile plaques with end-specific Aβ monoclonals: Evidence that an initially deposited species is Aβ42(43). *Neuron* **1994**, *13*, 45–53. [CrossRef]

47. De Strooper, B.; Saftig, P.; Craessaerts, K.; Vanderstichele, H.; Guhde, G.; Annaert, W.; Von Figura, K.; Van Leuven, F. Deficiency of presenilin-1 inhibits the normal cleavage of amyloid precursor protein. *Nature* **1998**, *391*, 387–390. [CrossRef] [PubMed]

48. Stokin, G.B.; Lillo, C.; Falzone, T.L.; Brusch, R.G.; Rockenstein, E.; Mount, S.L.; Raman, R.; Davies, P.; Masliah, E.; Williams, D.S.; et al. Axonopathy and transport deficits early in the pathogenesis of Alzheimer's disease. *Science* **2005**, *307*, 1282–1288. [CrossRef] [PubMed]

49. Lewis, H.; Beher, D.; Cookson, N.; Oakley, A.; Piggott, M.; Morris, C.M.; Jaros, E.; Perry, R.; Ince, P.; Kenny, R.A.; et al. Quantification of Alzheimer pathology in ageing and dementia: Age-related accumulation of amyloid-beta(42) peptide in vascular dementia. *Neuropathol. Appl. Neurobiol.* **2006**, *32*, 103–118. [CrossRef] [PubMed]

50. Sosa, L.J.; Caceres, A.; Dupraz, S.; Oksdath, M.; Quiroga, S.; Lorenzo, A. The physiological role of the amyloid precursor protein as an adhesion molecule in the developing nervous system. *J. Neurochem.* **2017**, *143*, 11–29. [CrossRef] [PubMed]

51. Goldgaber, D.; Lerman, M.I.; McBride, O.W.; Saffiotti, U.; Gajdusek, D.C. Characterization and chromosomal localization of a cDNA encoding brain amyloid of Alzheimer's disease. *Science* **1987**, *235*, 877–880. [CrossRef] [PubMed]

52. Tanzi, R.E.; Gusella, J.F.; Watkins, P.C.; Bruns, G.A.; St George-Hyslop, P.; Van Keuren, M.L.; Patterson, D.; Pagan, S.; Kurnit, D.M.; Neve, R.L. Amyloid beta protein gene: cDNA, mRNA distribution, and genetic linkage near the Alzheimer locus. *Science* **1987**, *235*, 880–884. [CrossRef] [PubMed]

53. Haass, C. Take five—BACE and the γ-secretase quartet conduct Alzheimer's amyloid β-peptide generation. *EMBO J.* **2004**, *23*, 483–488. [CrossRef] [PubMed]

54. Haass, C.; Schlossmacher, M.G.; Hung, A.Y.; Vigo-Pelfrey, C.; Mellon, A.; Ostaszewski, B.L.; Lieberburg, I.; Koo, E.H.; Schenk, D.; Teplow, D.B.; et al. Amyloid β-peptide is produced by cultured cells during normal metabolism. *Nature* **1992**, *359*, 322–325. [CrossRef] [PubMed]

55. Gouras, G.K.; Tsai, J.; Naslund, J.; Vincent, B.; Edgar, M.; Checler, F.; Greenfield, J.P.; Haroutunian, V.; Buxbaum, J.D.; Xu, H.; et al. Intraneuronal Aβ42 accumulation in human brain. *Am. J. Pathol.* **2000**, *156*, 15–20. [CrossRef]

56. Bitan, G.; Vollers, S.S.; Teplow, D.B. Elucidation of primary structure elements controlling early amyloid β-protein oligomerization. *J. Biol. Chem.* **2003**, *12*, 34882–34889. [CrossRef] [PubMed]

57. Suzuki, N.; Cheung, T.T.; Cai, X.-D.; Odaka, A.; Otvos, L.; Eckman, C.; Golde, T.E.; Younkin, S.G. An increased percentage of long amyloid beta protein secreted by familial amyloid beta protein precursor (beta APP717) mutants. *Science* **1994**, *264*, 1336–1340. [CrossRef] [PubMed]

58. Lovell, M.A.; Robertson, J.D.; Teesdale, W.J.; Campbell, J.L.; Markesbery, W.R. Copper, iron and zinc in Alzheimer's disease senile plaques. *J. Neurol. Sci.* **1998**, *158*, 47–52. [CrossRef]

59. Alexandrescu, A.T. Amyloid accomplices and enforcers. *Protein Sci.* **2005**, *14*, 1–12. [CrossRef] [PubMed]

60. Scheuner, D.; Eckman, C.; Jensen, M.; Song, X.; Citron, M.; Suzuki, N.; Bird, T.D.; Hardy, J.; Hutton, M.; Kukull, W.; et al. Secreted amyloid beta-protein similar to that in the senile plaques of Alzheimer's disease is increased in vivo by the presenilin 1 and 2 and APP mutations linked to familial Alzheimer's disease. *Nat. Med.* **1996**, *2*, 864–870. [CrossRef] [PubMed]

61. Citron, M.; Westaway, D.; Xia, W.; Carlson, G.; Diehl, T.; Levesque, G.; Johnson-Wood, K.; Lee, M.; Seubert, P.; Davis, A.; et al. Mutant presenilins of Alzheimer's disease increase production of 42-residue amyloid beta-protein in both transfected cells and transgenic mice. *Nat. Med.* **1997**, *3*, 67–72. [CrossRef] [PubMed]

62. Mullan, M.; Crawford, F.; Axelman, K.; Houlden, H.; Lilius, L.; Winblad, B.; Lannfelt, L. A pathogenic mutation for probable Alzheimer's disease in the APP gene at the N-terminus of β-amyloid. *Nat. Genet.* **1992**, *1*, 345–347. [CrossRef] [PubMed]

63. Sherrington, R.; Rogaev, E.I.; Liang, Y.; Rogaeva, E.A.; Levesque, G.; Ikeda, M.; Chi, H.; Lin, C.; Li, G.; Holman, K.; et al. Cloning of a gene bearing missense mutations in early-onset familial Alzheimer's disease. *Nature* **1995**, *375*, 754–760. [CrossRef] [PubMed]

64. Selkoe, D.J.; Hardy, J. The amyloid hypothesis of Alzheimer's disease at 25 years. *EMBO Mol. Med.* **2016**, *8*, 595–608. [CrossRef] [PubMed]

65. Lanoiselée, H.M.; Nicolas, G.; Wallon, D.; Rovelet-Lecrux, A.; Lacour, M.; Rousseau, S.; Richard, A.C.; Pasquier, F.; Rollin-Sillaire, A.; Martinaud, O.; et al. APP, PSEN1, and PSEN2 mutations in early-onset Alzheimer disease: A genetic screening study of familial and sporadic cases. *PLoS Med.* **2017**, *14*, e1002270. [CrossRef] [PubMed]

66. Simmons, L.K.; May, P.C.; Tomaselli, K.J.; Rydel, R.E.; Fuson, K.S.; Brigham, E.F.; Wright, S.; Lieberburg, I.; Becker, G.W.; Brems, D.N. Secondary structure of amyloid beta peptide correlates with neurotoxic activity in vitro. *Mol. Pharmacol.* **1994**, *45*, 373–379. [PubMed]

67. Kirkitadze, M.D.; Condron, M.M.; Teplow, D.B. Identification and characterization of key kinetic intermediates in amyloid beta-protein fibrillogenesis. *J. Mol. Biol.* **2001**, *312*, 1103–1119. [CrossRef] [PubMed]

68. Yan, Y.; Wang, C. Aβ42 is More Rigid than Aβ40 at the C Terminus: Implications for Aβ Aggregation and Toxicity. *J. Mol. Biol.* **2006**, *364*, 853–862. [CrossRef] [PubMed]

69. Petkova, A.T.; Ishii, Y.; Balbach, J.J.; Antzutkin, O.N.; Leapman, R.D.; Delaglio, F.; Tycko, R. A structural model for Alzheimer's beta-amyloid fibrils based on experimental constraints from solid state NMR. *Proc. Natl. Acad. Sci. USA* **2002**, *99*, 16742–16747. [CrossRef] [PubMed]

70. Petkova, A.T.; Leapman, R.D.; Guo, Z.; Yau, W.M.; Mattson, M.P.; Tycko, R. Self-propagating, molecular-level polymorphism in Alzheimer's β-amyloid fibrils. *Science* **2005**, *307*, 262–265. [CrossRef] [PubMed]

71. Bertini, I.; Gonnelli, L.; Luchinat, C.; Mao, J.; Nesi, A. A new structural model of Aβ40 fibrils. *J. Am. Chem. Soc.* **2011**, *133*, 16013–16022. [CrossRef] [PubMed]

72. Parthasarathy, S.; Long, F.; Miller, Y.; Xiao, Y.; McElheny, D.; Thurber, K.; Ma, B.; Nussinov, R.; Ishii, Y. Molecular-level examination of Cu^{2+} binding structure for amyloid fibrils of 40-residue Alzheimer's β by solid-state NMR spectroscopy. *J. Am. Chem. Soc.* **2011**, *133*, 3390–3400. [CrossRef] [PubMed]

73. Sgourakis, N.G.; Yau, W.M.; Qiang, W. Modeling an in-register, parallel "Iowa" Aβ fibril structure using solid-state NMR data from labeled samples with Rosetta. *Structure* **2015**, *23*, 216–227. [CrossRef] [PubMed]

74. Xiao, Y.; Ma, B.; McElheny, D.; Parthasarathy, S.; Long, F.; Hoshi, M.; Nussinov, R.; Ishii, Y. Aβ(1-42) fibril structure illuminates self-recognition and replication of amyloid in Alzheimer's disease. *Nat. Struct. Mol. Biol.* **2015**, *22*, 499–505. [CrossRef] [PubMed]

75. Colvin, M.T.; Silvers, R.; Ni, Q.Z.; Can, T.V.; Sergeyev, I.; Rosay, M.; Donovan, K.J.; Michael, B.; Wall, J.; Linse, S.; et al. Atomic Resolution Structure of Monomorphic Aβ42 Amyloid Fibrils. *J. Am. Chem. Soc.* **2016**, *138*, 9663–9674. [CrossRef] [PubMed]

76. Wälti, M.A.; Ravotti, F.; Arai, H.; Glabe, C.G.; Wall, J.S.; Böckmann, A.; Güntert, P.; Meier, B.H.; Riek, R. Atomic-resolution structure of a disease-relevant Aβ(1–42) amyloid fibril. *Proc. Natl. Acad. Sci. USA* **2016**, *113*, E4976–E4984. [CrossRef] [PubMed]

77. Masters, C.L.; Multhaup, G.; Simms, G.; Martins, R.N.; Beyreuther, K. Neuronal origin of a cerebral amyloid: Neurofibrilliary tangles of Alzheimer's disease contain the same protein as the amyloid of plaque cores and blood vessels. *EMBO J.* **1985**, *4*, 2757–2763. [CrossRef] [PubMed]

78. Portelius, E.; Bogdanovic, N.; Gustavsson, M.K.; Volkmann, I.; Brinkmalm, G.; Zetterberg, H.; Winblad, B.; Blennow, K. Mass spectrometric characterization of brain amyloid beta isoform signatures in familial and sporadic Alzheimer's disease. *Acta Neuropathol.* **2010**, *120*, 185–193. [CrossRef] [PubMed]

79. Weingarten, M.D.; Lockwood, A.H.; Hwo, S.Y.; Kirschner, M.W. A protein factor essential for microtubule assembly. *Proc. Natl. Acad. Sci. USA* **1975**, *72*, 1858–1862. [CrossRef] [PubMed]

80. Ebneth, A.; Godemann, R.; Stamer, K.; Illenberger, S.; Trinczek, B.; Mandelkow, E. Overexpression of tau protein inhibits kinesin-dependent trafficking of vesicles, mitochondria, and endoplasmic reticulum: Implications for Alzheimer's disease. *J. Cell Biol.* **1998**, *143*, 777–794. [CrossRef] [PubMed]

81. Borna, H.; Assadoulahei, K.; Riazi, G.; Harchegani, A.B.; Shahriary, A. Structure, Function and Interactions of Tau: Particular Focus on Potential Drug Targets for the Treatment of Tauopathies. *CNS Neurol. Disord Drug Targets* **2018**, *17*, 325–337. [CrossRef] [PubMed]

82. Bakota, L.; Ussif, A.; Jeserich, G.; Brandt, R. Systemic and network functions of the microtubule-associated protein tau: Implications for tau-based therapies. *Mol. Cell. Neurosci.* **2017**, *84*, 132–141. [CrossRef]

83. Kosik, K.S.; Joachim, C.L.; Selkoe, D.J. Microtubule-associated protein tau (tau) is a major antigenic component of paired helical filaments in Alzheimer disease. *Proc. Natl. Acad. Sci. USA* **1986**, *83*, 4044–4448. [CrossRef] [PubMed]

84. Andreadis, A.; Brown, W.M.; Kosik, K.S. Structure and novel exons of the human tau gene. *Biochemistry* **1992**, *31*, 10626–10633. [CrossRef] [PubMed]

85. Goedert, M.; Spillantini, M.G.; Jakes, R.; Rutherford, D.; Crowther, R.A. Multiple isoforms of human microtubule-associated protein tau: Sequences and localization in neurofibrillary tangles of Alzheimer's disease. *Neuron* **1989**, *3*, 519–526. [CrossRef]

86. Goedert, M.; Spillantini, M.G.; Potier, M.C.; Ulrich, J.; Crowther, R.A. Cloning and sequencing of the cDNA encoding an isoform of microtubule-associated protein tau containing four tandem repeats: Differential expression of tau protein mRNAs in human brain. *EMBO J.* **1989**, *8*, 393–399. [CrossRef] [PubMed]

87. Goedert, M.; Wischik, C.M.; Crowther, R.A.; Walker, J.E.; Klug, A. Cloning and sequencing of the cDNA encoding a core protein of the paired helical filament of Alzheimer disease: Identification as the microtubule-associated protein tau. *Proc. Natl. Acad. Sci. USA* **1988**, *85*, 4051–4055. [CrossRef] [PubMed]

88. Lee, V.M.; Balin, B.J.; Otvos, L., Jr.; Trojanowski, J.Q. A68: A major subunit of paired helical filaments and derivatized forms of normal Tau. *Science* **1991**, *251*, 675–678. [CrossRef] [PubMed]

89. Santa-Maria, I.; Varghese, M.; Ksiezak-Reding, H.; Dzhun, A.; Wang, J.; Pasinetti, G.M. Paired helical filaments from Alzheimer disease brain induce intracellular accumulation of Tau protein in aggresomes. *J. Biol. Chem.* **2012**, *287*, 20522–20533. [CrossRef] [PubMed]

90. Bhattacharya, K.; Rank, K.B.; Evans, D.B.; Sharma, S.K. Role of cysteine-291 and cysteine-322 in the polymerization of human tau into Alzheimer-like filaments. *Biochem. Biophys. Res. Commun.* **2001**, *285*, 20–26. [CrossRef] [PubMed]

91. Novak, P.; Cehlar, O.; Skrabana, R.; Novak, M. Tau Conformation as a Target for Disease-Modifying Therapy: The Role of Truncation. *J. Alzheimers Dis.* **2018**, *64*, S535–S546. [CrossRef] [PubMed]

92. Florenzano, F.; Veronica, C.; Ciasca, G.; Ciotti, M.T.; Pittaluga, A.; Olivero, G.; Feligioni, M.; Iannuzzi, F.; Latina, V.; Maria Sciacca, M.F.; et al. Extracellular truncated tau causes early presynaptic dysfunction associated with Alzheimer's disease and other tauopathies. *Oncotarget* **2017**, *8*, 64745–64778. [CrossRef] [PubMed]

93. Schedin-Weiss, S.; Winblad, B.; Tjernberg, L.O. The role of protein glycosylation in Alzheimer disease. *FEBS J.* **2014**, *281*, 46–62. [CrossRef] [PubMed]

94. Ligabue-Braun, R.; Carlini, C.R. Moonlighting Toxins: Ureases and Beyond. In *Plant Toxins*; Gopalakrishnakone, P., Carlini, C.R., Ligabue-Braun, R., Eds.; Springer: Dordrecht, The Netherlands, 2015; pp. 199–219.

95. Oliveira, J.M.; Henriques, A.G.; Martins, F.; Rebelo, S.; da Cruz e Silva, O.A. Amyloid-beta Modulates Both AbetaPP and Tau Phosphorylation. *J. Alzheimers Dis.* **2015**, *45*, 495–507. [CrossRef] [PubMed]

96. Ittner, L.M.; Ke, Y.D.; Delerue, F.; Bi, M.; Gladbach, A.; van Eersel, J.; Wölfing, H.; Chieng, B.C.; Christie, M.J.; Napier, I.A.; et al. Dendritic function of tau mediates amyloid-beta toxicity in Alzheimer's disease mouse models. *Cell* **2010**, *142*, 387–397. [CrossRef] [PubMed]

97. Prusiner, S.B. Novel proteinaceous infectious particles cause scrapie. *Science* **1982**, *216*, 136–144. [CrossRef] [PubMed]

98. Prusiner, S.B.; Mckinley, M.P.; Groth, D.F.; Bowman, K.A.; Mock, N.I.; Cochran, S.P.; Masiarz, F.R. Scrapie agent contains a hydrophobic protein. *Proc. Natl. Acad. Sci. USA* **1981**, *78*, 6675–6679. [CrossRef] [PubMed]

99. Prusiner, S.B.; Groth, D.F.; Bolton, D.C.; Kent, S.B.; Hood, L.E. Purification and structural studies of a major scrapie prion protein. *Cell* **1984**, *38*, 127–134. [CrossRef]

100. Westergard, L.; Christensen, H.M.; Harris, D.A. The cellular prion protein (PrP(C)): Its physiological function and role in Disease. *Biochim. Biophys. Acta* **2007**, *1772*, 629–644. [CrossRef] [PubMed]
101. Bremer, J.; Baumann, F.; Tiberi, C.; Wessig, C.; Fischer, H.; Schwarz, P.; Steele, A.D.; Toyka, K.V.; Nave, K.A.; Weis, J.; et al. Axonal prion protein is required for peripheral myelin maintenance. *Nat. Neurosci.* **2010**, *13*, 310–318. [CrossRef] [PubMed]
102. Chakravarty, A.K.; Jarosz, D.F. More than Just a Phase: Prions at the Crossroads of Epigenetic Inheritance and Evolutionary Change. *J. Mol. Biol.* **2018**, *430*, 4607–4618. [CrossRef] [PubMed]
103. Riek, R.; Hornemann, S.; Wider, G.; Glockshuber, R.; Wüthrich, K. NMR characterization of the full-length recombinant murine prion protein, mPrP(23-231). *FEBS Lett.* **1997**, *413*, 282–288. [CrossRef]
104. Donne, D.G.; Viles, J.H.; Groth, D.; Mehlhorn, I.; James, T.L.; Cohen, F.E.; Prusiner, S.B.; Wright, P.E.; Dyson, H.J. Structure of the recombinant full-length hamster prion protein PrP (29-231): The N terminus is highly flexible. *Proc. Natl. Acad. Sci. USA* **1997**, *94*, 13452–13457. [CrossRef] [PubMed]
105. Zahn, R.; Liu, A.; Luhrs, T.; Riek, R.; von Schroetter, C.; Lopez-Garcia, F.; Billeter, M.; Calzolai, L.; Wider, G.; Wuthrich, K. NMR solution structure of the human prion protein. *Proc. Natl. Acad. Sci. USA* **2000**, *97*, 145–150. [CrossRef] [PubMed]
106. Lopez-García, F.L.; Zahn, R.; Riek, R.; Wüthrich, K. NMR structure of the bovine prion protein. *Proc. Natl. Acad. Sci. USA* **2000**, *97*, 8334–8339. [CrossRef] [PubMed]
107. Sabate, R.; Rousseau, F.; Schymkowitz, J.; Ventura, S. What makes a protein sequence a prion? *PLoS Comput. Biol.* **2015**, *11*, e1004013. [CrossRef] [PubMed]
108. Cong, X.; Casiraghi, N.; Rossetti, G.; Mohanty, S.; Giachin, G.; Legname, G.; Carloni, P. Role of Prion Disease-Linked Mutations in the Intrinsically Disordered N-Terminal Domain of the Prion Protein. *J. Chem. Theory Comput.* **2013**, *9*, 5158–5167. [CrossRef] [PubMed]
109. Pan, K.-M.; Baldwin, M.; Nguyen, J.; Gasset, M.; Serban, A.N.A.; Groth, D.; Mehlhorn, I.; Huang, Z.; Fletterick, R.J.; Cohen, F.E. Conversion of alpha-helices into beta-sheets features in the formation of the scrapie prion proteins. *Proc. Natl. Acad. Sci. USA* **1993**, *90*, 10962–10966. [CrossRef] [PubMed]
110. Prusiner, S.B. Molecular biology of prion diseases. *Science* **1991**, *252*, 1515–1522. [CrossRef] [PubMed]
111. Prusiner, S.B. Chemistry and biology of prions. *Biochemistry* **1992**, *31*, 12277–12288. [CrossRef] [PubMed]
112. Brown, P.; Gajdusek, D.C. The Human Spongiform Encephalopathies: Kuru, Creutzfeldt-Jakob Disease, and the Gerstmann-Sträussler-Scheinker Syndrome. In *Transmissible Spongiform Encephalopathies: Current Topics in Microbiology and Immunology*; Chesebro, B.W., Ed.; Springer: Berlin, Germany, 1991; Volume 172, pp. 1–20.
113. Nathanson, N.; Wilesmith, J.; Griot, C. Bovine spongiform encephalopathy (BSE): Causes and consequences of a common source epidemic. *Am. J. Epidemiol.* **1997**, *145*, 959–969. [CrossRef] [PubMed]
114. Pattison, I.H. The relative susceptibility of sheep, goats and mice to two types of the goat scrapie agent. *Res. Vet. Sci.* **1966**, *7*, 207–212. [CrossRef]
115. Scott, M.; Foster, D.; Mirenda, C.; Serban, D.; Coufal, F.; Wälchli, M.; Torchia, M.; Groth, D.; Carlson, G.; DeArmond, S.J.; et al. Transgenic mice expressing hamster prion protein produce species-specific scrapie infectivity and amyloid plaques. *Cell* **1989**, *59*, 847–857. [CrossRef]
116. Prusiner, S.B.; Scott, M.; Foster, D.; Pan, K.-M.; Groth, D.; Mirenda, C.; Torchia, M.; Yang, S.-L.; Serban, D.; Carlson, G.A.; et al. Transgenetic studies implicate interactions between homologous PrP isoforms in scrapie prion replication. *Cell* **1990**, *63*, 673–686. [CrossRef]
117. Bartz, J.C.; McKenzie, D.I.; Bessen, R.A.; Marsh, R.F.; Aiken, J.M. Transmissible mink encephalopathy species barrier effect between ferret and mink: PrP gene and protein analysis. *J. Gen. Virol.* **1994**, *75*, 2947–2953. [CrossRef] [PubMed]
118. Bian, J.; Khaychuk, V.; Angers, R.C.; Fernández-Borges, N.; Vidal, E.; Meyerett-Reid, C.; Kim, S.; Calvi, C.L.; Bartz, J.C.; Hoover, E.A.; et al. Prion replication without host adaptation during interspecies transmissions. *Proc. Natl. Acad. Sci. USA* **2017**, *114*, 1141–1146. [CrossRef] [PubMed]
119. Prusiner, S.B. Molecular biology and pathogenesis of prion diseases. *Trends Biochem. Sci.* **1996**, *21*, 482–487. [CrossRef]
120. Colby, D.W.; Prusiner, S.B. Prions. *Cold Spring Harb. Perspect. Biol.* **2011**, *3*, a006833. [CrossRef] [PubMed]
121. Makarava, N.; Kovacs, G.G.; Bocharova, O.; Savtchenko, R.; Alexeeva, I.; Budka, H.; Rohwer, R.G.; Baskakov, I.V. Recombinant prion protein induces a new transmissible prion disease in wild-type animals. *Acta Neuropathol.* **2010**, *119*, 177–187. [CrossRef] [PubMed]

122. Stahl, N.; Baldwin, M.A.; Prusiner, S.B.; Teplow, D.B.; Hood, L.; Gibson, B.W.; Burlingame, A.L. Structural Studies of the Scrapie Prion Protein Using Mass Spectrometry and Amino Acid Sequencing. *Biochemistry* **1993**, *32*, 1991–2002. [CrossRef] [PubMed]

123. Meyer, R.K.; McKinley, M.P.; Bowman, K.A.; Braunfeld, M.B.; Barry, R.A.; Prusiner, S.B. Separation and properties of cellular and scrapie prion proteins. *Proc. Natl. Acad. Sci. USA* **1986**, *83*, 2310–2314. [CrossRef] [PubMed]

124. Saverioni, D.; Notari, S.; Capellari, S.; Poggiolini, I.; Giese, A.; Kretzschmar, H.A.; Parchi, P. Analyses of protease resistance and aggregation state of abnormal prion protein across the spectrum of human prions. *J. Biol. Chem.* **2013**, *288*, 27972–27985. [CrossRef] [PubMed]

125. Safar, J.; Wille, H.; Itri, V.; Groth, D.; Serban, H.; Torchia, M.; Cohen, F.E.; Prusiner, S.B. Eight prion strains have PrP Sc molecules with different conformations. *Nat. Med.* **1998**, *4*, 1157–1165. [CrossRef] [PubMed]

126. Tzaban, S.; Friedlander, G.; Schonberger, O.; Horonchik, L.; Yedidia, Y.; Shaked, G.; Gabizon, R.; Taraboulos, A. Protease-sensitive scrapie prion protein in aggregates of heterogeneous sizes. *Biochemistry* **2002**, *41*, 12868–12875. [PubMed]

127. DeArmond, S.J.; Sánchez, H.; Yehiely, F.; Qiu, Y.; Ninchak-Casey, A.; Daggett, V.; Camerino, A.P.; Cayetano, J.; Rogers, M.; Groth, D.; et al. Selective neuronal targeting in prion disease. *Neuron* **1997**, *19*, 1337–1348. [CrossRef]

128. Colby, D.W.; Zhang, Q.; Wang, S.; Groth, D.; Legname, G.; Riesner, D.; Prusiner, S.B. Prion detection by an amyloid seeding assay. *Proc. Natl. Acad. Sci. USA* **2007**, *104*, 20914–20919. [CrossRef] [PubMed]

129. Wille, H.; Prusiner, S.B.; Cohen, F.E. Scrapie infectivity is independent of amyloid staining properties of the N-Terminally truncated prion protein. *J. Struct. Biol.* **2000**, *130*, 323–338. [CrossRef] [PubMed]

130. Uversky, V.N.; Davé, V.; Iakoucheva, L.M.; Malaney, P.; Metallo, S.J.; Pathak, R.R.; Joerger, A.C. Pathological unfoldomics of uncontrolled chaos: Intrinsically disordered proteins and human diseases. *Chem. Rev.* **2014**, *114*, 6844–6879. [CrossRef] [PubMed]

131. Mol, P.R. Oncogenes as Therapeutic Targets in Cancer: A Review. *IOSR J. Dent. Med. Sci.* **2013**, *5*, 46–56. [CrossRef]

132. Dunker, A.K.; Uversky, V.N. Drugs for "protein clouds": Targeting intrinsically disordered transcription factors. *Curr. Opin. Pharmacol.* **2010**, *10*, 782–788. [CrossRef] [PubMed]

133. Uversky, V.N. p53 Proteoforms and Intrinsic Disorder: An Illustration of the Protein Structure-Function Continuum Concept. *Int. J. Mol. Sci.* **2016**, *17*, 1874. [CrossRef] [PubMed]

134. Levine, A.J. Targeting therapies for the p53 protein in cancer treatments. *Annu. Rev. Cancer Biol.* **2019**. [CrossRef]

135. Hollstein, M.; Sidransky, D.; Vogelstein, B.; Curtis, C. p53 Mutation Human Cancers. *Science* **1991**, *253*, 49–53. [CrossRef] [PubMed]

136. Muller, P.A.J.; Vousden, K.H. P53 mutations in cancer. *Nat. Cell Biol.* **2013**, *15*, 2–8. [CrossRef] [PubMed]

137. Dawson, R.; Müller, L.; Dehner, A.; Klein, C.; Kessler, H.; Buchner, J. The N-terminal domain of p53 is natively unfolded. *J. Mol. Biol.* **2003**, *332*, 1131–1141. [CrossRef] [PubMed]

138. Kubbutat, M.H.G.; Jones, S.N.; Vousden, K.H. Regulation of p53 stability by Mdm2. *Nature* **1997**, *387*, 299–303. [CrossRef] [PubMed]

139. Nag, S.; Qin, J.; Srivenugopal, K.S.; Wang, M.; Zhang, R. The MDM2-p53 pathway revisited. *J. Biomed. Res.* **2013**, *27*, 254–271. [PubMed]

140. Williams, A.B.; Schumacher, B. p53 in the DNA-damage-repair process. *Cold Spring Harb. Perspect. Med.* **2016**, *6*, a026070. [CrossRef] [PubMed]

141. Cho, Y.; Gorina, S.; Jeffrey, P.D.; Pavletich, N.P. Crystal structure of a p53 tumor suppressor-DNA complex: Understanding tumorigenic mutations. *Science* **1994**, *265*, 346–355. [CrossRef] [PubMed]

142. Clore, G.M.; Ernst, J.; Clubb, R.; Omichinski, J.G.; Kennedy, W.M.P.; Sakaguchi, K.; Appella, E.; Gronenborn, A.M. Refined solution structure of the oligomerization domain of the tumour suppressor p53. *Nat. Struct. Mol. Biol.* **1995**, *2*, 321–333. [CrossRef]

143. Fields, S.; Jang, S.K. Presence of a potent transcription activating sequence in the p53 protein. *Science* **1990**, *249*, 1046–1049. [CrossRef] [PubMed]

144. Haupt, Y.; Maya, R.; Kazaz, A.; Oren, M. Mdm2 promotes the rapid degradation of p53. *Nature* **1997**, *387*, 296–299. [CrossRef] [PubMed]

145. Honda, R.; Tanaka, H.; Yasuda, H. Oncoprotein MDM2 is a ubiquitin ligase E3 for tumor suppressor p53. *FEBS Lett.* **1997**, *420*, 25–27. [CrossRef]

146. Joerger, A.C.; Fersht, A.R. Structural Biology of the Tumor Suppressor p53. *Annu. Rev. Biochem.* **2008**, *77*, 557–582. [CrossRef] [PubMed]

147. Lee, W.; Harvey, T.S.; Yin, Y.; Yau, P.; Litchfield, D.; Arrowsmith, C.H. Solution structure of the tetrameric minimum transforming domain of p53. *Nat. Struct. Mol. Biol.* **1994**, *1*, 877–890. [CrossRef]

148. Uversky, A.V.; Xue, B.; Peng, Z.; Kurgan, L.; Uversky, V.N. On the intrinsic disorder status of the major players in programmed cell death pathways. *F1000Research* **2013**, *2*, 190. [CrossRef] [PubMed]

149. Kussie, P.H.; Gorina, S.; Marechal, V.; Elenbaas, B.; Moreau, J.; Levine, A.J.; Pavletich, N.P. Structure of the MDM2 oncoprotein bound to the p53 tumor suppressor transactivation domain. *Science* **1996**, *274*, 948–953. [CrossRef] [PubMed]

150. Marine, J.-C.; Jochemsen, A.G. Mdmx as an essential regulator of p53 activity. *Biochem. Biophys. Res. Commun.* **2005**, *331*, 750–760. [CrossRef] [PubMed]

151. Schon, O.; Friedler, A.; Bycroft, M.; Freund, S.M.V.; Fersht, A.R. Molecular mechanism of the interaction between MDM2 and p53. *J. Mol. Biol.* **2002**, *323*, 491–501. [CrossRef]

152. Borcherds, W.; Kashtanov, S.; Wu, H.; Daughdrill, G.W. Structural divergence is more extensive than sequence divergence for a family of intrinsically disordered proteins. *Proteins Struct. Funct. Bioinform.* **2013**, *81*, 1686–1698. [CrossRef] [PubMed]

153. Chi, S.W.; Lee, S.H.; Kim, D.H.; Ahn, M.J.; Kim, J.S.; Woo, J.Y.; Torizawa, T.; Kainosho, M.; Han, K.H. Structural details on mdm2-p53 interaction. *J. Biol. Chem.* **2005**, *280*, 38795–38802. [CrossRef] [PubMed]

154. Popowicz, G.M.; Czarna, A.; Rothweiler, U.; Szwagierczak, A.; Krajewski, M.; Weber, L.; Holak, T.A. Molecular basis for the inhibition of p53 by Mdmx. *Cell Cycle* **2007**, *6*, 2386–2392. [CrossRef] [PubMed]

155. Vise, P.D.; Baral, B.; Latos, A.J.; Daughdrill, G.W. NMR chemical shift and relaxation measurements provide evidence for the coupled folding and binding of the p53 transactivation domain. *Nucleic Acids Res.* **2005**, *33*, 2061–2077. [CrossRef] [PubMed]

156. Chene, P. The role of tetramerization in p53 function. *Oncogene* **2001**, *20*, 2611–2617. [CrossRef] [PubMed]

157. Pelengaris, S.; Khan, M.; Evan, G. c-MYC: More than just a matter of life and death. *Nat. Rev. Cancer* **2002**, *2*, 764–776. [CrossRef] [PubMed]

158. Dang, C.V. c-Myc Target Genes Involved in Cell Growth, Apoptosis, and Metabolism. *Mol. Cell. Biol.* **1999**, *19*, 1–11. [CrossRef] [PubMed]

159. Nesbit, C.E.; Tersak, J.M.; Prochownik, E.V. MYC oncogenes and human neoplastic disease. *Oncogene* **1999**, *18*, 3004–3016. [CrossRef] [PubMed]

160. Schlagbauer-Wadl, H.; Griffioen, M.; Van Elsas, A.; Schrier, P.I.; Pustelnik, T.; Eichler, H.; Wolff, K.; Pehamberger, H.; Jansen, B. Influence of Increased c-Myc Expression on the Growth Characteristics of Human Melanoma. *J. Investig. Dermatol.* **1999**, *112*, 332–336. [CrossRef] [PubMed]

161. Kumar, D.; Sharma, N.; Giri, R. Therapeutic interventions of cancers using intrinsically disordered proteins as drug targets: C-myc as model system. *Cancer Inform.* **2017**, *16*. [CrossRef] [PubMed]

162. Amati, B.; Dalton, S.; Brooks, M.W.; Littlewood, T.D.; Evan, G.I.; Land, H. Transcriptional activation by the human c-Myc oncoprotein in yeast requires interaction with Max. *Nature* **1992**, *359*, 423–426. [CrossRef] [PubMed]

163. Blackwell, T.K.; Kretzner, L.; Blackwood, E.M.; Eisenman, R.N.; Weintraub, H. Sequence-specific DNA binding by the c-Myc protein. *Science* **1990**, *250*, 1149–1151. [CrossRef] [PubMed]

164. Blackwood, E.M.; Eisenman, R.N. Max: A helix-loop-helix zipper protein that forms a sequence-specific DNA-binding complex with Myc. *Science* **1991**, *251*, 1211–1217. [CrossRef] [PubMed]

165. Kato, G.J.; Lee, W.M.; Chen, L.L.; Dang, C.V. Max: Functional domains and interaction with c-Myc. *Genes Dev.* **1992**, *6*, 81–92. [CrossRef] [PubMed]

166. Andresen, C.; Helander, S.; Lemak, A.; Farès, C.; Csizmok, V.; Carlsson, J.; Penn, L.Z.; Forman-Kay, J.D.; Arrowsmith, C.H.; Lundström, P.; et al. Transient structure and dynamics in the disordered c-Myc transactivation domain affect Bin1 binding. *Nucleic Acids Res.* **2012**, *40*, 6353–6366. [CrossRef] [PubMed]

167. Mao, D.Y.L.; Watson, J.D.; Yan, P.S.; Barsyte-Lovejoy, D.; Khosravi, F.; Wong, W.W.-L.; Farnham, P.J.; Huang, T.H.-M.; Penn, L.Z. Analysis of Myc bound loci identified by CpG island arrays shows that Max is essential for Myc-dependent repression. *Curr. Biol.* **2003**, *13*, 882–886. [CrossRef]

168. Clausen, D.M.; Guo, J.; Parise, R.A.; Beumer, J.H.; Egorin, M.J.; Lazo, J.S.; Prochownik, E.V.; Eiseman, J.L. In vitro cytotoxicity and in vivo efficacy, pharmacokinetics, and metabolism of 10074-G5, a novel small-molecule inhibitor of c-Myc/Max dimerization. *J. Pharmacol. Exp. Ther.* **2010**, *335*, 715–727. [CrossRef] [PubMed]

169. Raffeiner, P.; Röck, R.; Schraffl, A.; Hartl, M.; Hart, J.R.; Janda, K.D.; Vogt, P.K.; Stefan, E.; Bister, K. In vivo quantification and perturbation of Myc-Max interactions and the impact on oncogenic potential. *Oncotarget* **2014**, *5*, 8869–8878. [CrossRef]

170. Ferrannini, E. Insulin resistance versus insulin deficiency in non-insulin-dependent diabetes mellitus: Problems and prospects. *Endocr. Rev.* **1998**, *19*, 477–490. [CrossRef] [PubMed]

171. Cooper, G.J.; Willis, A.C.; Clark, A.; Turner, R.C.; Sim, R.B.; Reid, K.B. Purification and characterization of a peptide from amyloid-rich pancreases of type 2 diabetic patients. *Proc. Natl. Acad. Sci. USA* **1987**, *84*, 8628–8632. [CrossRef] [PubMed]

172. Westermark, P.; Wernstedt, C.; Wilander, E.; Hayden, D.W.; O'Brien, T.D.; Johnson, K.H. Amyloid fibrils in human insulinoma and islets of Langerhans of the diabetic cat are derived from a neuropeptide-like protein also present in normal islet cells. *Proc. Natl. Acad. Sci. USA* **1987**, *84*, 3881–3885. [CrossRef] [PubMed]

173. Mosselman, S.; Höppener, J.W.; Zandberg, J.; van Mansfeld, A.D.; Geurts van Kessel, A.H.; Lips, C.J.; Jansz, H.S. Islet amyloid polypeptide: Identification and chromosomal localization of the human gene. *FEBS Lett.* **1988**, *239*, 227–232. [CrossRef]

174. Kapurniotu, A. Amyloidogenicity and cytotoxicity of islet amyloid polypeptide. *Biopolymers* **2001**, *60*, 438–459. [CrossRef]

175. Moore, S.J.; Sonar, K.; Bharadwaj, P.; Deplazes, E.; Mancera, R.L. Characterisation of the Structure and Oligomerisation of Islet Amyloid Polypeptides (IAPP): A Review of Molecular Dynamics Simulation Studies. *Molecules* **2018**, *23*, 2142. [CrossRef] [PubMed]

176. Goldsbury, C.; Goldie, K.; Pellaud, J.; Seelig, J.; Frey, P.; Müller, S.A.; Kistler, J.; Cooper, G.J.; Aebi, U. Amyloid fibril formation from full-length and fragments of amylin. *J. Struct. Biol.* **2000**, *130*, 352–362. [CrossRef] [PubMed]

177. Yonemoto, I.T.; Kroon, G.J.; Dyson, H.J.; Balch, W.E.; Kelly, J.W. Amylin proprotein processing generates progressively more amyloidogenic peptides that initially sample the helical state. *Biochemistry* **2008**, *47*, 9900–9910. [CrossRef] [PubMed]

178. Reddy, A.S.; Wang, L.; Singh, S.; Ling, Y.L.; Buchanan, L.; Zanni, M.T.; Skinner, J.L.; de Pablo, J.J. Stable and metastable states of human amylin in solution. *Biophys. J.* **2010**, *99*, 2208–2216. [CrossRef] [PubMed]

179. Qiao, Q.; Bowman, G.R.; Huang, X. Dynamics of an intrinsically disordered protein reveal metastable conformations that potentially seed aggregation. *J. Am. Chem. Soc.* **2013**, *135*, 16092–16101. [CrossRef] [PubMed]

180. Höppener, J.W.; Ahrén, B.; Lips, C.J. Islet amyloid and type 2 diabetes mellitus. *N. Engl. J. Med.* **2000**, *343*, 411–419. [CrossRef] [PubMed]

181. Höppener, J.W.; Lips, C.J. Role of islet amyloid in type 2 diabetes mellitus. *Int. J. Biochem. Cell Biol.* **2006**, *38*, 726–736. [CrossRef] [PubMed]

182. Dong, X.; Qiao, Q.; Qian, Z.; Wei, G. Recent computational studies of membrane interaction and disruption of human islet amyloid polypeptide: Monomers, oligomers and protofibrils. *Biochim. Biophys. Acta Biomembr.* **2018**, *1860*, 1826–1839. [CrossRef] [PubMed]

183. Longhena, F.; Spano, P.; Bellucci, A. Targeting of Disordered Proteins by Small Molecules in Neurodegenerative Diseases. *Handb. Exp. Pharmacol.* **2018**, *245*, 85–110. [PubMed]

184. Babu, M.M.; van der Lee, R.; de Groot, N.S.; Gsponer, J. Intrinsically disordered proteins: Regulation and disease. *Curr. Opin. Struct. Biol.* **2011**, *21*, 432–440. [CrossRef] [PubMed]

185. Wright, P.E.; Dyson, H.J. Intrinsically disordered proteins in cellular signalling and regulation. *Nat. Rev. Mol. Cell Biol.* **2015**, *16*, 18–29. [CrossRef] [PubMed]

186. Bah, A.; Forman-Kay, J.D. Modulation of intrinsically disordered protein function by post-translational modifications. *J. Biol. Chem.* **2016**, *291*, 6696–6705. [CrossRef] [PubMed]

187. Dyson, H.J. Making Sense of Intrinsically Disordered Proteins. *Biophys. J.* **2016**, *110*, 1013–1016. [CrossRef] [PubMed]

188. Van Der Lee, R.; Buljan, M.; Lang, B.; Weatheritt, R.J.; Daughdrill, G.W.; Dunker, A.K.; Fuxreiter, M.; Gough, J.; Gsponer, J.; Jones, D.T.; et al. Classification of intrinsically disordered regions and proteins. *Chem. Rev.* **2014**, *114*, 6589–6631. [CrossRef] [PubMed]

189. Babu, M.M. The contribution of intrinsically disordered regions to protein function, cellular complexity, and human disease. *Biochem. Soc. Trans.* **2016**, *44*, 1185–1200. [CrossRef] [PubMed]

190. Strome, B.; Hsu, I.S.; Li Cheong Man, M.; Zarin, T.; Nguyen Ba, A.; Moses, A.M. Short linear motifs in intrinsically disordered regions modulate HOG signaling capacity. *BMC Syst. Biol.* **2018**, *12*, 75. [CrossRef] [PubMed]

191. Van Roey, K.; Uyar, B.; Weatheritt, R.J.; Dinkel, H.; Seiler, M.; Budd, A.; Gibson, T.J.; Davey, N.E. Short linear motifs: Ubiquitous and functionally diverse protein interaction modules directing cell regulation. *Chem. Rev.* **2014**, *114*, 6733–6778. [CrossRef] [PubMed]

192. Tompa, P. Unstructural biology coming of age. *Curr. Opin. Struct. Biol.* **2011**, *21*, 419–425. [CrossRef] [PubMed]

193. Hu, G.; Wu, Z.; Uversky, V.N.; Kurgan, L. Functional analysis of human hub proteins and their interactors involved in the intrinsic disorder-enriched interactions. *Int. J. Mol. Sci.* **2017**, *18*, 2761. [CrossRef] [PubMed]

194. Tompa, P.; Kovacs, D. Intrinsically disordered chaperones in plants and animals. *Biochem. Cell Biol.* **2010**, *88*, 167–174. [CrossRef] [PubMed]

195. Sharma, R.; Raduly, Z.; Miskei, M.; Fuxreiter, M. Fuzzy complexes: Specific binding without complete folding. *FEBS Lett.* **2015**, *589*, 2533–2542. [CrossRef] [PubMed]

196. Weiss, M.A.; Ellenberger, T.; Wobbe, C.R.; Lee, J.P.; Harrison, S.C.; Struhl, K. Folding transition in the DMA-binding domain of GCN4 on specific binding to DNA. *Nature* **1990**, *347*, 575–578. [CrossRef] [PubMed]

197. Bracken, C.; Carr, P.A.; Cavanagh, J.; Palmer, A.G. Temperature dependence of intramolecular dynamics of the basic leucine zipper of GCN4: Implications for the entropy of association with DNA. *J. Mol. Biol.* **1999**, *285*, 2133–2146. [CrossRef] [PubMed]

198. Shammas, S.L.; Crabtree, M.D.; Dahal, L.; Wicky, B.I.M.; Clarke, J. Insights into coupled folding and binding mechanisms from kinetic studies. *J. Biol. Chem.* **2016**, *291*, 6689–6695. [CrossRef] [PubMed]

199. Hammes, G.G.; Chang, Y.-C.; Oas, T.G. Conformational selection or induced fit: A flux description of reaction mechanism. *Proc. Natl. Acad. Sci. USA* **2009**, *106*, 13737–13741. [CrossRef] [PubMed]

200. DeForte, S.; Uversky, V.N. Order, disorder, and everything in between. *Molecules* **2016**, *21*, 1090. [CrossRef] [PubMed]

201. Xie, H.; Vucetic, S.; Iakoucheva, L.M.; Oldfield, C.J.; Dunker, A.K.; Obradovic, Z.; Uversky, V.N. Functional anthology of intrinsic disorder. 3. Ligands, post-translational modifications, and diseases associated with intrinsically disordered proteins. *J. Proteome Res.* **2007**, *6*, 1917–1932. [CrossRef] [PubMed]

202. Schwalbe, M.; Biernat, J.; Bibow, S.; Ozenne, V.; Jensen, M.R.; Kadavath, H.; Blackledge, M.; Mandelkow, E.; Zweckstetter, M. Phosphorylation of human tau protein by microtubule affinity-regulating kinase 2. *Biochemistry* **2013**, *52*, 9068–9079. [CrossRef] [PubMed]

203. Ou, L.; Ferreira, A.M.; Otieno, S.; Xiao, L.; Bashford, D.; Kriwacki, R.W. Incomplete folding upon binding mediates Cdk4/cyclin D complex activation by tyrosine phosphorylation of inhibitor p27 protein. *J. Biol. Chem.* **2011**, *286*, 30142–30151. [CrossRef] [PubMed]

204. Zeng, Y.; He, Y.; Yang, F.; Mooney, S.M.; Getzenberg, R.H.; Orban, J.; Kulkarni, P. The cancer/testis antigen prostate-associated gene 4 (PAGE4) is a highly intrinsically disordered protein. *J. Biol. Chem.* **2011**, *286*, 13985–13994. [CrossRef] [PubMed]

205. Coskuner-Weber, O.; Uversky, V.N. Insights into the molecular mechanisms of Alzheimer's and Parkinson's diseases with molecular simulations: Understanding the roles of artificial and pathological missense mutations in intrinsically disordered proteins related to pathology. *Int. J. Mol. Sci.* **2018**, *19*, 336. [CrossRef] [PubMed]

206. Edwards, Y.J.K.; Lobley, A.E.; Pentony, M.M.; Jones, D.T. Insights into the regulation of intrinsically disordered proteins in the human proteome by analyzing sequence and gene expression data. *Genome Biol.* **2009**, *10*, R50. [CrossRef] [PubMed]

207. Gsponer, J.; Futschik, M.E.; Teichmann, S.A.; Babu, M.M. Tight regulation of unstructured proteins: From transcript synthesis to protein degradation. *Science* **2008**, *322*, 1365–1368. [CrossRef] [PubMed]

208. Forbes, J.G.; Jin, A.J.; Ma, K.; Gutierrez-Cruz, G.; Tsai, W.L.; Wang, K. Titin PEVK segment: Charge-driven elasticity of the open and flexible polyampholyte. *J. Muscle Res. Cell Motil.* **2005**, *26*, 291–301. [CrossRef] [PubMed]

209. Boothby, T.C.; Tapia, H.; Brozena, A.H.; Piszkiewicz, S.; Smith, A.E.; Giovannini, I.; Rebecchi, L.; Pielak, G.J.; Koshland, D.; Goldstein, B. Tardigrades Use Intrinsically Disordered Proteins to Survive Desiccation. *Mol. Cell* **2017**, *65*, 975.e5–984.e5. [CrossRef] [PubMed]

210. Russo, J.; Olivas, W.M. Conditional regulation of Puf1p, Puf4p, and Puf5p activity alters YHB1 mRNA stability for a rapid response to toxic nitric oxide stress in yeast. *Mol. Biol. Cell* **2015**, *26*, 1015–1029. [CrossRef] [PubMed]

211. Wang, M.; Ogé, L.; Perez-Garcia, M.D.; Hamama, L.; Sakr, S. The PUF protein family: Overview on PUF RNA targets, biological functions, and post transcriptional regulation. *Int. J. Mol. Sci.* **2018**, *19*, 410. [CrossRef] [PubMed]

212. Tsvetkov, P.; Reuven, N.; Shaul, Y. The nanny model for IDPs. *Nat. Chem. Biol.* **2009**, *5*, 778–781. [CrossRef] [PubMed]

213. Inobe, T.; Matouschek, A. Paradigms of protein degradation by the proteasome. *Curr. Opin. Struct. Biol.* **2014**, *24*, 156–164. [CrossRef] [PubMed]

214. Hagai, T.; Azia, A.; Tóth-Petróczy, Á.; Levy, Y. Intrinsic disorder in ubiquitination substrates. *J. Mol. Biol.* **2011**, *412*, 319–324. [CrossRef] [PubMed]

215. Wenzel, T.; Baumeister, W. Conformational constraints in protein degradation by the 20S proteasome. *Nat. Struct. Biol.* **1995**, *2*, 199–204. [CrossRef] [PubMed]

216. Theillet, F.-X.; Binolfi, A.; Frembgen-Kesner, T.; Hingorani, K.; Sarkar, M.; Kyne, C.; Li, C.; Crowley, P.B.; Gierasch, L.; Pielak, G.J.; et al. Physicochemical Properties of Cells and Their Effects on Intrinsically Disordered Proteins (IDPs). *Chem. Rev.* **2014**, *114*, 6661–6714. [CrossRef] [PubMed]

217. Dou, F.; Netzer, W.J.; Tanemura, K.; Li, F.; Hartl, F.U.; Takashima, A.; Gouras, G.K.; Greengard, P.; Xu, H. Chaperones increase association of tau protein with microtubules. *Proc. Natl. Acad. Sci. USA* **2003**, *100*, 721–726. [CrossRef] [PubMed]

218. Wang, H.; Tan, M.S.; Lu, R.C.; Yu, J.T.; Tan, L. Heat shock proteins at the crossroads between cancer and Alzheimer's disease. *BioMed Res. Int.* **2014**, *2014*, 239164. [CrossRef] [PubMed]

219. Tortosa, E.; Santa-Maria, I.; Moreno, F.; Lim, F.; Perez, M.; Avila, J. Binding of Hsp90 to tau promotes a conformational change and aggregation of tau protein. *J. Alzheimers Dis.* **2009**, *17*, 319–325. [CrossRef] [PubMed]

220. Dou, F.; Chang, X.; Ma, D. Hsp90 maintains the stability and function of the tau phosphorylating kinase GSK3β. *Int. J. Mol. Sci* **2007**, *8*, 51–60. [CrossRef]

221. Shelton, L.B.; Baker, J.D.; Zheng, D.; Sullivan, L.E.; Solanki, P.K.; Webster, J.M.; Sun, Z.; Sabbagh, J.J.; Nordhues, B.A.; Koren, J.; et al. Hsp90 activator Aha1 drives production of pathological tau aggregates. *Proc. Natl. Acad. Sci. USA* **2017**, *114*, 9707–9712. [CrossRef] [PubMed]

222. Mayer, M.P. Hsp70 chaperone dynamics and molecular mechanism. *Trends Biochem. Sci.* **2013**, *38*, 507–514. [CrossRef] [PubMed]

223. Young, Z.T.; Rauch, J.N.; Assimon, V.A.; Jinwal, U.K.; Ahn, M.; Li, X.; Dunyak, B.M.; Ahmad, A.; Carlson, G.A.; Srinivasan, S.R.; et al. Stabilizing the Hsp70-Tau complex promotes turnover in models of Tauopathy. *Cell Chem. Biol.* **2016**, *23*, 992–1001. [CrossRef] [PubMed]

224. Westhoff, B.; Chapple, J.P.; van der Spuy, J.; Höhfeld, J.; Cheetham, M.E. HSJ1 is a neuronal shuttling factor for the sorting of chaperone clients to the proteasome. *Curr. Biol.* **2005**, *15*, 1058–1064. [CrossRef] [PubMed]

225. Shin, Y.; Klucken, J.; Patterson, C.; Hyman, B.T.; McLean, P.J. The co-chaperone carboxyl terminus of Hsp70-interacting protein (CHIP) mediates alpha-synuclein degradation decisions between proteasomal and lysosomal pathways. *J. Biol. Chem.* **2005**, *280*, 23727–23734. [CrossRef] [PubMed]

226. Lang, K.; Schmid, F.X.; Fischer, G. Catalysis of protein folding by prolyl isomerase. *Nature* **1987**, *329*, 268–270. [CrossRef] [PubMed]

227. Nigro, P.; Pompilio, G.; Capogrossi, M.C. Cyclophilin A: A key player for human disease. *Cell Death Dis.* **2013**, *4*, e888. [CrossRef] [PubMed]

228. Torbeev, V.Y.; Hilvert, D. Both the cis-trans equilibrium and isomerization dynamics of a single proline amide modulate β2-microglobulin amyloid assembly. *Proc. Natl. Acad. Sci. USA* **2013**, *110*, 20051–20056. [CrossRef] [PubMed]

229. Baker, J.D.; Shelton, L.B.; Zheng, D.; Favretto, F.; Nordhues, B.A.; Darling, A.; Sullivan, L.E.; Sun, Z.; Solanki, P.K.; Martin, M.D.; et al. Human cyclophilin 40 unravels neurotoxic amyloids. *PLoS Biol.* **2017**, *15*, e2001336. [CrossRef] [PubMed]

230. Tsvetkov, P.; Reuven, N.; Shaul, Y. Ubiquitin-independent p53 proteasomal degradation. *Cell Death Differ.* **2010**, *17*, 103–108. [CrossRef] [PubMed]

231. Cheng, Y.; LeGall, T.; Oldfield, C.J.; Mueller, J.P.; Van, Y.Y.J.; Romero, P.; Cortese, M.S.; Uversky, V.N.; Dunker, A.K. Rational drug design via intrinsically disordered protein. *Trends Biotechnol.* **2006**, *24*, 435–442. [CrossRef] [PubMed]

232. Metallo, S.J. Intrinsically disordered proteins are potential drug targets. *Curr. Opin. Chem. Biol.* **2010**, *14*, 481–488. [CrossRef] [PubMed]

233. Wang, J.H.; Cao, Z.X.; Zhao, L.L.; Li, S.Q. Novel strategies for drug discovery based on intrinsically disordered proteins (IDPs). *Int. J. Mol. Sci.* **2011**, *12*, 3205–3219. [CrossRef] [PubMed]

234. Zhang, Y.; Cao, H.; Liu, Z. Binding cavities and druggability of intrinsically disordered proteins. *Protein Sci.* **2015**, *24*, 688–705. [CrossRef] [PubMed]

235. Liu, Z.; Huang, Y. Advantages of proteins being disordered. *Protein Sci.* **2014**, *23*, 539–550. [CrossRef] [PubMed]

236. Zhu, M.; De Simone, A.; Schenk, D.; Toth, G.; Dobson, C.M.; Vendruscolo, M. Identification of small-molecule binding pockets in the soluble monomeric form of the Aβ42 peptide. *J. Chem. Phys.* **2013**, *139*, 035101. [CrossRef] [PubMed]

237. Fokkens, M.; Schrader, T.; Klärner, F.G. A molecular tweezer for lysine and arginine. *J. Am. Chem. Soc.* **2005**, *127*, 14415–14421. [CrossRef] [PubMed]

238. Sinha, S.; Lopes, D.H.; Du, Z.; Pang, E.; Shanmugam, A.; Lomakin, A.; Talbiersky, P.; Tennstaedt, A.; McDaniel, K.; Bakshi, R.; et al. Lysine-specific molecular tweezers are broad-spectrum inhibitors of assembly and toxicity of amyloid proteins. *J. Am. Chem. Soc.* **2011**, *133*, 16958–16969. [CrossRef] [PubMed]

239. O'Hare, E.; Scopes, D.I.; Kim, E.M.; Palmer, P.; Spanswick, D.; McMahon, B.; Amijee, H.; Nerou, E.; Treherne, J.M.; Jeggo, R. Novel 5-aryloxypyrimidine SEN1576 as a candidate for the treatment of Alzheimer's disease. *Int. J. Neuropsychopharmacol.* **2014**, *17*, 117–126. [CrossRef] [PubMed]

240. Prabhudesai, S.; Sinha, S.; Attar, A.; Kotagiri, A.; Fitzmaurice, A.G.; Lakshmanan, R.; Ivanova, M.I.; Loo, J.A.; Klärner, F.G.; Schrader, T.; et al. A novel "molecular tweezer" inhibitor of α-synuclein neurotoxicity in vitro and in vivo. *Neurotherapeutics* **2012**, *9*, 464–476. [CrossRef] [PubMed]

241. Sievers, S.A.; Karanicolas, J.; Chang, H.W.; Zhao, A.; Jiang, L.; Zirafi, O.; Stevens, J.T.; Münch, J.; Baker, D.; Eisenberg, D. Structure-based design of non-natural amino-acid inhibitors of amyloid fibril formation. *Nature* **2011**, *475*, 96–100. [CrossRef] [PubMed]

242. Frenkel-Pinter, M.; Tal, S.; Scherzer-Attali, R.; Abu-Hussien, M.; Alyagor, I.; Eisenbaum, T.; Gazit, E.; Segal, D. Naphthoquinone-Tryptophan Hybrid Inhibits Aggregation of the Tau-Derived Peptide PHF6 and Reduces Neurotoxicity. *J. Alzheimers Dis.* **2016**, *51*, 165–178. [CrossRef] [PubMed]

243. Jones, C.L.; Njomen, E.; Sjögren, B.; Dexheimer, T.S.; Tepe, J.J. Small Molecule Enhancement of 20S Proteasome Activity Targets Intrinsically Disordered Proteins. *ACS Chem. Biol.* **2017**, *12*, 2240–2247. [CrossRef] [PubMed]

244. Joerger, A.C.; Fersht, A.R. The tumor suppressor p53: From structures to drug discovery. *Cold Spring Harb. Perspect. Biol.* **2010**, *2*, a000919. [CrossRef] [PubMed]

245. Li, Z.Y.; Ni, M.; Li, J.K.; Zhang, Y.P.; Ouyang, Q.; Tang, C. Decision making of the p53 network: Death by integration. *J. Theor. Biol.* **2011**, *271*, 205–211. [CrossRef] [PubMed]

246. Huang, Y.Q.; Liu, Z.R. Anchoring intrinsically disordered proteins to multiple targets: Lessons from N terminus of the p53 protein. *Int. J. Mol. Sci.* **2011**, *12*, 1410–1430. [CrossRef] [PubMed]

247. Vassilev, L.T.; Vu, B.T.; Graves, B.; Carvajal, D.; Podlaski, F.; Filipovic, Z.; Kong, N.; Kammlott, U.; Lukacs, C.; Klein, C.; et al. In vivo activation of the p53 pathway by small-molecule antagonists of MDM2. *Science* **2004**, *303*, 844–848. [CrossRef] [PubMed]

248. Tovar, C.; Rosinski, J.; Filipovic, Z.; Higgins, B.; Kolinsky, K.; Hilton, H.; Zhao, X.L.; Vu, B.T.; Qing, W.G.; Packman, K.; et al. Small-molecule MDM2 antagonists reveal aberrant p53 signaling in cancer: Implications for therapy. *Proc. Natl. Acad. Sci. USA* **2006**, *103*, 1888–1893. [CrossRef] [PubMed]

249. Yu, X.; Narayanan, S.; Vazquez, A.; Carpizo, D.R. Small molecule compounds targeting the p53 pathway: Are we finally making progress? *Apoptosis* **2014**, *19*, 1055–1068. [CrossRef] [PubMed]

250. Burgess, A.; Chia, K.M.; Haupt, S.; Thomas, D.; Haupt, Y.; Lim, E. Clinical Overview of MDM2/X-Targeted Therapies. *Front. Oncol.* **2016**, *6*, 7. [CrossRef] [PubMed]

251. Hammoudeh, D.I.; Follis, A.V.; Prochownik, E.V.; Metallo, S.J. Multiple independent binding sites for small molecule inhibitors on the oncoprotein c-Myc. *J. Am. Chem. Soc.* **2009**, *131*, 7390–7401. [CrossRef] [PubMed]

252. Berg, T.; Cohen, S.B.; Desharnais, J.; Sonderegger, C.; Maslyar, D.J.; Goldberg, J.; Boger, D.L.; Vogt, P.K. Small-molecule antagonists of Myc/Max dimerization inhibit Myc-induced transformation of chicken embryo fibroblasts. *Proc. Natl. Acad. Sci. USA* **2002**, *99*, 3830–3835. [CrossRef] [PubMed]

253. Shi, J.; Stover, J.S.; Whitby, L.R.; Vogt, P.K.; Boger, D.L. Small molecule inhibitors of Myc/Max dimerization and Myc-induced cell transformation. *Bioorg. Med. Chem. Lett.* **2009**, *19*, 6038–6041. [CrossRef] [PubMed]

254. Yin, X.Y.; Giap, C.; Lazo, J.S.; Prochownik, E.V. Low molecular weight inhibitors of Myc-Max interaction and function. *Oncogene* **2003**, *22*, 6151–6159. [CrossRef] [PubMed]

255. Zirath, H.; Frenzel, A.; Oliynyk, G.; Segerstrom, L.; Westermark, U.K.; Larsson, K.; Persson, M.M.; Hultenby, K.; Lehtio, J.; Einvik, C.; et al. MYC inhibition induces metabolic changes leading to accumulation of lipid droplets in tumor cells. *Proc. Natl. Acad. Sci. USA* **2013**, *110*, 10258–10263. [CrossRef] [PubMed]

256. Fletcher, S.; Prochownik, E.V. Small-molecule inhibitors of the Myc oncoprotein. *Biochim. Biophys. Acta* **2015**, *1849*, 525–543. [CrossRef] [PubMed]

257. Yu, C.; Niu, X.; Jin, F.; Liu, Z.; Jin, C.; Lai, L. Structure-based Inhibitor Design for the Intrinsically Disordered Protein c-Myc. *Sci. Rep.* **2016**, *6*, 22298. [CrossRef] [PubMed]

258. Erkizan, H.V.; Kong, Y.L.; Merchant, M.; Schlottmann, S.; Barber-Rotenberg, J.S.; Yuan, L.S.; Abaan, O.D.; Chou, T.H.; Dakshanamurthy, S.; Brown, M.L.; et al. A small molecule blocking oncogenic protein EWS-FLI1 interaction with RNA helicase A inhibits growth of Ewing's sarcoma. *Nat. Med.* **2009**, *15*, 750–757. [CrossRef] [PubMed]

259. Hong, S.H.; Youbi, S.E.; Hong, S.P.; Kallakury, B.; Monroe, P.; Erkizan, H.V.; Barber-Rotenberg, J.S.; Houghton, P.; Uren, A.; Toretsky, J.A. Pharmacokinetic modeling optimizes inhibition of the 'undruggable' EWS-FLI1 transcription factor in Ewing Sarcoma. *Oncotarget* **2014**, *5*, 338–350. [CrossRef] [PubMed]

260. Hegyi, H.; Buday, L.; Tompa, P. Intrinsic structural disorder confers cellular viability on oncogenic fusion proteins. *PLoS Comput. Biol.* **2009**, *5*, e1000552. [CrossRef] [PubMed]

261. Huang, Y.Q.; Liu, Z.R. Do intrinsically disordered proteins possess high specificity in protein-protein interactions? *Chem.-Eur. J.* **2013**, *19*, 4462–4467. [CrossRef] [PubMed]

262. Srinivasan, R.S.; Nesbit, J.B.; Marrero, L.; Erfurth, F.; LaRussa, V.F.; Hemenway, C.S. The synthetic peptide PFWT disrupts AF4-AF9 protein complexes and induces apoptosis in t(4;11) leukemia cells. *Leukemia* **2004**, *18*, 1364–1372. [CrossRef] [PubMed]

263. Palermo, C.M.; Bennett, C.A.; Winters, A.C.; Hemenway, C.S. The AF4-mimetic peptide, PFWT, induces necrotic cell death in MV4-11 leukemia cells. *Leuk. Res.* **2008**, *32*, 633–642. [CrossRef] [PubMed]

264. Watson, V.G.; Drake, K.M.; Peng, Y.; Napper, A.D. Development of a high-throughput screening-compatible assay for the discovery of inhibitors of the AF4-AF9 interaction using AlphaScreen technology. *Assay Drug Dev. Technol.* **2013**, *11*, 253–268. [CrossRef] [PubMed]

265. Johnson, T.O.; Ermolieff, J.; Jirousek, M.R. Protein tyrosine phosphatase 1B inhibitors for diabetes. *Nat. Rev. Drug Discov.* **2002**, *1*, 696–709. [CrossRef] [PubMed]

266. Krishnan, N.; Koveal, D.; Miller, D.H.; Xue, B.; Akshinthala, S.D.; Kragelj, J.; Jensen, M.R.; Gauss, C.M.; Page, R.; Blackledge, M.; et al. Targeting the disordered C terminus of PTP1B with an allosteric inhibitor. *Nat. Chem. Biol.* **2014**, *10*, 558–566. [CrossRef] [PubMed]

267. Bieschke, J.; Russ, J.; Friedrich, R.P.; Ehrnhoefer, D.E.; Wobst, H.; Neugebauer, K.; Wanker, E.E. EGCG remodels mature alpha-synuclein and amyloid-beta fibrils and reduces cellular toxicity. *Proc. Natl. Acad. Sci. USA* **2010**, *107*, 7710–7715. [CrossRef] [PubMed]

268. Martin-Bastida, A.; Ward, R.J.; Newbould, R.; Piccini, P.; Sharp, D.; Kabba, C.; Patel, M.C.; Spino, M.; Connelly, J.; Tricta, F.; et al. Brain iron chelation by deferiprone in a phase 2 randomised double-blinded placebo controlled clinical trial in Parkinson's disease. *Sci. Rep.* **2017**, *7*, 1398. [CrossRef] [PubMed]

269. Hung, L.W.; Villemagne, V.L.; Cheng, L.; Sherratt, N.A.; Ayton, S.; White, A.R.; Crouch, P.J.; Lim, S.; Leong, S.L.; Wilkins, S.; et al. The hypoxia imaging agent CuII(atsm) is neuroprotective and improves motor and cognitive functions in multiple animal models of Parkinson's disease. *J. Exp. Med.* **2012**, *209*, 837–854. [CrossRef] [PubMed]

270. Savolainen, M.H.; Richie, C.T.; Harvey, B.K.; Männistö, B.T.; Maguire-Zeiss, K.A.; Myöhänen, T.T. The beneficial effect of a prolyl oligopeptidase inhibitor, KYP-2047, on alpha-synuclein clearance and autophagy in A30P transgenic mouse. *Neurobiol. Dis.* **2014**, *68*, 1–15. [CrossRef] [PubMed]

271. Svarcbahs, R.; Julku, U.H.; Myöhänen, T.T. Inhibition of Prolyl Oligopeptidase Restores Spontaneous Motor Behavior in the α-Synuclein Virus Vector-Based Parkinson's Disease Mouse Model by Decreasing α-Synuclein Oligomeric Species in Mouse Brain. *J. Neurosci.* **2016**, *36*, 12485–12497. [CrossRef] [PubMed]

272. Levin, J.; Schmidt, F.; Boehm, C.; Prix, C.; Bötzel, K.; Ryazanov, S.; Leonov, A.; Griesinger, C.; Giese, A. The oligomer modulator anle138b inhibits disease progression in a Parkinson mouse model even with treatment started after disease onset. *Acta Neuropathol.* **2014**, *127*, 779–780. [CrossRef] [PubMed]

273. Price, D.L.; Koike, M.A.; Khan, A.; Wrasidlo, W.; Rockenstein, E.; Masliah, E.; Bonhaus, D. The small molecule alpha-synuclein misfolding inhibitor, NPT200-11, produces multiple benefits in an animal model of Parkinson's disease. *Sci. Rep.* **2018**, *8*, 16165. [CrossRef] [PubMed]

274. Krishnan, R.; Hefti, F.; Tsubery, H.; Lulu, M.; Proschitsky, M.; Fisher, R. Conformation as the Therapeutic Target for Neurodegenerative Diseases. *Curr. Alzheimer Res.* **2017**, *14*, 393–402. [CrossRef] [PubMed]

275. Levenson, J.M.; Schroeter, S.; Carroll, J.C.; Cullen, V.; Asp, E.; Proschitsky, M.; Chung, C.H.; Gilead, S.; Nadeem, M.; Dodiya, H.B.; et al. NPT088 reduces both amyloid-β and tau pathologies in transgenic mice. *Alzheimers Dement.* **2016**, *2*, 141–155. [CrossRef] [PubMed]

276. Yuan, B.; Sierks, M.R. Intracellular targeting and clearance of oligomeric alpha-synuclein alleviates toxicity in mammalian cells. *Neurosci. Lett.* **2009**, *459*, 16–18. [CrossRef] [PubMed]

277. Bhatt, M.A.; Messer, A.; Kordower, J.H. Can intrabodies serve as neuroprotective therapies for Parkinson's disease? Beginning thoughts. *J. Parkinsons Dis.* **2013**, *3*, 581–591. [PubMed]

278. Emadi, S.; Liu, R.; Yuan, B.; Schulz, P.; McAllister, C.; Lyubchenko, Y.; Messer, A.; Sierks, M.R. Inhibiting aggregation of alpha-synuclein with human single chain antibody fragments. *Biochemistry* **2004**, *43*, 2871–2878. [CrossRef] [PubMed]

279. Butler, D.C.; Joshi, S.N.; Genst, E.; Baghel, A.S.; Dobson, C.M.; Messer, A. Bifunctional Anti-Non-Amyloid Component α-Synuclein Nanobodies Are Protective In Situ. *PLoS ONE* **2016**, *11*, e0165964. [CrossRef] [PubMed]

280. Chatterjee, D.; Bhatt, M.; Butler, D.; De Genst, E.; Dobson, C.M.; Messer, A.; Kordower, J.H. Proteasome-targeted nanobodies alleviate pathology and functional decline in an α-synuclein-based Parkinson's disease model. *NPJ Parkinsons Dis.* **2018**, *4*, 25. [CrossRef] [PubMed]

281. Dehay, B.; Bourdenx, M.; Gorry, P.; Przedborski, S.; Vila, M.; Hunot, S.; Singleton, A.; Olanow, C.; Merchant, K.; Bezard, E.; et al. Targeting alpha-synuclein for treatment of Parkinson's disease: Mechanistic and therapeutic considerations. *Lancet Neurol.* **2015**, *14*, 855–866. [CrossRef]

282. Schenk, D.B.; Koller, M.; Ness, D.K.; Griffith, S.G.; Grundman, M.; Zago, W.; Soto, J.; Atiee, G.; Ostrowitzki, S.; Kinney, G.G. First-in-human assessment of PRX002, an anti-α-synuclein monoclonal antibody, in healthy volunteers. *Mov. Disord.* **2017**, *32*, 211–218. [CrossRef] [PubMed]

283. Jankovic, J.; Goodman, I.; Safirstein, B.; Marmon, T.K.; Schenk, D.B.; Koller, M.; Zago, W.; Ness, D.K.; Griffith, S.G.; Grundman, M.; et al. Safety and Tolerability of Multiple Ascending Doses of PRX002/RG7935, an Anti-α-Synuclein Monoclonal Antibody, in Patients With Parkinson Disease: A Randomized Clinical Trial. *JAMA Neurol.* **2018**, *75*, 1206–1214. [CrossRef] [PubMed]

284. Mandler, M.; Valera, E.; Rockenstein, E.; Weninger, H.; Patrick, C.; Adame, A.; Santic, R.; Meindl, S.; Vigl, B.; Smrzka, O.; et al. Next-generation active immunization approach for synucleinopathies: Implications for Parkinson's disease clinical trials. *Acta Neuropathol.* **2014**, *127*, 861–879. [CrossRef] [PubMed]

285. O'Hara, D.M.; Kalia, S.K.; Kalia, L.V. Emerging disease-modifying strategies targeting alpha-synuclein for the treatment of Parkinson's disease. *Br. J. Pharmacol.* **2018**, *175*, 3080–3089. [CrossRef] [PubMed]

286. Kiss, R.; Csizmadia, G.; Solti, K.; Keresztes, A.; Zhu, M.; Pickhardt, M.; Mandelkow, E.; Toth, G. Structural Basis of Small Molecule Targetability of Monomeric Tau Protein. *ACS Chem. Neurosci.* **2018**, *9*, 2997–3006. [CrossRef] [PubMed]

287. Jouanne, M.; Rault, S.; Voisin-Chiret, A.S. Tau protein aggregation in Alzheimer's disease: An attractive target for the development of novel therapeutic agents. *Eur. J. Med. Chem.* **2017**, *139*, 153–167. [CrossRef] [PubMed]

288. Pickhardt, M.; Neumann, T.; Schwizer, D.; Callaway, K.; Vendruscolo, M.; Schenk, D.; St George-Hyslop, P.; Mandelkow, E.M.; Dobson, C.M.; McConlogue, L.; et al. Identification of Small Molecule Inhibitors of Tau Aggregation by Targeting Monomeric Tau As a Potential Therapeutic Approach for Tauopathies. *Curr. Alzheimer Res.* **2015**, *12*, 814–828. [CrossRef] [PubMed]

289. Baggett, D.W.; Nath, A. The Rational Discovery of a Tau Aggregation Inhibitor. *Biochemistry* **2018**, *57*, 6099–6107. [CrossRef] [PubMed]

290. Shiryaev, N.; Pikman, R.; Giladi, E.; Gozes, I. Protection against tauopathy by the drug candidates NAP (davunetide) and D-SAL: Biochemical, cellular and behavioral aspects. *Curr. Pharm. Des.* **2011**, *17*, 2603–2612. [CrossRef] [PubMed]

291. Ivashko-Pachima, Y.; Gozes, I. NAP protects against Tau hyperphosphorylation through GSK3. *Curr. Pharm. Des.* **2018**, *24*, 3868–3877. [CrossRef] [PubMed]

292. Dammers, C.; Yolcu, D.; Kukuk, L.; Willbold, D.; Pickhardt, M.; Mandelkow, E.; Horn, A.H.; Sticht, H.; Malhis, M.N.; Will, N.; et al. Selection and Characterization of Tau Binding -Enantiomeric Peptides with Potential for Therapy of Alzheimer Disease. *PLoS ONE* **2016**, *11*, e0167432. [CrossRef] [PubMed]

293. Kim, J.H.; Kim, E.; Choi, W.H.; Lee, J.; Lee, J.H.; Lee, H.; Kim, D.E.; Suh, Y.H.; Lee, M.J. Inhibitory RNA Aptamers of Tau Oligomerization and Their Neuroprotective Roles against Proteotoxic Stress. *Mol. Pharm.* **2016**, *13*, 2039–2048. [CrossRef] [PubMed]

294. Rafiee, S.; Asadollahi, K.; Riazi, G.; Ahmadian, S.; Saboury, A.A. Vitamin B12 Inhibits Tau Fibrillization via Binding to Cysteine Residues of Tau. *ACS Chem. Neurosci.* **2017**, *8*, 2676–2682. [CrossRef] [PubMed]

295. Yoshitake, J.; Soeda, Y.; Ida, T.; Sumioka, A.; Yoshikawa, M.; Matsushita, K.; Akaike, T.; Takashima, A. Modification of Tau by 8-Nitroguanosine 3,5-Cyclic Monophosphate (8-Nitro-cGMP): Effects of nitric oxide-linked chemical modification on tau aggregation. *J. Biol. Chem.* **2016**, *291*, 22714–22720. [CrossRef] [PubMed]

296. Sun, W.; Lee, S.; Huang, X.; Liu, S.; Inayathullah, M.; Kim, K.-M.; Tang, H.; Ashford, J.W.; Rajadas, J. Attenuation of synaptic toxicity and MARK4/PAR1-mediated Tau phosphorylation by methylene blue for Alzheimer's disease treatment. *Sci. Rep.* **2016**, *6*, 34784. [CrossRef] [PubMed]

297. George, R.C.; Lew, J.; Graves, D.J. Interaction of Cinnamaldehyde and Epicatechin with Tau: Implications of Beneficial Effects in Modulating Alzheimer's Disease Pathogenesis. *J. Alzheimers Dis.* **2013**, *36*, 21–40. [CrossRef] [PubMed]

298. Gandini, A.; Bartolini, M.; Tedesco, D.; Martinez-Gonzalez, L.; Roca, C.; Campillo, N.E.; Zaldivar-Diez, J.; Perez, C.; Zuccheri, G.; Miti, A.; et al. Tau-Centric Multitarget Approach for Alzheimer's Disease: Development of First-in-Class Dual Glycogen Synthase Kinase 3 beta and Tau-Aggregation Inhibitors. *J. Med. Chem.* **2018**, *61*, 7640–7656. [CrossRef] [PubMed]

299. Llorach-Pares, L.; Nonell-Canals, A.; Avila, C.; Sanchez-Martinez, M. Kororamides, Convolutamines, and Indole Derivatives as Possible Tau and Dual-Specificity Kinase Inhibitors for Alzheimer's Disease: A Computational Study. *Mar. Drugs* **2018**, *16*, 386. [CrossRef] [PubMed]

300. Moussa, C.E. Beta-secretase inhibitors in phase I and phase II clinical trials for Alzheimer's disease. *Expert Opin. Investig. Drugs* **2017**, *26*, 1131–1136. [CrossRef] [PubMed]

301. Mead, E.; Kestoras, D.; Gibson, Y.; Hamilton, L.; Goodson, R.; Jones, S.; Eversden, S.; Davies, P.; O'Neill, M.; Hutton, M.; et al. Halting of Caspase Activity Protects Tau from MC1-Conformational Change and Aggregation. *J. Alzheimers Dis.* **2016**, *54*, 1521–1538. [CrossRef] [PubMed]

302. Rao, M.V.; McBrayer, M.K.; Campbell, J.; Kumar, A.; Hashim, A.; Sershen, H.; Stavrides, P.H.; Ohno, M.; Hutton, M.; Nixon, R.A. Specific calpain inhibition by calpastatin prevents tauopathy and neurodegeneration and restores normal lifespan in tau P301L mice. *J. Neurosci.* **2014**, *34*, 9222–9234. [CrossRef] [PubMed]

303. Blair, L.J.; Sabbagh, J.J.; Dickey, C.A. Targeting Hsp90 and its co-chaperones to treat Alzheimer's disease. *Expert Opin. Ther. Targets* **2014**, *18*, 1219–1232. [CrossRef] [PubMed]

304. Kontsekova, E.; Zilka, N.; Kovacech, B.; Novak, P.; Novak, M. First-in-man tau vaccine targeting structural determinants essential for pathological tau-tau interaction reduces tau oligomerisation and neurofibrillary degeneration in an Alzheimer's disease model. *Alzheimers Res. Ther.* **2014**, *6*, 44. [CrossRef] [PubMed]

305. Novak, P.; Kontsekova, E.; Zilka, N.; Novak, M. Ten Years of Tau-Targeted Immunotherapy: The Path Walked and the Roads Ahead. *Front. Neurosci.* **2018**, *12*, 798. [CrossRef] [PubMed]

306. Shahpasand, K.; Sepehri Shamloo, A.; Nabavi, S.M.; Lu, K.P.; Zhou, X.Z. Tau immunotherapy: Hopes and hindrances. *Hum. Vaccin Immunother.* **2018**, *14*, 277–284. [CrossRef] [PubMed]

307. Cehlar, O.; Skrabana, R.; Kovac, A.; Kovacech, B.; Novak, M. Crystallization and preliminary X-ray diffraction analysis of tau protein microtubule-binding motifs in complex with Tau5 and DC25 antibody Fab fragments. *Acta Crystallogr. Sect. F Struct. Biol. Cryst. Commun.* **2012**, *68*, 1181–1185. [CrossRef] [PubMed]

308. Panza, F.; Solfrizzi, V.; Seripa, D.; Imbimbo, B.P.; Lozupone, M.; Santamato, A.; Tortelli, R.; Galizia, I.; Prete, C.; Daniele, A.; et al. Tau-based therapeutics for Alzheimer's disease: Active and passive immunotherapy. *Immunotherapy* **2016**, *8*, 1119–1134. [CrossRef] [PubMed]

309. Novak, P.; Zilka, N.; Zilkova, M.; Kovacech, B.; Skrabana, R.; Ondrus, M.; Fialova, L.; Kontsekova, E.; Otto, M.; Novak, M. AADvac1, an Active Immunotherapy for Alzheimer's Disease and Non Alzheimer Tauopathies: An Overview of Preclinical and Clinical Development. *J. Prev. Alzheimers Dis.* **2019**, *6*, 63–69. [PubMed]

310. Novak, P.; Schmidt, R.; Kontsekova, E.; Zilka, N.; Kovacech, B.; Skrabana, R.; Vince-Kazmerova, Z.; Katina, S.; Fialova, L.; Prcina, M.; et al. Safety and immunogenicity of the tau vaccine AADvac1 in patients with Alzheimer's disease: A randomised, double-blind, placebo-controlled, phase 1 trial. *Lancet Neurol.* **2017**, *16*, 123–134. [CrossRef]

311. Novak, P.; Schmidt, R.; Kontsekova, E.; Kovacech, B.; Smolek, T.; Katina, S.; Fialova, L.; Prcina, M.; Parrak, V.; Dal-Bianco, P.; et al. FUNDAMANT: An interventional 72-week phase 1 follow-up study of AADvac1, an active immunotherapy against tau protein pathology in Alzheimer's disease. *Alzheimers Res. Ther.* **2018**, *10*, 108. [CrossRef] [PubMed]

312. Theunis, C.; Crespo-Biel, N.; Gafner, V.; Pihlgren, M.; Lopez-Deber, M.; Reis, P.; Hickman, D.; Adolfsson, O.; Chuard, N.; Ndao, D.; et al. Efficacy and Safety of A Liposome-Based Vaccine against Protein Tau, Assessed in Tau.P301L Mice That Model Tauopathy. *PLoS ONE* **2013**, *8*, e72301. [CrossRef] [PubMed]

313. Yanamandra, K.; Jiang, H.; Mahan, T.; Maloney, S.; Wozniak, D.; Diamond, M.; Holtzmanm, D. Anti-tau antibody reduces insoluble tau and decreases brain atrophy. *Ann. Clin. Trans. Neurol.* **2016**, *2*, 278–288. [CrossRef] [PubMed]

314. West, T.; Hu, Y.; Verghese, P.B.; Bateman, R.J.; Braunstein, J.B.; Fogelman, I.; Budur, K.; Florian, H.; Mendonca, N.; Holtzman, D.M. Preclinical and Clinical Development of ABBV-8E12, a Humanized Anti-Tau Antibody, for Treatment of Alzheimer's Disease and Other Tauopathies. *J. Prev. Alzheimers Dis.* **2017**, *4*, 236–241. [PubMed]

315. Rosenberg, R.N.; Fu, M.; Lambracht-Washington, D. Active full-length DNA Aβ_{42} immunization in 3xTg-AD mice reduces not only amyloid deposition but also tau pathology. *Alzheimers Res. Ther.* **2018**, *10*, 115. [CrossRef] [PubMed]

316. Dai, C.L.; Tung, Y.C.; Liu, F.; Gong, C.X.; Iqbal, K. Tau passive immunization inhibits not only tau but also Aβ pathology. *Alzheimers Res. Ther.* **2017**, *9*, 1. [CrossRef] [PubMed]

317. Hol, W.G.J. Protein Crystallography and Computer Graphics—Toward Rational Drug Design. *Angew. Chem.* **1986**, *25*, 767–778. [CrossRef]

318. Roberts, N.A.; Martin, J.A.; Kinchington, D.; Broadhurst, A.V.; Craig, J.C.; Duncan, I.B.; Galpin, S.A.; Handa, B.K.; Kay, J.; Kröhn, A.; et al. Rational design of peptide-based HIV proteinase inhibitors. *Science* **1990**, *248*, 358–361. [CrossRef] [PubMed]

319. Von Itzstein, M.; Wu, W.Y.; Kok, G.B.; Pegg, M.S.; Dyason, J.C.; Jin, B.; Van Phan, T.; Smythe, M.L.; White, H.F.; Oliver, S.W.; et al. Rational design of potent sialidase-based inhibitors of influenza virus replication. *Nature* **1993**, *363*, 418–423. [CrossRef] [PubMed]

320. Sliwoski, G.; Kothiwale, S.; Meiler, J.; Lowe, E.W., Jr. Computational methods in drug discovery. *Pharmacol. Rev.* **2013**, *66*, 334–395. [CrossRef] [PubMed]

321. Joshi, P.; Vendruscolo, M. Druggability of Intrinsically Disordered Proteins. *Adv. Exp. Med. Biol.* **2015**, *870*, 383–400. [PubMed]

322. Marasco, D.; Scognamiglio, P.L. Identification of inhibitors of biological interactions involving intrinsically disordered proteins. *Int. J. Mol. Sci.* **2015**, *16*, 7394–7412. [CrossRef] [PubMed]

323. Tsafou, K.; Tiwari, P.B.; Forman-Kay, J.D.; Metallo, S.J.; Toretsky, J.A. Targeting Intrinsically Disordered Transcription Factors: Changing the Paradigm. *J. Mol. Biol.* **2018**, *430*, 2321–2341. [CrossRef] [PubMed]

324. Rezaei-Ghaleh, N.; Blackledge, M.; Zweckstetter, M. Intrinsically disordered proteins: From sequence and conformational properties toward drug discovery. *ChemBioChem* **2012**, *13*, 930–950. [CrossRef] [PubMed]

325. Uversky, V.N. Dancing Protein Clouds: The Strange Biology and Chaotic Physics of Intrinsically Disordered Proteins. *J. Biol. Chem.* **2016**, *291*, 6681–6688. [CrossRef] [PubMed]

326. Jin, F.; Yu, C.; Lai, L.; Liu, Z. Ligand clouds around protein clouds: A scenario of ligand binding with intrinsically disordered proteins. *PLoS Comput. Biol.* **2013**, *9*, e1003249. [CrossRef] [PubMed]

327. Bier, D.; Thiel, P.; Briels, J.; Ottmann, C. Stabilization of Protein-Protein Interactions in chemical biology and drug discovery. *Prog. Biophys. Mol. Biol.* **2015**, *119*, 10–19. [CrossRef] [PubMed]

328. Arkin, M.R.; Wells, J.A. Small-molecule inhibitors of protein-protein interactions: Progressing towards the dream. *Nat. Rev. Drug Discov.* **2004**, *3*, 301–317. [CrossRef] [PubMed]

329. Wells, J.A.; McClendon, C.L. Reaching for high-hanging fruit in drug discovery at protein-protein interfaces. *Nature* **2007**, *450*, 1001–1009. [CrossRef] [PubMed]

330. Hopkins, A.L.; Groom, C.R. The druggable genome. *Nat. Rev. Drug Discov.* **2002**, *1*, 727–730. [CrossRef] [PubMed]

331. Drews, J.; Ryser, S. The role of innovation in drug development. *Nat. Biotechnol.* **1997**, *15*, 1318–1319. [CrossRef] [PubMed]

332. Drews, J. Drug discovery: A historical perspective. *Science* **2000**, *287*, 1960–1964. [CrossRef] [PubMed]

333. Li, J.; Feng, Y.; Wang, X.; Li, J.; Liu, W.; Rong, L.; Bao, J. An Overview of Predictors for Intrinsically Disordered Proteins over 2010–2014. *Int. J. Mol. Sci.* **2015**, *16*, 23446–23562. [CrossRef] [PubMed]

334. Heller, G.T.; Aprile, F.A.; Vendruscolo, M. Methods of probing the interactions between small molecules and disordered proteins. *Cell. Mol. Life Sci.* **2017**, *74*, 3225–3243. [CrossRef] [PubMed]

335. Ambadipudi, S.; Zweckstetter, M. Targeting intrinsically disordered proteins in rational drug discovery. *Expert Opin. Drug Discov.* **2016**, *11*, 65–77. [CrossRef] [PubMed]

336. Henriques, J.; Cragnell, C.; Skepö, M. Molecular Dynamics Simulations of Intrinsically Disordered Proteins: Force Field Evaluation and Comparison with Experiment. *J. Chem. Theory Comput.* **2015**, *11*, 3420–3431. [CrossRef] [PubMed]

337. Robustelli, P.; Piana, S.; Shaw, D.E. Developing a molecular dynamics force field for both folded and disordered protein states. *Proc. Natl. Acad. Sci. USA* **2018**, *115*, E4758–E4766. [CrossRef] [PubMed]

338. Uversky, V.N. Unusual biophysics of intrinsically disordered proteins. *Biochim. Biophys. Acta* **2013**, *1834*, 932–951. [CrossRef] [PubMed]

339. Dogan, J.; Gianni, S.; Jemth, P. The binding mechanisms of intrinsically disordered proteins. *Phys. Chem. Chem. Phys.* **2014**, *16*, 6323–6331. [CrossRef] [PubMed]

340. Shirai, N.C.; Kikuchi, M. Structural flexibility of intrinsically disordered proteins induces stepwise target recognition. *J. Chem. Phys.* **2013**, *139*, 225103. [CrossRef] [PubMed]

341. Sammak, S.; Zinzalla, G. Targeting protein-protein interactions (PPIs) of transcription factors: Challenges of intrinsically disordered proteins (IDPs) and regions (IDRs). *Prog. Biophys. Mol. Biol.* **2015**, *119*, 41–46. [CrossRef] [PubMed]

342. Hausrath, A.C.; Kingston, R.L. Conditionally disordered proteins: Bringing the environment back into the fold. *Cell. Mol. Life Sci.* **2017**, *74*, 3149–3162. [CrossRef] [PubMed]

343. Hultqvist, G.; Åberg, E.; Camilloni, C.; Sundell, G.N.; Andersson, E.; Dogan, J.; Chi, C.N.; Vendruscolo, M.; Jemth, P. Emergence and evolution of an interaction between intrinsically disordered proteins. *eLife* **2017**, *6*, e16059. [CrossRef] [PubMed]

344. De Cássia Ruy, P.; Torrieri, R.; Toledo, J.S.; de Souza Alves, V.; Cruz, A.K.; Ruiz, J.C. Intrinsically disordered proteins (IDPs) in trypanosomatids. *BMC Genom.* **2014**, *15*, 1100.

345. Longhi, S. Structural disorder within paramyxoviral nucleoproteins. *FEBS Lett.* **2015**, *589*, 2649–2659. [CrossRef] [PubMed]

346. Russo, A.; Manna, S.L.; Novellino, E.; Malfitano, A.M.; Marasco, D. Molecular signaling involving intrinsically disordered proteins in prostate cancer. *Asian J. Androl.* **2016**, *18*, 673–681. [PubMed]

347. Ruan, H.; Sun, Q.; Zhang, W.; Liu, Y.; Lai, L. Targeting intrinsically disordered proteins at the edge of chaos. *Drug Discov. Today* **2019**, *24*, 217–227. [CrossRef] [PubMed]

8

Repeats in S1 Proteins: Flexibility and Tendency for Intrinsic Disorder

Andrey Machulin [1], Evgenia Deryusheva [2], Mikhail Lobanov [3] and Oxana Galzitskaya [3,*]

[1] Skryabin Institute of Biochemistry and Physiology of Microorganisms, Russian Academy of Sciences, Federal Research Center "Pushchino Scientific Center for Biological Research of the Russian Academy of Sciences, 142290 Pushchino, Russia; and.machul@gmail.com

[2] Institute for Biological Instrumentation, Federal Research Center "Pushchino Scientific Center for Biological Research of the Russian Academy of Sciences, 142290 Pushchino, Russia; evgenia.deryusheva@gmail.com

[3] Institute of Protein Research, Russian Academy of Sciences, 142290 Pushchino, Russia; mlobanov@phys.protres.ru

* Correspondence: ogalzit@vega.protres.ru

Abstract: An important feature of ribosomal S1 proteins is multiple copies of structural domains in bacteria, the number of which changes in a strictly limited range from one to six. For S1 proteins, little is known about the contribution of flexible regions to protein domain function. We exhaustively studied a tendency for intrinsic disorder and flexibility within and between structural domains for all available UniProt S1 sequences. Using charge–hydrophobicity plot cumulative distribution function (CH-CDF) analysis we classified 53% of S1 proteins as ordered proteins; the remaining proteins were related to molten globule state. S1 proteins are characterized by an equal ratio of regions connecting the secondary structure within and between structural domains, which indicates a similar organization of separate S1 domains and multi-domain S1 proteins. According to the FoldUnfold and IsUnstruct programs, in the multi-domain proteins, relatively short flexible or disordered regions are predominant. The lowest percentage of flexibility is in the central parts of multi-domain proteins. Our results suggest that the ratio of flexibility in the separate domains is related to their roles in the activity and functionality of S1: a more stable and compact central part in the multi-domain proteins is vital for RNA interaction, terminals domains are important for other functions.

Keywords: ribosomal proteins S1; structural domains; intrinsically flexibility; FoldUnfold program; IsUnstruct program

1. Introduction

It is known that multi-domain proteins are frequently characterized by the occurrence of domain repeats in proteomes across the three domains of life: Bacteria, Archaea, and Eukaryotes [1,2]. Proteins with repeats participate in nearly every cellular process from transcriptional regulation in the nucleus to cell adhesion at the plasma membrane [3]. In addition, due to their flexibility, domain repeats can be found in cytoskeleton proteins, proteins responsible for transport and cell cycle control [4]. Proteins with structural repeats are believed to be ancient folds.

One such unique protein family is a family of bacterial ribosomal proteins S1 in which structural domain S1 (one of the oligonucleotide/oligosaccharide-binding fold (OB-fold) options) repeats and changes in a strictly limited range from one to six [5]. As demonstrated in our recent paper [5], the family of polyfunctional ribosomal proteins S1 contains about 20% of all bacterial proteins, including the S1 domain. This fold also could be found in different eukaryotic protein families and protein complexes in different number variations. Such multiple copies of the structure increase the affinity and/or specificity of the protein binding to nucleic acid molecules.

Recently we have shown that the sequence alignments of S1 proteins between separate domains in each group reveal a rather low percentage of identity. In addition, the verification of the equivalence of the domain characteristics showed that for long S1 proteins (five- and six-domain containing S1 proteins) the central part of the proteins (the third domain) is more conservative than the terminal domains and apparently is vital for the activity and functionality of S1. Data obtained indicated that for general functioning of these proteins, the structure scaffold (OB-fold) is obviously more important than the amino acid sequence [6]. This statement is in good agreement with the fact that there is a high degree of conservatism and topology position of the binding site on the OB-fold surface in others proteins, as well as "fold resistance" to mutations and the ability to adapt to a wide range of ligands, which allows us to consider this fold as one of the ancient protein folds. For example, the author of article [7] proposed considering this core structure of inorganic pyrophosphatase as the evolutionary precursor of all other superfamilies.

At present, the structure of S1 from *Escherichia coli* was obtained only with a very low resolution of 11.5 Å using cryo-electron microscopy [8]. In the Protein Data Bank, there are only 3D structures of separate domains of ribosomal S1 from *E. coli* obtained by NMR [9,10]. Recently, protein S1 on the 70S ribosome was visualized by ensemble cryo-electron microscopy [11]. It was shown that S1 cooperates with other ribosomal proteins (S2, S3, S6, and S18) to form a dynamic mesh near the mRNA exit and entrance channels to modulate the binding, folding and movement of mRNA. The cryo-electron microscopy was also used to obtain the structure of the inactive conformation of the S1 protein as part of a hibernating 100S ribosome [12].

A separate S1 domain from the ribosomal proteins S1 [9] and other bacterial proteins containing an S1 domain [13–16] represents a β-barrel with an additional α-helix between the third and fourth β-sheets. As shown in the articles [13–16], the S1 domain as a part of different bacterial proteins (as well as in eukaryotic proteins) itself is quite compact, therefore it crystallizes and is visualized very well.

At the same time, there are currently no determined structures for full-length, intact ribosomal S1 proteins containing a different number of structural domains (six in *E. coli*, five in *Thermus thermophilus*, etc.). This may be due to the increased flexibility of multi-domain proteins as was noted in [17]. In addition, some biochemical studies suggest that in solution and on the ribosome, S1 can have an elongated shape stretching over 200 Å long [17–20].

Moreover, recently it was shown that the prediction of intrinsic disorder within proteins with the tandem repeats supports the conclusion that the level of repetition correlates with their tendency to be unstructured and the chance to find natural structured proteins in the Protein Data Bank (PDB) increases with a decrease in the level of repeat perfection. Also, the authors suggested that in general, the repeat perfection is a sign of recent evolutionary events rather than of exceptional structural and/or functional importance of the repeat residues [21].

Despite all these observations, the flexibility of S1 proteins, their tendency for intrinsic disorder, and the structural characteristics of this family have not been studied as of yet. To fill this gap, we have analyzed here the flexibility of the bacterial S1 proteins within and between structural domains, as well as the tendency for intrinsic disorder of the S1 protein family.

2. Results and Discussion

2.1. Analysis of Tendency for Intrinsic Disorder of the Bacterial S1 Proteins

Binary disorder analysis using the charge–hydrophobicity plot cumulative distribution function (CH-CDF) plot [22] showed that most of the bacterial S1 proteins (1374 sequences) (53%) are expected to be mostly ordered (or folded, 'F') (Figure 1a).

Mixed or molten globular ('MG') forms comprised the remaining 47% of the bacterial S1 proteins. Major protein states for separate groups of the S1 proteins (different number of structural domains) according to the CH-CDF analysis are shown in Figure 1b. In the case of S1 proteins containing one, two or six structural domains (1S1, 2S1, 6S1) the ordered state prevailed (83%, 78% and 67%, respectively). S1 proteins containing three, four and five domains were classified as molten globule state according to the CH-CDF analysis in 69%, 74% and 56% cases, respectively. It was seen that with an increase in the number of structural domains (starting from the three-domain containing proteins), the MG state prevailed, but for six-domain proteins only 34% of the records belonged to this area. Despite the fact that one-domain and two-domain containing proteins were the least represented in our dataset, the data obtained for these groups results are in good agreement with the fact that the separate S1 domain is stable and has rather rigid structure [13–16]. Note that for other structural variants of the OB-fold (for example, CSD domain [23], inorganic pyrophosphatase [24], *MOP-like* [25], etc.) there are available structures that also have only one or two (repeated) domains [5].

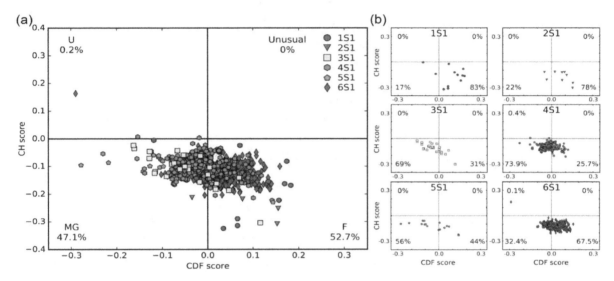

Figure 1. (**a**) Binary disorder analysis (charge–hydrophobicity plot cumulative distribution function (CH-CDF) plots [22]) of 1374 S1 proteins; (**b**) separate S1 proteins groups containing different numbers of structural domains.

2.2. Analysis of Intrinsic Flexibility and Disorder of the Bacterial S1 Proteins and Its Domains.

For analysis of intrinsic flexibility and disorder of the full length bacterial S1 proteins and its separate structural domains we used the FoldUnfold (average window 11 aa and 5 aa) and IsUnstruct programs; their possibilities and accuracy were described in [26–29]. The obtained results are given in Table 1.

Analysis of the percentage of disorder in the full length S1 proteins and in their separate domains by the FoldUnfold (average window 11 aa and 5 aa) and IsUnstruct programs revealed their close similarity (Table 1).

For full-length proteins, the highest percentage of disorder was detected for four- (30%) and five-domain (30%) containing proteins using the FoldUnfold program (average window 5 aa). The smallest percentage was in the six-domain proteins (13%) when using the FoldUnfold program (average window 11 aa). This indicates the predominance of relatively short flexible or unstructured regions in the considered sequences of the proteins of this group, consistent with the fact that the binary predictor of the CH-CDF plot revealed the ordered states for 67% of proteins in this group.

Most of the separate S1 domains exhibited disorder values around 20%. The lowest percentage of disorder (except the third domain in three-domain containing proteins and the separate domains in the one-domain containing proteins) predicted by the FoldUnfold program (average window 5 aa) was the third domain in six-domain containing proteins (13%). Using the FoldUnfold program (average window 11 aa) and IsUnstruct for this domain also revealed a relatively low percentage of intrinsically disorder compared with other domains in this group and other groups (by the number of domains), 19% and 21%, respectively. The largest percentage of disorder predicted by the IsUnstruct program belonged to the sixth domain in the six-domain containing proteins (45%). Using the FoldUnfold program for six-domain containing proteins, a propensity for a more disordered state in the terminal domains was also identified. Note that, earlier, we have shown that for long S1 proteins (six-domain S1 proteins) the central part of the proteins (the third domain) is more conservative (as a percent of identity between separate domains) than the terminal domains, and apparently is vital for the activity and functionality of S1 proteins [6].

The concept of order and disorder in protein segments has often been investigated in correlation with the presence or absence of protein repeats at the sequence level. It is noticed that intrinsically disordered proteins often correspond to regions of low compositional complexity (low sequence entropy) and sometimes to repetitive sub-sequences, for example, in fibrillar proteins [30]. Also in some special cases, protein repeats (for example, in the PEVK ((Pro-Glu-Val-Lys) domain) regions of human titin, the prion proteins, or the CTD domain of RNA polymerase) are discussed in detail [31]. However, these findings on specific instances are hard to generalize. A general property observed is that a higher level of repeat perfection correlates positively with the disordered state of protein sub-chains [21].

S1 proteins, having a low degree of conservatism (not perfect repeats) [6], in addition to the found low degree of disorder within and between the domains, demonstrate the unique structural organization of proteins of this family. Apparently, the organization is closer to the formation of the quaternary structure of globular proteins, with the same structural organization of individual structural domains.

Table 1. Intrinsic flexibility and disorder of S1 protein family and its structural domains. The largest and smallest values are highlighted in bold.

Number of Structural S1 Domains	FoldUnfold (11 aa)		FoldUnfold (5 aa)		IsUnstruct	
	% Disorder for Each Domain	Full Length Proteins	% Disorder for Each Domain	Full Length Proteins	% Disorder for Each Domain	Full Length Proteins
1S1	20 ± 3	25 ± 10	17 ± 11	22 ± 13	17 ± 11	24 ± 17
2S1	1　16 ± 1	20 ± 11	1　13 ± 6	20 ± 5	1　18 ± 5	19 ± 10
	2　24 ± 10		2　20 ± 10		2　28 ± 11	
3S1	1　17 ± 1	15 ± 9	1　20 ± 6	26 ± 7	1　36 ± 13	26 ± 9
	2　21 ± 7		2　21 ± 7		2　36 ± 16	
	3　0		3　13 ± 6		3　20 ± 4	
4S1	1　21 ± 5	18 ± 5	1　25 ± 7	**30 ± 4**	1　24 ± 9	22 ± 5
	2　18 ± 1		2　13 ± 5		2　24 ± 5	
	3　21 ± 6		3　17 ± 8		3　28 ± 10	
	4　18 ± 3		4　16 ± 7		4　23 ± 7	
5S1	1　21 ± 3	17 ± 13	1　22 ± 12	**30 ± 11**	1　28 ± 16	21 ± 15
	2　21 ± 5		2　15 ± 8		2　23 ± 13	
	3　20 ± 3		3　22 ± 12		3　28 ± 13	
	4　24 ± 1		4　22 ± 8		4　35 ± 16	
	5　18 ± 2		5　22 ± 5		5　28 ± 10	
6S1	1　24 ± 9	**13 ± 4**	1　22 ± 8	27 ± 3	1　27 ± 12	16 ± 4
	2　18 ± 3		2　14 ± 8		2　22 ± 7	
	3　18 ± 4		3　**12 ± 6**		3　21 ± 3	
	4　19 ± 3		4　19 ± 6		4　24 ± 4	
	5　20 ± 5		5　27 ± 7		5　25 ± 5	
	6　22 ± 7		6　32 ± 9		6　**45 ± 19**	

2.3. Flexibility of S1 Domain in the Bacterial Proteins

Besides the ribosomal proteins, S1 domains are identified in different quantities in different archaeal, bacterial and eukaryotic proteins [5]. As we recently showed, archaeal proteins contain one copy of the S1 domain, while the number of repeats in the eukaryotic proteins varies between 1 and 15 and correlates with the protein size. In the bacterial proteins, the number of repeats is no more than 6, regardless of the protein size. To compare the obtained data on the flexibility of ribosomal proteins S1, S1 domains from some bacterial proteins [5] were investigated using the approaches described above (Table 2).

Table 2. Intrinsic flexibility and disorder of S1 domains in some bacterial proteins.

Protein Name	Source Organism	UniProt Code	Percent of Flexibility/Disorder		
			FoldUnfold (11 aa)	FoldUnfold (5 aa)	IsUnstruct
S1 domain PNPase	*E. coli*	P05055	0	17	17
Protein YhgF	*E. coli*	P46837	0	0	11
Antitermination protein NusA	*E. coli*	P0AFF6	0	36	26
Ribonuclease R	*E. coli*	P21499	13	6	27
Ribonuclease E	*E. coli*	P21513	0	20	26
Tex-like protein N-terminal domain protein	*Kingella denitrificans*	F0F1S0	0	0	13

In all proteins (Table 2, Figure 2), one S1 domain was identified and had a low degree of disorder (about 20%). It can be seen that when the size of average window of the FoldUnfold program decreases, this percentage increases, indicating the presence of flexible sections of short length in the considered proteins. This is consistent with the fact that S1 domains in these proteins are well determined by various methods (Figure 2).

(a)   (b)   (c)

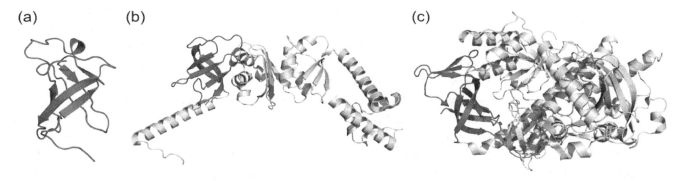

Figure 2. Protein structures with the S1 domain from different bacterial proteins. The S1 domain in each structure is highlighted with red color. (**a**) S1 domain PNPase, PDB code: 1sro; (**b**) antitermination protein NusA, PDB code: 5ml9; (**c**) Ribonuclease R, PDB code: 5xgu.

However, structures of proteins containing three or more S1 domains have not been determined yet. In the eukaryotic proteins containing more than two S1 domain (from 7 to 15) determined structures also are not available. Note that in these proteins, functions of separate S1 domains are not defined, for example, Rrp5p [32], Prp22p [33].

2.4. Analysis of the Ratio of Secondary Structures in the Bacterial S1 Proteins and Its Domains

Obtained ratios of regions connecting secondary structure according to the JPred predictions are shown in Table 3.

Table 3. Ratio of regions connecting elements of the secondary structure according to the JPred predictions.

Number of Structural S1 Domains	JPred			
	% aa in the Regions Connecting Secondary Structures (Separate Domains)		% aa in the Regions Connecting Secondary Structures (Full Length Proteins)	% aa of the Linkers between Structural Domains (Full Length Proteins)
1S1	41 ± 6		49 ± 7	45 ± 13
2S1	1	50 ± 6	47 ± 3	33 ± 13
	2	41 ± 4		
3S1	1	48 ± 3	46 ± 4	38 ± 7
	2	44 ± 10		
	3	43 ± 3		
4S1	1	47 ± 2	51 ± 2	38 ± 7
	2	47 ± 6		
	3	51 ± 4		
	4	44 ± 2		
5S1	1	49 ± 3	53 ± 4	33 ± 8
	2	51 ± 4		
	3	51 ± 5		
	4	48 ± 3		
	5	48 ± 6		
6S1	1	47 ± 2	52 ± 2	27 ± 3
	2	52 ± 5		
	3	49 ± 3		
	4	50 ± 4		
	5	51 ± 4		
	6	47 ± 3		

It can be seen that the ratio of regions connecting the secondary structure in separate domains was approximately the same and equal to about 50%, which in addition to conservative secondary structure indicates about the same organization of separate S1 domains. For full length proteins this ratio (linkers and regions connecting secondary structures within domains) was also about 50%, indicating about the same organization of multi-domains containing S1 proteins. The average percent of linkers between structural domains was about 30–40%. The obtained results are in a good agreement with the predictions of the FoldUnfold and IsUnstruct programs and CH-CDF plots, and characterized the family of S1 proteins as proteins with relatively short flexible regions within domains and between them that apparently prefer to be in the folded or MG state. In addition to the aforementioned lower conservatism between separate domains in each group, it can be argued that the unique S1 protein family is different in the classical sense from a protein with tandem repeats, such as the ANK family, leucine-rich-repeat proteins, etc. [4]. This family having repeats (separate structural domains) with 70 residues is close to a "beads-on-a-string" organization with each repeat being folded into a globular domain, for example, Zn-finger domains [34], Ig-domains [35] and the human matrix metalloproteinase [36]. Thus, one of the reasons for the absence of allowed three-dimensional structures of multi-domain S1 proteins may be the mobility of domains relative to each other due to the flexibility of interdomain linkers.

In fact, the biochemical experimental study of various fragments allowed establishing the functions of individual protein domains and parts only for the well-studied 30S ribosomal protein S1 with six S1 domain repeats from *E. coli*. For example, it has been shown that cutting one S1 domain from the C-terminus or two S1 domains from the N-terminus of the protein reduces only the effectiveness of protein functions but not its functional abilities; the sixth domain is bound with the process of autoregulation of synthesis, thus cutting off the fifth and sixth domain leads to effective participation of the remaining part of protein only in synthetic mRNA translation [37,38]. Our results indicated about the same organization of separate S1 domains and full-length proteins (conservative secondary structure, ratio of linkers and regions connecting secondary structures within domains). In addition, the percent of intrinsic flexibility is less for the central domains in the multi-domain proteins. These facts allowed us to assume that for all multi-domain S1 proteins more stable and compact domain are located in the central part and are vital for RNA interaction, while more flexible terminals domains are for other functions. The obtained results will be used as a base for investigation of the proposed theories on the evolutionary development of proteins with structural repeats: From the multi-repeat assemblies to single repeat or vice versa.

3. Materials and Methods

3.1. Construction of Ribosomal Proteins S1 Dataset

To make a representative dataset of records for the family of ribosomal proteins S1 from the UniProt database, all records for the bacteria containing any one of the keywords «30s ribosomal protein s1», «ribosomal protein s1», «30s ribosomal protein s1 (ec 1.17.1.2)», «30s ribosomal protein s1 (ribosomal protein s1)», «ribosomal protein s1 domain protein», «rna binding protein s1», «rna binding s1 domain protein», «s1 rna binding domain protein» in the protein name were selected (UniProt release 2018_04). Then the obtained array of data was used to choose only proteins encoded by the rpsA gene or its analog; for example, rpsA_1, rpsA_2, rpsA_3, etc. Only this gene, coding the ribosomal protein S1, in the European nucleotide archive (ENA, http://www.ebi.ac.uk/ena) is affiliated to the STD class, that is, the class of standard annotated sequences. From the obtained dataset records, those with six-digital identification numbers (annotated records in the UniProt database) were selected. All data were collected in one file that was the basis for further analysis, namely for collection of data on the number of structural domains and for phylogenetic grouping in the main bacterial phyla (http://bioinfo.protres.ru/other/uniprot_S1.xlsx). Records characterized by the presence of the word "candidate" were removed from our dataset. The automated advanced exhaustive analysis allowed us to choose 1374 records corresponding to these search parameters.

3.2. Number and Identification of Structural Domains in Protein Sequences

The values of the number of S1 domains corresponding to the SMART database (about 1200 domains), were selected for each analyzed record. If no data on the number of domains in one of the analyzed bases was available (None), this number was taken to be zero (these records were removed from investigated dataset). Accurate borders for each S1 domain for each record were taken from the UniProt database (position, domain and repeats field).

3.3. Prediction of Disordered Regions and Tendency for Intrinsic Disorder

3.3.1. FoldUnfold and IsUnstruct Programs

The FoldUnfold program is accessible at http://bioinfo.protres.ru/ogu/. The principle of its operation is described elsewhere [26,27]. Such a property of residues as the observed average number of contacts in a globular state, closed at a given distance, was used. To predict IDRs (intrinsically disordered regions) in the protein chain using the amino acid sequence, every residue was given an expected number of contacts in the globular state. Then averaging was done by the residue equal to the window width. The obtained average value of expected contacts was ascribed to the central residue in the chosen window. After that the window was shifted by one residue, and the procedure was repeated. On the profile of expected contacts, a boundary was marked that separated structured and unstructured residues. The mean expected number of closed residues, estimated from the sequence, was equal to the sum of expected contact residues divided by the number of amino acid residues in the protein. According to the algorithm of the program, the size of disordered (flexible) regions in such a protein must be equal to or greater than the size of the averaged window. Therefore, the number of predicted regions depended on the window size. The window size in 11 amino acid residues was optimal for the search for relatively short disordered regions in the polypeptide chain. In the case of searching for long disordered regions in partially disordered proteins, the window size must be increased to several tens of amino acid resides. At the same time, for searching for short loops one should use the averaged window size of five amino acid residues, which is optimal for this task.

The IsUnstruct program (v.2.02) is accessible at http://bioinfo.protres.ru/IsUnstruct/. The algorithm of the IsUnstruct program is based on the Ising model. For estimation the energy of any state, the energy of the border between ordered and disordered residues and the energies of initiation of disordered state at the ends were used [39]. After the optimization procedure [28], 20 energetic potentials for residues were obtained which were considered to be in a disordered state, the energy of border, and the energies of initiation of disordered state at the ends. The energy of the completely ordered state was taken to be zero.

3.3.2. CH-CDF Analysis

The charge–hydrophobicity plots (CH-plots) [40] and the cumulative distribution function (CDF) analysis [41] were used for binary prediction of protein stability based of its amino acid sequence.

The Y-coordinate in the CH-CDF plot corresponded to the distance from the obtained ordinate value to the correlation line separating the structured and unstructured conformational state of the protein on the CH (charge-hydrophobicity) plot. The X-coordinate on the CH-CDF plot corresponded to the distance from the obtained ordinate value to the correlation line separating the structured and unstructured conformational state of the protein in the CDF. Thus, in the coordinates of CH-CDF plot it was possible to assign the sequence to one of four quadrants (four conformational states). I quadrant (CH > 0, CDF > 0) were rare proteins for which it was impossible to determine accurately the state (unusual/rare); II quadrant (CH > 0, CDF < 0) were unfolded proteins (U), III quadrant (CH < 0, CDF < 0) was the state of the molten globule (MG), IV quadrant (CH < 0, CDF > 0) were structured proteins (F) [22]. Calculation of the Y-coordinate (CH-coordinate) was performed automatically. The CH coordinate values were calculated as a distance between the CH values calculated using PONDR® online service (http://www.pondr.com/) and the linear border between IDPs and structured proteins ($y = 2.743 \times x - 1.109$) [41]. Values of the X-coordinate (CDF) were the average of the vertical distances from the CDF curve to the seven boundary points. To obtain CDF-values, the version VSL2 PONDR was used [42].

3.4. Prediction of Secondary Structure

Jpred4 (http://www.compbio.dundee.ac.uk/jpred/) was used for prediction of secondary structure for each sequence in our dataset [43].

3.5. Analysis and Visualization

Algorithms of search, collection, representation and analysis by the described methods of the data were realized using the freely available programming language Python 3 (https://www.python.org/). The result of the obtained two-dimensional array of data (for CH-CDF plots) was visualized using the Matplotlib library.

4. Conclusions

In this work, we show that S1 proteins belong to a unique family, which differs in the classical sense from proteins with tandem repeats. We found that the one-domain and two-domain containing S1 proteins apparently have more stable and rigid structure. An increase in the number of structural domains contributes to the possible transition of a portion of proteins from the folded state to the MG state. For example, for three- and four-domain containing proteins, the ratio of predicted MG state is about 70%. A relatively small percentage of internal flexibility/disorder within individual structural domains could be seen as an indicator of the stability of the S1 domain as one of the OB-fold in this family. At the same time the ratio of flexibility in the separate domains apparently is related to their roles in the activity and functionality of S1. A more stable, compact and conservative central part in the multi-domain proteins is vital for RNA interaction, while terminals domains are for other functions. At the same time, an equal ratio of regions connecting the secondary structure in separate domains and between structural domains indicates about the same organization of multi-domains containing S1 proteins, as well as position and ratio of the secondary structures within separate domains. Reasons for the lack of intact 3D structure of full-length ribosomal protein S1 is not well-understood Perhaps this is due to the high mobility of domains relative to each other in the multi-domain proteins. Further investigation of the flexibility of the available 3D structures for separate S1 domains and the full length S1 domain from *E. coli* in complex with 70S ribosomal subunit will allow finding an accurate explanation.

Author Contributions: Conceptualization and experiment design, O.G., A.M., E.D.; software, A.M., M.L.; formal analysis, A.M., M.L., E.D., O.G.; data analysis, A.M., E.D., O.G.; visualization, A.M.; writing—original draft preparation, E.D. and O.G.; writing—review & editing, O.G.; supervision, O.G.

Abbreviations

OB-fold	Oligonucleotide/oligosaccharide-binding fold
CH-plot	Charge–hydrophobicity plot
CDF	Cumulative distribution function
U	Unfolded
MG	Molten globule
F	Folded
SMART	Simple Modular Architecture Research Tool
ANK	Ankyrin
Ig	Immunoglobulin

References

1. Björklund, A.K.; Ekman, D.; Elofsson, A. Expansion of protein domain repeats. *PLoS Comput. Biol.* **2006**, *2*, e114. [CrossRef] [PubMed]
2. Jernigan, K.K.; Bordenstein, S.R. Tandem-repeat protein domains across the tree of life. *PeerJ* **2015**, *3*, e732. [CrossRef] [PubMed]
3. Andrade, M.A.; Petosa, C.; O'Donoghue, S.I.; Müller, C.W.; Bork, P. Comparison of ARM and HEAT protein repeats. *J. Mol. Biol.* **2001**, *309*, 1–18. [CrossRef] [PubMed]
4. Andrade, M.A.; Perez-Iratxeta, C.; Ponting, C.P. Protein Repeats: Structures, Functions, and Evolution. *J. Struct. Biol.* **2001**, *134*, 117–131. [CrossRef] [PubMed]
5. Deryusheva, E.I.; Machulin, A.V.; Selivanova, O.M.; Galzitskaya, O. V Taxonomic distribution, repeats, and functions of the S1 domain-containing proteins as members of the OB-fold family. *Proteins* **2017**, *85*, 602–613. [CrossRef]
6. Machulin, A.; Deryusheva, E.; Selivanova, O.; Galzitskaya, O. Phylogenetic bacterial grouping by numbers of structural domains in the family of ribosomal proteins S1. *Sci. Rep.* under review.
7. Arcus, V. OB-fold domains: A snapshot of the evolution of sequence, structure and function. *Curr. Opin. Struct. Biol.* **2002**, *12*, 794–801. [CrossRef]
8. Sengupta, J.; Agrawal, R.K.; Frank, J. Visualization of protein S1 within the 30S ribosomal subunit and its interaction with messenger RNA. *Proc. Natl. Acad. Sci. USA* **2001**, *98*, 11991–11996. [CrossRef]
9. Salah, P.; Bisaglia, M.; Aliprandi, P.; Uzan, M.; Sizun, C.; Bontems, F. Probing the relationship between gram-negative and gram-positive S1 proteins by sequence analysis. *Nucleic Acids Res.* **2009**, *37*, 5578–5588. [CrossRef]
10. Giraud, P.; Créchet, J.-B.; Uzan, M.; Bontems, F.; Sizun, C. Resonance assignment of the ribosome binding domain of E. coli ribosomal protein S1. *Biomol. NMR Assign.* **2015**, *9*, 107–111. [CrossRef]

11. Loveland, A.B.; Korostelev, A.A. Structural dynamics of protein S1 on the 70S ribosome visualized by ensemble cryo-EM. *Methods* **2018**, *137*, 55–66. [CrossRef]

12. Beckert, B.; Turk, M.; Czech, A.; Berninghausen, O.; Beckmann, R.; Ignatova, Z.; Plitzko, J.M.; Wilson, D.N. Structure of a hibernating 100S ribosome reveals an inactive conformation of the ribosomal protein S1. *Nat. Microbiol.* **2018**, *3*, 1115–1121. [CrossRef]

13. Bycroft, M.; Hubbard, T.J.; Proctor, M.; Freund, S.M.; Murzin, A.G. The solution structure of the S1 RNA binding domain: a member of an ancient nucleic acid-binding fold. *Cell* **1997**, *88*, 235–242. [CrossRef]

14. Schubert, M.; Edge, R.E.; Lario, P.; Cook, M.A.; Strynadka, N.C.J.; Mackie, G.A.; McIntosh, L.P. Structural characterization of the RNase E S1 domain and identification of its oligonucleotide-binding and dimerization interfaces. *J. Mol. Biol.* **2004**, *341*, 37–54. [CrossRef]

15. Beuth, B.; Pennell, S.; Arnvig, K.B.; Martin, S.R.; Taylor, I.A. Structure of a Mycobacterium tuberculosis NusA-RNA complex. *EMBO J.* **2005**, *24*, 3576–3587. [CrossRef]

16. Battiste, J.L.; Pestova, T.V.; Hellen, C.U.; Wagner, G. The eIF1A solution structure reveals a large RNA-binding surface important for scanning function. *Mol. Cell* **2000**, *5*, 109–119. [CrossRef]

17. Giri, L.; Subramanian, A.R. Hydrodynamic properties of protein S1 from Escherichia coli ribosome. *FEBS Lett.* **1977**, *81*, 199–203. [CrossRef]

18. Laughrea, M.; Moore, P.B. Physical properties of ribosomal protein S1 and its interaction with the 30 S ribosomal subunit of Escherichia coli. *J. Mol. Biol.* **1977**, *112*, 399–421. [CrossRef]

19. Labischinski, H.; Subramanian, A.R. Protein S1 from Escherichia coli ribosomes: an improved isolation procedure and shape determination by small-angle X-ray scattering. *Eur. J. Biochem.* **1979**, *95*, 359–366. [CrossRef]

20. Sillers, I.Y.; Moore, P.B. Position of protein S1 in the 30 S ribosomal subunit of Escherichia coli. *J. Mol. Biol.* **1981**, *153*, 761–780. [CrossRef]

21. Jorda, J.; Xue, B.; Uversky, V.N.; Kajava, A. V Protein tandem repeats - the more perfect, the less structured. *FEBS J.* **2010**, *277*, 2673–2682. [CrossRef]

22. Huang, F.; Oldfield, C.; Meng, J.; Hsu, W.L.; Xue, B.; Uversky, V.N.; Romero, P.; Dunker, A.K. Subclassifying disordered proteins by the CH-CDF plot method. *Pac. Symp. Biocomput.* **2012**, 128–139. [CrossRef]

23. Schindelin, H.; Jiang, W.; Inouye, M.; Heinemann, U. Crystal structure of CspA, the major cold shock protein of Escherichia coli. *Proc. Natl. Acad. Sci. USA* **1994**, *91*, 5119–5123. [CrossRef]

24. Heikinheimo, P.; Tuominen, V.; Ahonen, A.K.; Teplyakov, A.; Cooperman, B.S.; Baykov, A.A.; Lahti, R.; Goldman, A. Toward a quantum-mechanical description of metal-assisted phosphoryl transfer in pyrophosphatase. *Proc. Natl. Acad. Sci. USA* **2001**, *98*, 3121–3126. [CrossRef] [PubMed]

25. Delarbre, L.; Stevenson, C.E.; White, D.J.; Mitchenall, L.A.; Pau, R.N.; Lawson, D.M. Two crystal structures of the cytoplasmic molybdate-binding protein ModG suggest a novel cooperative binding mechanism and provide insights into ligand-binding specificity. *J. Mol. Biol.* **2001**, *308*, 1063–1079. [CrossRef] [PubMed]

26. Galzitskaya, O.V.; Garbuzynskiy, S.O.; Lobanov, M.Y. Prediction of amyloidogenic and disordered regions in protein chains. *PLoS Comput. Biol.* **2006**, *2*, 10. [CrossRef]

27. Galzitskaya, O.V.; Garbuzynskiy, S.O.; Lobanov, M.Y. FoldUnfold: Web server for the prediction of disordered regions in protein chain. *Bioinformatics* **2006**, *22*, 2948–2949. [CrossRef] [PubMed]

28. Lobanov, M.Y.; Galzitskaya, O.V. The Ising model for prediction of disordered residues from protein sequence alone. *Phys. Biol.* **2011**, *8*, 035004. [CrossRef]

29. Deryusheva, E.; Machulin, A.; Nemashkalova, E.; Glyakina, A.; Galzitskaya, O. Search for functional flexible regions in the G-protein family: new reading of the FoldUnfold program. *Protein Pept. Lett.* **2018**, *25*, 589–598. [CrossRef]

30. Dunker, A.K.; Lawson, J.D.; Brown, C.J.; Williams, R.M.; Romero, P.; Oh, J.S.; Oldfield, C.J.; Campen, A.M.; Ratliff, C.M.; Hipps, K.W.; et al. Intrinsically disordered proteins. *J. Mol. Graph. Model.* **2001**, *19*, 26–59. [CrossRef]

31. Tompa, P.; Fersht, A. *Structure and Function of Intrinsically Disordered Proteins*; Chapman and Hall/CRC: New York, NY, USA, 2010.

32. Hierlmeier, T.; Merl, J.; Sauert, M.; Perez-Fernandez, J.; Schultz, P.; Bruckmann, A.; Hamperl, S.; Ohmayer, U.; Rachel, R.; Jacob, A.; et al. Rrp5p, Noc1p and Noc2p form a protein module which is part of early large ribosomal subunit precursors in S. cerevisiae. *Nucleic Acids Res.* **2013**, *41*, 1191–1210. [CrossRef] [PubMed]

33. Mayas, R.M.; Maita, H.; Staley, J.P. Exon ligation is proofread by the DExD/H-box ATPase Prp22p. *Nat. Struct. Mol. Biol.* **2006**, *13*, 482–490. [CrossRef]

34. Lee, M.S.; Gippert, G.P.; Soman, K.V.; Case, D.A.; Wright, P.E. Three-dimensional solution structure of a single zinc finger DNA-binding domain. *Science* **1989**, *245*, 635–637. [CrossRef]

35. Sawaya, M.R.; Wojtowicz, W.M.; Andre, I.; Qian, B.; Wu, W.; Baker, D.; Eisenberg, D.; Zipursky, S.L. A double S shape provides the structural basis for the extraordinary binding specificity of Dscam isoforms. *Cell* **2008**, *134*, 1007–1018. [CrossRef]

36. Elkins, P.A.; Ho, Y.S.; Smith, W.W.; Janson, C.A.; D'Alessio, K.J.; McQueney, M.S.; Cummings, M.D.; Romanic, A.M. Structure of the C-terminally truncated human ProMMP9, a gelatin-binding matrix metalloproteinase. *Acta Crystallogr. D Biol. Crystallogr.* **2002**, *58*, 1182–1192. [CrossRef] [PubMed]

37. Amblar, M.; Barbas, A.; Gomez-Puertas, P.; Arraiano, C.M. The role of the S1 domain in exoribonucleolytic activity: substrate specificity and multimerization. *Rna* **2007**, *13*, 317–327. [CrossRef]

38. Boni, I.V.; Artamonova, V.S.; Dreyfus, M. The last RNA-binding repeat of the Escherichia coli ribosomal protein S1 is specifically involved in autogenous control. *J. Bacteriol.* **2000**, *182*, 5872–5879. [CrossRef] [PubMed]

39. Lobanov, M.Y.; Sokolovskiy, I.V.; Galzitskaya, O.V. IsUnstruct: prediction of the residue status to be ordered or disordered in the protein chain by a method based on the Ising model. *J. Biomol. Struct. Dyn.* **2013**, *31*, 1034–1043. [CrossRef]

40. Kyte, J.; Doolittle, R.F. A simple method for displaying the hydropathic character of a protein. *J. Mol. Biol.* **1982**, *157*, 105–132. [CrossRef]

41. Xue, B.; Oldfield, C.J.; Dunker, A.K.; Uversky, V.N. CDF it all: Consensus prediction of intrinsically disordered proteins based on various cumulative distribution functions. *FEBS Lett.* **2009**, *583*, 1469–1474. [CrossRef]

42. Pace, C.N.; Vajdos, F.; Fee, L.; Grimsley, G.; Gray, T. How to measure and predict the molar absorption coefficient of a protein. *Protein Sci.* **1995**, *4*, 2411–2423. [CrossRef] [PubMed]

43. Drozdetskiy, A.; Cole, C.; Procter, J.; Barton, G.J. JPred4: a protein secondary structure prediction server. *Nucleic Acids Res.* **2015**, *43*, W389–394. [CrossRef] [PubMed]

9

Investigation into Early Steps of Actin Recognition by the Intrinsically Disordered N-WASP Domain V

Maud Chan-Yao-Chong [1,2], Dominique Durand [2] and Tâp Ha-Duong [1,*]

[1] BioCIS, University Paris-Sud, CNRS UMR 8076, University Paris-Saclay, 92290 Châtenay-Malabry, France; maud.chan-yao-chong@u-psud.fr

[2] Institute for Integrative Biology of the Cell (I2BC), CEA, CNRS, University Paris-Sud, University Paris-Saclay, 91190 Gif-sur-Yvette, France; dominique.durand@i2bc.paris-saclay.fr

* Correspondence: tap.ha-duong@u-psud.fr

Abstract: Cellular regulation or signaling processes are mediated by many proteins which often have one or several intrinsically disordered regions (IDRs). These IDRs generally serve as binders to different proteins with high specificity. In many cases, IDRs undergo a disorder-to-order transition upon binding, following a mechanism between two possible pathways, the induced fit or the conformational selection. Since these mechanisms contribute differently to the kinetics of IDR associations, it is important to investigate them in order to gain insight into the physical factors that determine the biomolecular recognition process. The verprolin homology domain (V) of the Neural Wiskott–Aldrich Syndrome Protein (N-WASP), involved in the regulation of actin polymerization, is a typical example of IDR. It is composed of two WH2 motifs, each being able to bind one actin molecule. In this study, we investigated the early steps of the recognition process of actin by the WH2 motifs of N-WASP domain V. Using docking calculations and molecular dynamics simulations, our study shows that actin is first recognized by the N-WASP domain V regions which have the highest propensity to form transient α-helices. The WH2 motif consensus sequences "LKKV" subsequently bind to actin through large conformational changes of the disordered domain V.

Keywords: intrinsically disordered protein; protein–protein interaction; molecular docking; molecular dynamics

1. Introduction

Intrinsically disordered proteins (IDPs) play important roles in the regulation of many biological processes, such as cell growth, cell signaling, and cell survival. To exert these functions, their intrinsically disordered regions (IDRs) often bind to different proteins with high specificity and low affinity [1–4]. In many cases, it is observed that IDRs adopt well structured conformations when bound to their partners [5]. Segments that undergo such a disorder-to-order transition upon binding are frequently called Molecular Recognition Features (MoRFs) in the literature [4,6–10].

A typical IDR with a MoRF is the WASP-homology 2 (WH2) motif, which is found in about 50 proteins [11]. WH2 motifs are actin-binding modules of about 30–50 residues that are key players in regulation of the cytoskeleton actin polymerization, dynamics, and organization [11–13]. Proteins of the WH2 family can contain one to four WH2 motifs, each being able to bind one G-actin monomer (Table S1). In unbound state, WH2 motifs are intrinsically disordered, and, in complex with actin, they all share a similar binding mode: their N-terminal part folds into an α-helix which interacts with the barbed face of actin, between subdomains 1 and 3, while their central consensus sequence "LKKV" has an extended conformation which lies on the actin's surface, between subdomains 1 and 2 [11,14,15] (see Figure 2B). Although these actin–WH2 motif structures were determined by X-ray diffraction, the common folding of different WH2 motifs upon binding to actin indicates, with reasonable confidence, that it is probably similar to the one adopted in solution.

It should be noted that, when a WH2 motif or a peptide construct encompassing a WH2 motif is co-crystallized with actin, only the coordinates of about 20 residues, generally from the beginning of the helical segment to the consensus sequence "LKKV", were resolved in most crystallographic complexes (Table S1). Only the crystallographic structures 2A41, 2D1K, and 5YPU contain almost all residue coordinates of the co-crystallized WH2 motifs. The absence in most crystallographic structures of atomic coordinates for regions after the consensus sequence "LKKV" indicates that they probably keep a highly flexible and disordered conformation upon binding to actin, forming so-called fuzzy complexes. Questions that could be raised here are: What is the conformational dynamics of these invisible regions? Are they interacting with actin, and, if so, with which residues?

A more general and still debated question regarding IDRs concerns the mechanism of their specific binding to their partners. The formation of IDP–protein complexes can indeed follow a pathway between two possible mechanisms [16]: the "induced fit" pathway, in which the disordered region binds to its partner and folds into an ordered structure on its surface, and the "conformational selection" mechanism, in which the folded structure preexists among the ensemble of conformations of the unbound IDP and is recognized by the protein partner. However, the observation of preexisting structured segments in IDRs does not necessarily prove that the binding proceeds by a direct conformational selection [17]. For example, an alternative mechanism could be that the protein partner first binds to any IDR region and slides to the specific binding site which has the correct complementary conformation [18]. Thus, closer investigations are required to gain insight into the early events and pathways of the IDP–protein recognition mechanism.

In this report, we address these issues in the case of the verprolin homology domain (V) of the Neural Wiskott–Aldrich Syndrome Protein (N-WASP), which has two WH2 motifs. With the Arp2/3 complex, N-WASP stimulates actin filament branching and the formation of dendritic networks of filaments that shape or deform cell membranes in several cellular processes, such as cell motility or endocytosis [19,20]. The 505-residue sequence of the human N-WASP can be decomposed into seven domains: a primary WASP homology domain WH1 (segment 1–150), a basic domain B (186–200), a GTPase-binding domain GBD (203–274), a proline-rich domain PRD (277–392), a verprolin homology domain V (405–450), a cofilin homology domain C (451–485), and an acidic domain A (486–505) [21,22]. N-WASP domain V binds and recruits G-actin monomers, while domains CA are attached to the Arp2/3 complex. These associations allow the nucleation of new branch filaments [19,23,24]. N-WASP domain V is composed of two WH2 motifs (Table S1), each being able to bind one G-actin [25–27]. Interestingly, the presence of two WH2 motifs in N-WASP domain V induces more rapid actin polymerization than the other proteins of the WASP family which have only one WH2 motif [28]. However, the structural mechanism by which a tandem of WH2 motifs binds two actin monomers and accelerates polymerization and branching is not completely elucidated.

Two crystallographic structures of the N-WASP WH2 tandem in complex with actin are available in the Protein Data Bank: a 1:1 actin–domain VC (2VCP [27]) and a 2:1 actin–WH2 tandem (3M3N [26]). Nevertheless, in both 2VCP and 3M3N structures, we emphasize again that only about 20 residues of each WH2 motif, from the helical N-terminal part to the consensus sequence "LKKV", could be resolved by X-ray experiments (Table S1). It should be noted that the actin dimer in 3M3N complex has an overall longitudinal arrangement similar to that one in actin filament [26]. This suggests that N-WASP domain V might favor the formation of actin dimers in a longitudinal filament-like conformation, which might accelerate actin polymerization. However, to confirm this scenario, a detailed description of the formation of the 1:1 and 2:1 actin–domain V complexes in solution is required.

Previously, we structurally characterized the unbound state of a construct encompassing N-WASP domain V (Figure S1) by combining various biophysical techniques [29]. Multiple molecular dynamics (MD) simulations allowed generating a conformational ensemble of this construct (which we continue to call "N-WASP domain V" for simplicity) in very good agreement with both NMR chemical shifts and SAXS intensity measurements. In this ensemble, several conformations were identified with transient α-helices in the WH2 motifs, suggesting that these secondary structures might be selected by actin during the recognition process. We query here the validity of this hypothesis and, more generally, investigate the early events of actin recognition by these α-MoRFs, using protein–protein docking calculations and multiple MD simulations. In addition, since N-WASP has a tandem of WH2 motifs, we examine the possible molecular pathways leading to the ternary complex of domain V with two actins.

2. Results

NMR experiments and MD simulations previously showed that unbound N-WASP domain V has two transient α-helical structures (one per WH2 domain) at regions 10–15 and 37–43 corresponding to residues 407–412 and 434–440 in the whole protein sequence (Figure S1) [29].

2.1. Monomeric Actin–Domain V Encounter Complexes Generated by Docking Calculations

To examine whether these two helical MoRFs are preferential recognition sites for actin, we blindly docked the 527 most populated clusters of N-WASP domain V conformational ensemble (derived from MD simulations with the A03ws force field [29]) onto the actin chain B extracted from the PDB structure 2VCP [27]. Each docking generated about 1300 different poses of domain V on actin, yielding a total number of 702,920 encounter complexes. The likeliness of these complexes was evaluated with the scoring function $2/3B^{best}$ InterEvScore [30]. We delineated the 1% of complexes (i.e., 7030 conformers) having the highest $2/3B^{best}$ score as the most probable actin–domain V structures. It could be noted that, when compared to the 527 cluster representative structures, the domain V conformations that are retrieved in the 7030 most probable complexes are sightly more compact, as indicated by the radius of gyration distributions (Figure S2), indicating that extended conformations of domain V did not particularly favor their binding to actin. At the local level, the difference in probability for residues to be in α-helix, between the two ensembles of 527 clusters and of 7030 ligands, appears quite small and may not be significant (Figure S2).

We first analyzed the residues at the protein–protein interface in the 7030 most probable complex structures. The probability of N-WASP domain V residues to be in contact with actin was computed, as plotted in Figure 1. Clearly, it can be observed that actin preferentially recognizes two regions of domain V which can be delimited by residues 8–18 and 37–50. The first binding site is shorter than the second one, which might be related to the difference in propensity of the two WH2 motifs to form α-helical structures (Figure S2). Nevertheless, when the two regions with high probability to be contacted by actin are compared, a consensus sequence can be identified as the most probable recognition site for actin: [9]KAALLDQIRE[18] and [37]RDALLDQIRQ[46] in the first and second WH2 motif, respectively. It is worth noting that both recognition segments exhibit a similar pattern in which a positively charged residue (K9 or R37) precedes two moderate probability residues (A10/A11 or D38/A39), followed by two high probability hydrophobic residues (L12/L13 or L40/L41) and again two moderate probability ones (D14/Q15 or D42/Q43), before two other high probability residues (I16/R17 or I44/R45). This pattern suggests that the domain V recognized regions are rather α-helical structures than short linear motifs (SLiMs) in coil or extended conformations. The chemical nature of the mentioned residues also indicates that the central parts of the recognized segments are amphiphilic helices with their hydrophobic faces in contact with actin.

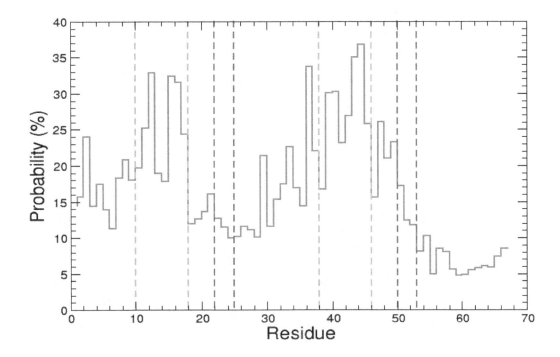

Figure 1. Probability of the N-WASP domain V residues to be distant by less than 4 Å from actin. Orange and magenta dashed lines indicate the protein regions in α-helix (as revealed by the X-ray structure 2VCP [27]) and the consensus sequences "LKKV" [14,31], respectively.

Besides, it could be noted that, among the most probable complexes, the conserved residues [22]LKKV[25] and [50]LKSV[53] have significantly lower probability to be in contact with actin than the two previous binding sites (Figure 1). This suggests that, after the recognition of regions 9–18 or 37–46 by actin, the N-WASP consensus sequences "LKKV" should move and anchor to the actin's surface in a second step. This scenario was further examined using MD simulations, as presented in the next section.

Before that, we investigated the preferential location of the two N-WASP regions 9–18 and 37–46 on actin's surface. To that end, the probability that actin residues are contacted by one of these two segments was computed over the 7030 most probable complexes predicted by docking, as plotted in Figure 2A. Among the actin residues which are frequently contacted by regions 9–18 and 37–46, we retrieved those (Y143, G146, T148, G168, Y169, L349, T351, M355, and F375) which make contacts with the N-WASP segment 37–46 in structure 2VCP [27]. However, we also observed that segments 9–18 or 37–46 can bind to other patches of the actin's surface with high probability, notably residues 171–173 and 283–290, which are not close to the cognate binding site (Figure 2). These observations could arise from various factors, including limitations of the rigid-body docking procedure and imperfections of the coarse-grained scoring function. This could be also related to the fact that, in most selected conformations of N-WASP domain V used in docking calculations, segments 9–18 and 37–46 were not fully helical, unlike in the crystallographic complex (Figure S2). This might favor the binding to pockets of the actin's surface with no particular shape, to the detriment of the groove that is expected to accommodate the WH2 motif helices. In these cases, the conformational transition of these N-WASP regions toward full α-helices might not lead to stable complexes. Besides, it could be noted that these non-specific binding sites on actin monomer also extend over the actin–actin interface in longitudinal dimers and, therefore, might be less observed in such actin assemblies.

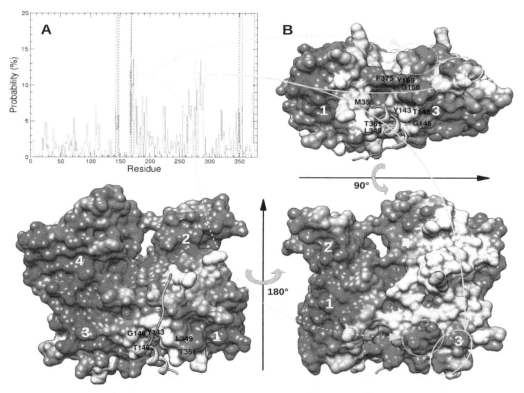

Figure 2. (**A**) Probability of actin residues to be distant by less than 4 Å from domain V regions 9–18 or 37–46. Red dashed lines indicate actin residues in contact with N-WASP helical segment in structure 2VCP [27]. (**B**) Views of actin's surface colored proportionally to previous probabilities. Blue, white, and red colors indicate actin residues with low, intermediate, and high probabilities to be contacted by domain V, respectively. As a reference, yellow and green ribbons represent the second WH2 motif helical region and conserved sequence LKSV as observed in 2VCP [27].

Overall, docking calculations of representative conformations of free domain V on actin monomer yielded many encounter complexes in which N-WASP segments 9–18 and 37–46 are preferentially bound to actin, but to both specific and non-specific sites. In these encounter complexes, consensus sequences "LKKV" have low probability to be in contact with actin, whereas they are found attached to actin in all available crystallographic complex structures. This suggests a two-step association mechanism involving large conformational rearrangements of domain V after the formation of a productive encounter complex with either segment 9–18 or 37–46 in cognate binding site of actin.

2.2. Identification and MD Simulations of Productive Actin–Domain V Encounter Complexes

The binding mechanism of N-WASP domain V to actin was further investigated using MD simulations of productive encounter complexes selected on the basis of the position and orientation of regions 9–18 or 37–46 in the cognate actin binding groove. More specifically, among the 7030 most probable complexes generated by docking, we identified those with residues 9–18 or 37–46 contacting at least six actin residues over the nine observed in contact with the N-WASP region 37–46 in the X-ray structure (Y143, G146, T148, G168, Y169, L349, T351, M355, and F375). We found a total of 194 complexes which have one of the two recognized segments in contact with at least six of the nine actin hot-spot residues. However, in a large number of these complexes, the segment 9–18 or 37–46 is oriented in the opposite direction of the crystallographic helix, so that the consensus sequence "LKKV" would not be able to reach its cognate binding site. Thus, we further filtered the 194 complexes based on the angle between the principal axis of segment 9–18 or 37–46 and that one of the helical region 37–46 in crystal. We obtained 16 and 18 complexes in which this angle is lower than 30° for N-WASP regions 9–18 and 37–46, respectively (Tables S2 and S3).

In these 34 productive actin–domain V encounter complexes, the recognized regions 9–18 and 37–46 are surprisingly not completely folded in α-helix, but can have various local conformations with 0–6 over 10 residues in helical structures. Nevertheless, it should be noted that the lack of helical residues is often balanced by several residues with a turn motif. This is notably the case for four over the five complexes which have region 9–18 or 37–46 RMSD lower than 5 Å relative to the crystallographic structure (Tables S2 and S3). In the 34 actin–domain V complexes, the consensus segments "LKKV" are variously far off from their cognate binding site on actin, as indicated by their RMSD values ranging from 8.7 to 37.7 Å. To study the complete association process of N-WASP WH2 motifs, we performed MD simulations of actin–bound domain V conformational changes starting from the two structures which have region 9–18 or 37–46 with the lowest RMSD relative the structure 2VCP (Figure 3). These selected productive encounter complexes are hereafter denoted CplxA and CplxB.

CplxA   CplxB

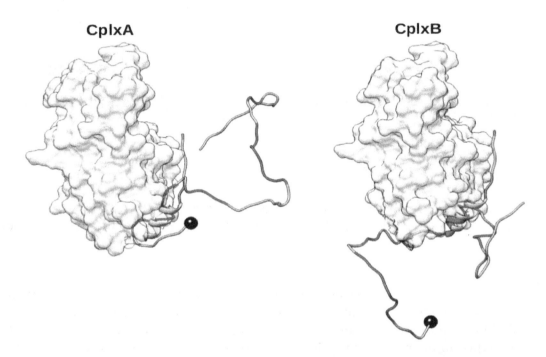

Figure 3. Side view of the two best 1:1 actin–domain V encounter complexes with N-WASP segment 9–18 (**left**) or 37–46 (**right**) located and oriented as in structure 2VCP. Black balls are N-terminal Cα-atoms of domain V. Red and magenta ribbons represent its regions 9–18 or 37–46 and consensus sequences "LKKV", respectively. As a reference, yellow and green ribbons indicate the helical and [50]LKSV[53] regions of domain VC in 2VCP.

For each selected encounter complex, two MD simulations of about 350 ns were performed from the same coordinates but with different initial velocities. These four simulations will be referred to as CplxA_MD1, CplxA_MD2, CplxB_MD1, and CplxB_MD2. In all complex trajectories, the actin tertiary structure remains stable, with RSMD relative to structure 2VCP fluctuating below 5.2 Å (Figure 4). Regarding the N-WASP regions 9–18 and 37–46 (which are bound to actin in CplxA and CplxB, respectively), their position and orientation are maintained in the actin binding site in three over four simulations (CplxA_MD1, CplxA_MD2, and CplxB_MD1), as indicated by their average RMSD values relative to the complex 2VCP (4.4, 4.4, and 2.7 Å, respectively). A visual inspection of the CplxB_MD2 trajectory showed that segment 37–46 slid toward the bottom of actin, explaining its higher RMSD (8.2 Å on average). For the three other simulations, the N-WASP regions 9–18 and 37–46 remain attached to their binding site after the formation of productive encounter complexes.

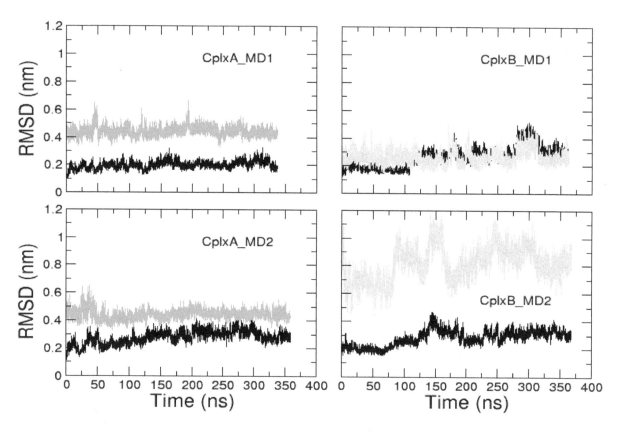

Figure 4. Time evolutions of RMSD relative to structure 2VCP, after fitting MD trajectories on crystallographic actin, for actin (black) and segments 9–18 (orange) and 37–46 (cyan) of N-WASP domain V.

Next, we monitored the dynamics of residues [22]LKKV[25] and [50]LKSV[53] relative to their cognate binding site on actin. As shown in Figure 5, segments [22]LKKV[25] and [50]LKSV[53] had large amplitude motions in all four simulations, without reaching stable bound positions on actin. Strikingly, the minimal distance to actin of these residues and their RMSD relative to structure 2VCP seem to be highly correlated, which can be explained as follows: Once N-WASP domain V helical region 9–18 or 37–46 is correctly positioned and oriented in its cognate binding site, if segment [22]LKKV[25] or [50]LKSV[53] is detached from actin's surface, it is largely free to move in solvent, accounting for large RMSD values. However, when it is bound to actin, its accessible space is narrowed down to a region close to the cognate site on actin, decreasing the RMSD relative to X-ray structure. However, in none of simulations, these segments were observed to persistently bind to their cognate binding site: In simulations CplxB_MD1 and CplxB_MD2, RMSD of residues [50]LKSV[53] relative to the crystallographic structure never decreased below 13.8 Å. The observed large RMSD values are mainly due to the fact that segment [50]LKSV[53] is, most of the time, detached from actin's surface in simulations of CplxB. In simulations of CplxA, segment [22]LKKV[25] was able to reach its cognate site, with minimal RMSD of 2.4 and 4.3 Å in CplxA_MD1 and CplxA_MD2, respectively, but these associations were only transient (Figure 5). Overall, in three over four simulations, residues [22]LKKV[25] or [50]LKSV[53] were observed to bind the actin's surface during quite long periods, but not necessarily at their cognate locations, confirming that these N-WASP segments are not primary recognition sites for actin. Finally, we should point out that the auto-correlation functions of minimal distances to actin of residues [22]LKKV[25] or [50]LKSV[53] are characterized by relaxation times of 102, 126, 164, and 133 ns for simulations CplxA_MD1, CplxA_MD2, CplxB_MD1, and CplxB_MD2, respectively. This notably indicates that the two short simulations of CplxA still provide reliable information about the dynamics of segment [22]LKKV[25].

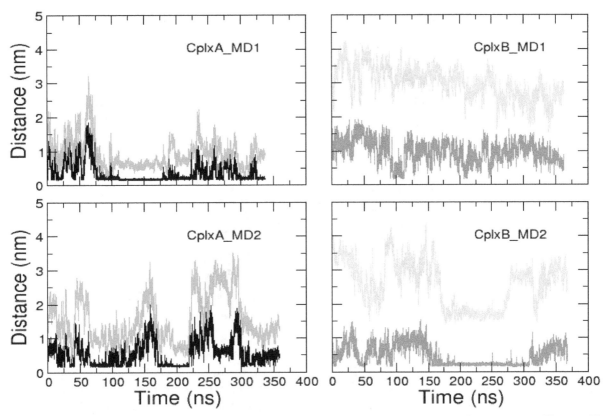

Figure 5. Time evolutions of minimal distance to actin of segments ^{22}LKKV^{25} (black) and ^{50}LKSV^{53} (brown) of N-WASP domain V. RMSD relative to structure 2VCP, after fitting trajectories on actin, are also displayed as a function of time for segments ^{22}LKKV^{25} (orange) and ^{50}LKSV^{53} (cyan).

The actin residues that have high probabilities to be contacted by these segments are shown in Figure 6. In both simulations of CplxA, segment ^{22}LKKV^{25} was found in contact with several actin residues close to the cognate binding site. In contrast, due to the sliding of region 37–46 toward the bottom of actin in simulation CplxB_MD2, the segment ^{50}LKSV^{53} is too far to reach and bind its cognate site on actin. All together, despite their limited number and length, our simulations suggest that CplxA (which has the N-WASP helical region 9–18 recognized by actin) is likely a productive encounter complex that can lead to a subsequent binding of segment ^{22}LKKV^{25} to its specific site on actin. In contrast, simulations of CplxB suggest that the complete binding of N-WASP second WH2 motif is less favorable than for the first WH2 motif. Beyond the limited statistics, this could result from the fact that segment ^{50}LKSV^{53} is less positively charged than ^{22}LKKV^{25}, whereas their cognate binding site on actin has two negatively charged residues (D24 and D25). Another possible explanation is that N-WASP region 37–46 has a higher propensity to form α-helices than segment 9–18. This would increase the stiffness of the second WH2 motif that might restrict the motion of residues ^{50}LKSV^{53} and their ability to reach their cognate binding site on actin.

Finally, we studied the dynamics of domain V regions ^{28}NSRPVS^{33} and ^{56}GQESTP^{61} following the conserved sequences ^{22}LKKV^{25} and ^{50}LKSV^{53}, respectively. Indeed, as mentioned in the introduction, most crystallographic structures of actin–WH2 motif lack atomic coordinates for regions after the consensus sequence "LKKV", indicating that they are highly flexible in their bound state. We thus characterized the preferential location of these two regions on actin's surface in our MD simulations. Figure 7 plots the minimal distance of regions ^{28}NSRPVS^{33} and ^{56}GQESTP^{61} to actin as a function of time in CplxA and CplxB simulations, respectively. It can be observed that these two regions mostly contact the actin's surface when the preceding conserved sequences ^{22}LKKV^{25} or ^{50}LKSV^{53} are already attached to actin, except in CplxB_MD1. In the latter, residues ^{56}GQESTP^{61} make frequent contacts with actin when segment ^{50}LKSV^{53} is not bound to actin.

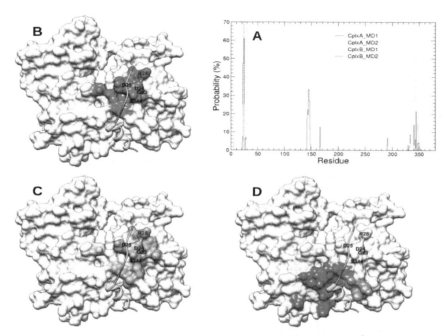

Figure 6. (**A**) Probability of actin residues to be distant by less than 4 Å from N-WASP segments [22]LKKV[25] or [50]LKSV[53] in CplxA_MD1 (red), CplxA_MD2 (orange), CplxB_MD1 (cyan), and CplxB_MD2 (blue). Brown dashed lines indicate the actin residues (G23, D24, D25, R28, and S344) in contact with N-WASP segment [50]LKSV[53] in structure 2VCP [27]). (**B–D**) Front views of the actin's surface colored proportionally to the previous probabilities. Red, orange, and blue colors indicate actin residues with high probabilities to be contacted by N-WASP segments [22]LKKV[25] or [50]LKSV[53] in simulations CplxA_MD1 (**B**), CplxA_MD2 (**C**), and CplxB_MD2 (**D**), respectively. As a reference, yellow and green ribbons represent the helical region and the conserved sequence LKSV of the second WH2 motif observed in structure 2VCP [27].

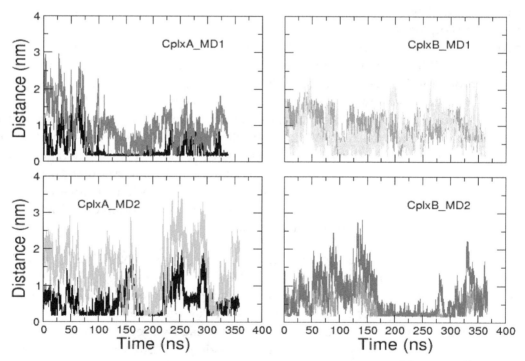

Figure 7. Time evolutions of minimal distances between actin and segment [28]NSRPVS[33] in simulations of CplxA (red and orange lines) and segment [56]GQESTP[61] in simulations of CplxB (cyan and blue lines). For comparison, time evolutions of minimal distances between actin and segments [22]LKKV[25] and [50]LKSV[53] are displayed with black and brown lines, respectively.

The actin residues that have high probabilities to be contacted by regions [28]NSRPVS[33] and [56]GQESTP[61] are displayed in Figure 8. In both simulations of CplxB, segment [56]GQESTP[61] was mostly found in contact with residues of the actin subdomain 3. In CplxB_MD1, this might be the reason the conserved segment [50]LKSV[53] cannot reach its cognate binding site on actin. In CplxB_MD2, this is probably because the helix 37–46 slid toward the bottom of actin and that segment [50]LKSV[53] is improperly located between actin subdomains 1 and 3 (Figure 6). Strikingly, in simulations of CplxA in which the helical segment 9–18 and conserved sequence [22]LKKV[25] are both satisfactorily positioned on actin's surface, the region [28]NSRPVS[33] is observed to contact several separated patches on actin's surface, mainly located on subdomains 2 and 4. This might explain why these disordered regions cannot crystallize in one homogeneous conformation and, therefore, are not visible in most crystallographic actin–WH2 complexes.

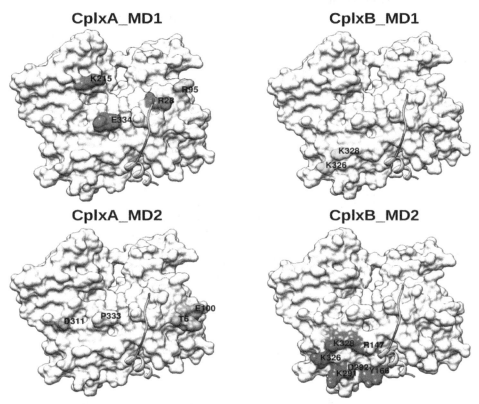

Figure 8. Actin residues distant by less than 4 Å from N-WASP segments [28]NSRPVS[33] or [56]GQESTP[61] in CplxA_MD1 (red), CplxA_MD2 (orange), CplxB_MD1 (cyan), and CplxB_MD2 (blue). As a reference, yellow and green ribbons represent the helical region and the conserved sequence LKSV of the second WH2 motif observed in structure 2VCP [27].

2.3. Dimeric Actin–Domain V Encounter Complexes Generated by Docking Calculations

As reported in the literature, a tandem of WH2 motifs, such as N-WASP domain V, can form a ternary complex with two actin molecules [26,32]. Rebowski et al. notably reported a 2:1 actin–domain V complex, in which two actins are assembled into a longitudinal filament-like dimer (PDB structure 3M3N) [26]. In this section, we investigate the early steps of formation of these ternary encounter complexes. As for actin monomer, we blindly docked the 527 most populated clusters of the MD-derived N-WASP domain V conformational ensemble [29], but here, onto the longitudinal actin dimer structure extracted from the PDB file 3M3N [26]. It should be noted that each chain of the 3M3N dimer is structurally very similar to actin in 2VCP (RMSD over Cα atoms being equal to 0.99 and 0.66 Å for chain A and B, respectively). Moreover, unlike in 2VCP structure, both chains of actin dimer 3M3N lack the coordinates of their last residue F375. A total number of 754,118 complex structures were generated. The likeliness of these complexes was evaluated with the scoring function 2/3B^{best}

InterEvScore [30]. We delineated the 1% complexes (that is 7540 conformers) having the highest $2/3B^{best}$ score as the most probable actin dimer-domain V structures. As for actin monomer, when compared to the 527 cluster representative structures, the domain V conformations that are retrieved in the most probable complexes with actin dimer are in average more compact as indicated by the radius of gyration distributions (Figure S3). The dimeric state of actin did not favor the binding of extended conformations of domain V.

We then analyzed the probability of domain V residues to be in contact with each chain of actin dimer. We observed again that actin preferentially recognizes the domain V regions [9]KAALLDQIRE[18] and [37]RDALLDQIRQ[46], with a similar pattern as for actin monomer (compare Figure 9 with Figure 1), indicating that the N-WASP recognized regions are rather in (partial) α-helical structures. It is also confirmed that the conserved sequences [22]LKKV[25] and [50]LKSV[53] have low probability to be contacted by actin dimer in the encounter complexes, suggesting again that they should move and anchor to the actin's surface after the recognition of the previously mentioned regions 9–18 and 37–46.

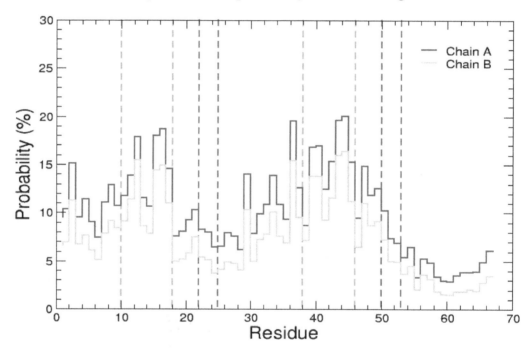

Figure 9. Probability of the N-WASP domain V residues to be distant by less than 4 Å from actin dimer. Orange and magenta dashed lines indicate the N-WASP regions in α-helix (as revealed by the X-ray structure 2VCP [27]) and the consensus sequences "LKKV" [14,31], respectively.

Finally, we determined the preferential location of the domain V regions 9–18 and 37–46 on actin dimer surface by computing over the 7540 most probable complexes the probability that actin residues are contacted by one of these segments (Figure 10). The N-WASP regions 9–18 and 37–46 can be retrieved in the cognate binding site of actin chain A but not of chain B. The presence of chain A at the bottom of chain B probably hinders the approach and accommodation of domain V in the binding site of chain B. As for actin monomer, we also observed that N-WASP segments 9–18 and 37–46 can bind to other patches of the actin's surface with high probability, notably at residues K191, E195, R256 and F266 which are located at the top of the back of actin dimer (Figure 10). It is not clear for us if these non-productive associations are artifacts or not. Nevertheless, since the consensus sequences "LKKV" have low probabilities to contact actin, large conformational changes of domain V are likely to occur after the formation of the encounter complexes. Only a productive encounter complex in which the cognate binding site of actin accommodates N-WASP segment 9–18 or 37–46 will be able to form the correct quaternary structure.

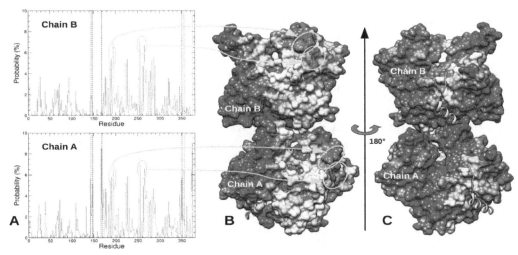

Figure 10. (**A**) Residue-specific probability of actin dimer chain A (bottom) and chain B (top) to be distant by less than 4 Å from N-WASP domain V regions 9–18 or 37–46 in the ensemble of 7540 ternary complexes generated by docking. Red dashed lines indicate the actin residues in contact with the N-WASP helical segment in the X-ray structure 2VCP [27]. (**B,C**) Back and front views of the actin dimer surface colored proportionally to the previous probabilities. Blue, white, and red colors indicate actin residues with low, intermediate, and high probabilities to be contacted by N-WASP domain V regions 9–18 or 37–46, respectively. As a reference, yellow and green ribbons represent helical regions and consensus sequences "LKKV" of the two WH2 motifs observed in the X-ray structure 3M3N [26].

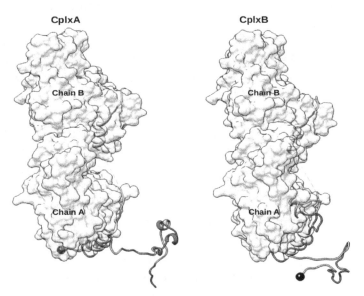

Figure 11. Side view of the two best 2:1 actin–domain V encounter complexes with N-WASP segment 9–18 (**left**) or 37–46 (**right**) located and oriented as in structure 3M3N. Black balls are N-terminal Cα-atoms of domain V. Red and magenta ribbons represent its regions 9–18 or 37–46 and consensus sequences "LKKV", respectively. As a reference, yellow and green ribbons indicate the helical and LKKV regions of domain V in 3M3N.

These productive actin–domain V encounter complexes were identified among the 7540 most probable complexes as those with segment 9–18 or 37–46 making contacts to at least 6 over the 8 hot-spot residues of 3M3N actin chain A (Y143, G146, T148, G168, Y169, L349, T351, and M355), and correctly oriented so that the conserved sequence ^{22}LKKV^{25} or ^{50}LKSV^{53} can reach their cognate binding site. We found 10 and 13 productive encounter complexes in which N-WASP segments 9–18 and 37–46 are bound to actin chain A, respectively (Tables S4 and S5). The two complexes for which the regions 9–18 or 37–46 have the lowest RMSD relative to structure 3M3N are displayed in Figure 11.

In all found productive encounter complexes, regions ^{22}LKKV^{25} or ^{50}LKSV^{53} are detached from actin, and actin chain B is not contacted by other parts of N-WASP domain V. The presence of chain B in the actin dimer does not seem to influence the recognition of N-WASP segments 9–18 or 37–46 by actin chain A. Besides, several representative structures of domain V conformational ensemble (clusters 105, 145, 230, 333, 407, and 411) were retrieved in the most probable encounter complexes on both the monomeric (2VCP) and dimeric (3M3N) states of actin. Nevertheless, as previously seen, the subsequent binding of residues ^{22}LKKV^{25} or ^{50}LKSV^{53} to actin was not persistent in our MD simulation of complexes with actin monomer, but this association might be stabilized by the presence of a second chain in complexes with actin dimer. This hypothesis can be assessed using extensive MD simulations. Unfortunately, our limited computational resources for this project did not allow us to perform these calculations.

3. Discussion

The characterization of the early events of protein–protein recognitions involving intrinsically disordered proteins is important for better understanding the molecular bases of regulation and signaling processes occurring in cells. This task is very challenging using current experimental techniques and can be fruitfully complemented by molecular modeling. However, MD simulations of encounter complexes starting from separated proteins are computationally very demanding and require extremely long trajectories in cases of IDPs. In this study, we propose a less expensive approach consisting, first, in discretizing the IDP large conformational ensemble into representative structures of the most populated clusters; secondly, in generating the protein–protein encounter complexes by rigid coarse-grained protein–protein docking; and, finally, in performing MD calculations of few selected productive complex conformations.

This approach was used to study the recognition by actin of the two WH2 motifs of N-WASP domain V, which is largely disordered in free state. Several crystallographic structures of actin–WH2 motif complexes show that the WH2 motif N-terminal part is folded into an amphiphilic α-helix located in a cleft at the bottom of actin, and that its consensus sequence "LKKV" has a rather extended conformation lying on the actin front surface (Figures 2 and 6). The pathway leading to these bound states remains largely unknown, especially in the case of tandems of WH2 motifs which bind two actins.

Previously, we identified several structures with transient α-helices at regions 9–18 and 37–46 in the unbound domain V conformational ensemble [29]. Our present docking calculations showed that these two regions are effectively preferential binding sites for actin (Figure 1). Our results also suggest that conformations with regions 9–18 or 37–46 completely structured in α-helix are not preferably recognized, but less folded conformations can be equally accommodated in the cognate binding site on actin (Tables S2–S5). Knowing the binding location on actin's surface of the conserved segments ^{22}LKKV^{25} or ^{50}LKSV^{53}, it is apparent that non-specific association and orientation of regions 9–18 and 37–46 on actin's surface cannot produce the observed quaternary structure of actin–WH2 motif complexes. Our MD simulations of a productive encounter complex even showed that, when the recognized helical region 37–46 of N-WASP is initially correctly located and oriented in the actin cognate binding site, a slight displacement of this region toward the bottom of actin prevents the segment ^{50}LKSV^{53} to reach and bind its specific site on actin (simulation CplxB_MD2).

In our modeling procedure, it could be noted that only the 7030 encounter complexes with the highest 2/3B^{best} score among the 702,920 generated by docking were deemed as probable and subsequently analyzed. Although this limited number could lead to possible missed relevant structures, it is much larger than the number of docking solutions that are usually analyzed to find near-native protein–protein interfaces (up to 1000) [30]. This provides reasonable confidence that our modeling generated relevant quaternary structures. Besides, the 7030 analyzed structures can be considered as representative of both the productive and non-productive encounter complexes (Figure 2), as they probably appear in vitro or in vivo. Strikingly, in all productive encounter complexes, the consensus sequence "LKKV" of WH2 motifs is found distant from actin's surface (Figure 3). This indicates that large amplitude motions of these segments are likely to occur in a second step to enable the formation of the final quaternary structure, as illustrated in our MD simulations of CplxA (Figures 5 and 7). Thus, we think that our modeling study has allowed going beyond the prediction of the actin–N-WASP complex quaternary structure and has also gained insight into its mechanism of formation. To sum up, our study of actin monomer recognition by N-WASP domain V indicates that actin first binds domain V regions 9–18 or 37–46 which are partially folded into amphiphilic helical structures, mainly through hydrophobic interactions. Then, the charged segments ^{22}LKKV^{25} or ^{50}LKSV^{53}, driven by electrostatic forces, move and attach to their cognate site on actin's surface.

When the binding of domain V to a longitudinal actin dimer was considered, our docking calculations showed that N-WASP helical regions 9–18 and 37–46 can bind their cognate binding sites, but preferentially on actin chain A, the access of the specific binding site on chain B being more restricted (Figure 10). Nevertheless, this result might depend on the quaternary structure of the actin dimer, particularly on the actin–actin interface, which can significantly vary, as observed in various crystallographic structures of actin oligomers (3M3N [26], 4JHD [32], and 6FHL [33]). All together, our results allow us to propose the following model for the early events of association of N-WASP domain V to two actins and the formation of a ternary complex with a longitudinal filament-like actin dimer, as observed in structure 3M3N (Figure 12): From isolated actin chains and N-WASP domain V, three possible binary complexes can be formed (States II-a, II-b, and II-c). In State II-a, the second WH2 motif attached to actin chain B prevents the approach and binding of chain A [11,15,34] and thus disfavors the formation of intermediate State III-a. When the actin dimer is already formed, our docking calculations indicate that the binding of N-WASP second WH2 motif to actin chain B is not favorable. Thus, the direct formation of the ternary State III-a from a preformed actin dimer or the evolution of intermediate State III-b toward the final complex are very unlikely. These considerations imply that the final state is likely formed through an intermediate ternary complex in which the two WH2 motifs are bound to two loosely interacting actin chains (State III-c). Then, this highly flexible assembly evolves toward the final state through the association of the two actin chains into a longitudinal dimer. This model suggests that the binding of N-WASP domain V to an actin dimer would not be a cooperative process, in line with fluorescence titration experiments reported by Gaucher et al. [27].

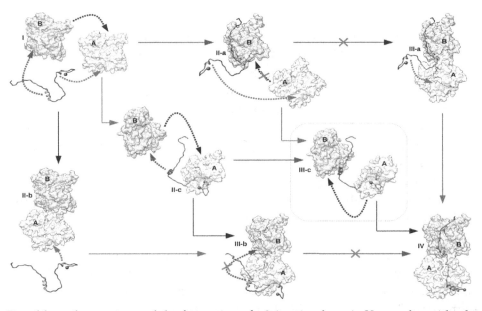

Figure 12. Possible pathways toward the formation of a 2:1 actin–domain V complex with a longitudinal actin dimer as observed in 3M3N. Starting from two actin chains and one N-WASP domain V (State I), three possible binary encounter complexes can be formed (States II-a, II-b, and II-c), leading to three possible intermediate ternary complexes (States III-a, III-b, and III-c) just before the final structure (State IV). Cyan and red arrows indicate the binding of the N-WASP first and second WH2 motif to actin chain A and B, respectively. Dark grey arrows represent the binding of two actins into a longitudinal dimer.

During this process, it is not clear whether the binding of the conserved sequences [22]LKKV[25] and [50]LKSV[53] to their cognate sites occurs before the formation of the longitudinal dimer. In crystallographic structure 2VCP, the four residues [50]LKSV[53] are found attached to the actin's surface, but our MD simulations in explicit water indicate that this binding is rather transient in 1:1 actin–domain V complexes. We speculate that the interactions between the consensus sequences and actin might guide the dynamics of dimerization into longitudinal assemblies. All together, our model for the early events of domain V association to two actins might explain how the two WH2 motifs of N-WASP favor the formation of longitudinal filament-like conformation of actin dimer and why they induce more rapid actin polymerization than proteins of the WASP family with only one WH2 motif [28].

4. Methods

4.1. Conformational Clustering

The conformational ensemble of the studied construct encompassing N-WASP domain V and previously generated by MD simulations with the Amber-03ws force field [29] was clustered with the GROMACS tool *gmx cluster* using the *gromos* method [35] and a RMSD cutoff of 0.5 nm (computed over the mainchain atoms). We obtained 2467 clusters and decided, for subsequent protein–protein docking calculations, to keep only the 527 most populated ones, which represent 50% of the 40,000 conformations sampled by MD simulations. To verify that the 527 clusters are representative of the overall conformational ensemble, we compared the residue probabilities to be in α-helix and the distributions of gyration averaged over the 40,000 conformations or the 527 cluster structures. As shown in Figure S2, the probabilities to form α-helices of the 527 clusters and of the whole conformational ensemble are almost identical, and the protein radius of gyration has similar distributions when computed over the sub-ensemble of representative structures or over the 40,000 conformations. This indicates that the selected 527 conformers are locally and globally representative of the whole conformational ensemble of N-WASP domain V.

4.2. Protein–Protein Docking

The 527 representative conformations of N-WASP domain V were docked into two crystallographic structures of actin (PDB ID: 2VCP [27] and 3M3N [26]), using the molecular modeling library PTools [36]. This toolbox performs rigid-body docking of coarse-grained proteins by multiple energy minimizations, starting from regularly distributed initial positions and orientations of the ligand around the receptor surface. It should be emphasized that no conformational change was allowed during these docking calculations for both protein partners, notably the intrinsically disordered domain V. The energy function minimized here is the physics-based pairwise protein–protein interaction energy SCORPION [37,38]. Then, to better discriminate the near-native interface between actin and domain V, the complexes previously generated with PTools were rescored using a knowledge-based scoring function which additionally takes into account three-body interactions. We used in this study the 2/3B^{best} InterEvScore, without any evolutionary information from the actin or N-WASP domain V sequences [30].

The performance of 2/3B^{best} InterEvScore was positively evaluated on an ensemble of 131 protein–protein complexes which, as far as we know, did not include IDP case [30]. Thus, to assess the validity of our approach to study the actin–domain V recognition, we performed the redocking of the folded segment 433–451 of N-WASP domain V into actin structure 2VCP [27] and checked if the X-ray structure of the complex can be retrieved. The results of this test are reported in Figure S4, which displays the actin–ligand interaction 2/3B^{best} score as a function of the RMSD relative to the peptide conformation in the crystallographic structure. It can be seen that the coarse-grained protein–protein redocking is able to retrieve the experimental structure with a RMSD calculated over the Cα atoms of only 0.5 Å. In this particular case, the modeled complex structure, which is the closest to the experimental one is ranked first (the higher is the score, the more native-like is the interface). This benchmark led us to adopt this two-step approach consisting in generating complex structures with PTools and rescoring them with InterEvScore.

4.3. MD Simulations

From the docking results, several probable structures of the actin–domain V complex were selected and submitted to extensive MD simulations performed with the GROMACS software (versions 5.0.2 and 2016.1) [39]. Each selected complex initial conformation was put and solvated in a dodecahedral rhombic box of 14.0 nm edge, then neutralized by adding 175 sodium and 176 chloride ions to reach the salt concentration of 150 mM. The non-bonded interactions were treated using the smooth PME method [40] for the electrostatic terms and a cutoff distance of 1.2 nm for the van der Waals potentials. The solute and water covalent bond lengths were kept constant using the LINCS [41] and SETTLE [42] algorithms, respectively, allowing to integrate the equations of motion with a 2 fs time step. All simulations were run in the NPT ensemble, at T = 310 K and P = 1 bar, using the Nose–Hoover and Parrinello–Rahman algorithms [43–45] with the time coupling constants $\tau_T = 0.5$ ps and $\tau_P = 2.5$ ps.

In our previous study of the free state N-WASP domain V, short preliminary MD simulations indicated that the force field AMBER-03w [46] combined with the modified water model TIP4P/2005s [47] (a combination referred to as A03ws) allowed correctly exploring the protein conformational space. For consistency, we kept this force field for the study of its complex with actin. Each selected complex was submitted to about 350 ns MD simulations within the general conditions previously described. Data collected every 20 ps were kept for subsequent analyses. The latter were made using mostly the GROMACS tools, such as *gmx mindist* or *gmx cluster* for computing specific distances or structural clusters, respectively. The program STRIDE [48] was used to assign secondary structures to the protein residues.

Supplementary Materials:
Table S1: List of proteins with WH2 motifs which were co-crystallized with actin; Figure S1: Alignment of the studied construct sequence with those of N-WASP in 2VCP and 3M3N structures; Figure S2: N-WASP domain V residue probabilities to be in α-helix and distributions of gyration of the 7030 conformations in the most probable 1:1 actin–domain V complexes; Table S2: Most probable 1:1 actin–domain V encounter complexes in which domain V segment 9–18 is in contact with at least six over nine actin hot-spot residues; Table S3: Most probable 1:1 actin–domain V encounter complexes in which domain V segment 37–46 is in contact with at least six over nine actin hot-spot residues; Figure S3: N-WASP domain V residue probabilities to be in α-helix and distributions of gyration of the 7540 conformations in the most probable 2:1 actin–domain V complexes; Table S4: Most probable 2:1 actin–domain V encounter complexes in which segment 9–18 is in contact with at least six over eight actin hot-spot residues; Table S5: Most probable 2:1 actin–domain V encounter complexes in which segment 37–46 is in contact with at least six over eight actin hot-spot residues; Figure S4: $2/3B^{best}$ score of N-WASP segment 433–451 redocked into actin as a function of the ligand RMSD relative to the conformation found in structure 2VCP.

Author Contributions: Conceptualization, M.C.-Y.-C., D.D. and T.H.-D.; Formal analysis, M.C.-Y.-C. and T.H.-D.; Funding acquisition, D.D. and T.H.-D.; Investigation, M.C.-Y.-C. and T.H.-D.; Methodology, M.C.-Y.-C., D.D. and T.H.-D.; Project administration, D.D. and T.H.-D.; Resources, D.D. and T.H.-D.; Supervision, D.D. and T.H.-D.; Validation, M.C.-Y.-C. and T.H.-D.; Visualization, M.C.-Y.-C., D.D. and T.H.-D.; Writing—original draft, M.C.-Y.-C., D.D. and T.H.-D.; and Writing—review and editing, M.C.-Y.-C., D.D. and T.H.-D.

Acknowledgments: We are grateful to L. Renault for fruitful discussions about actin and WH2 motifs.

Abbreviations

The following abbreviations are used in this manuscript:

IDP   Intrinsically Disordered Protein
IDR   Intrinsically Disordered Region
PDB   Protein Data Bank
N-WASP   Neural Wiskott–Aldrich Syndrome Protein
MoRF   Molecular Recognition Feature
NMR   Nuclear Magnetic Resonance
SAXS   Small-Angle X-ray Scattering
MD   Molecular Dynamics
RMSD   Root Mean Square Deviation

References

1.   Wright, P.E.; Dyson, H.J. Intrinsically unstructured proteins: Re-assessing the protein structure-function paradigm. *J. Mol. Biol.* **1999**, *293*, 321–331. [CrossRef]
2.   Dunker, A.K.; Lawson, J.D.; Brown, C.J.; Williams, R.M.; Romero, P.; Oh, J.S.; Oldfield, C.J.; Campen, A.M.; Ratliff, C.M.; Hipps, K.W.; et al. Intrinsically disordered protein. *J. Mol. Graph. Model.* **2001**, *19*, 26–59. [CrossRef]
3.   Dyson, H.J.; Wright, P.E. Intrinsically unstructured proteins and their functions. *Nat. Rev. Mol. Cell Biol.* **2005**, *6*, 197–208. [CrossRef] [PubMed]
4.   Dunker, A.K.; Silman, I.; Uversky, V.N.; Sussman, J.L. Function and structure of inherently disordered proteins. *Curr. Opin. Struct. Biol.* **2008**, *18*, 756–764. [CrossRef]
5.   Zea, D.J.; Monzon, A.M.; Gonzalez, C.; Fornasari, M.S.; Tosatto, S.C.E.; Parisi, G. Disorder transitions and conformational diversity cooperatively modulate biological function in proteins. *Protein Sci.* **2016**, *25*, 1138–1146. [CrossRef]
6.   Oldfield, C.J.; Cheng, Y.; Cortese, M.S.; Romero, P.; Uversky, V.N.; Dunker, A.K. Coupled Folding and Binding with α-Helix-Forming Molecular Recognition Elements. *Biochemistry* **2005**, *44*, 12454–12470. [CrossRef] [PubMed]
7.   Mohan, A.; Oldfield, C.J.; Radivojac, P.; Vacic, V.; Cortese, M.S.; Dunker, A.K.; Uversky, V.N. Analysis of Molecular Recognition Features (MoRFs). *J. Mol. Biol.* **2006**, *362*, 1043–1059. [CrossRef]
8.   Vacic, V.; Oldfield, C.J.; Mohan, A.; Radivojac, P.; Cortese, M.S.; Uversky, V.N.; Dunker, A.K. Characterization of Molecular Recognition Features, MoRFs, and Their Binding Partners. *J. Proteome Res.* **2007**, *6*, 2351–2366. [CrossRef] [PubMed]
9.   Cheng, Y.; Oldfield, C.J.; Meng, J.; Romero, P.; Uversky, V.N.; Dunker, A.K. Mining α-helix-forming molecular recognition features with cross species sequence alignments. *Biochemistry* **2007**, *46*, 13468–13477. [CrossRef]
10. Lee, C.; Kalmar, L.; Xue, B.; Tompa, P.; Daughdrill, G.W.; Uversky, V.N.; Han, K.H. Contribution of proline to the pre-structuring tendency of transient helical secondary structure elements in intrinsically disordered proteins. *Biochim. Biophys. Acta Gen. Subj.* **2014**, *1840*, 993–1003. [CrossRef] [PubMed]
11. Carlier, M.F.; Husson, C.; Renault, L.; Didry, D. Chapter Two–Control of Actin Assembly by the WH2 Domains and Their Multifunctional Tandem Repeats in Spire and Cordon-Bleu. In *International Review of Cell and Molecular Biology*; Jeon, K.W., Ed.; Academic Press: Cambridge, MA, USA, 2011; Volume 290, pp. 55–85.
12. Derry, J.M.J.; Ochs, H.D.; Francke, U. Isolation of a novel gene mutated in Wiskott-Aldrich syndrome. *Cell* **1994**, *78*, 635–644. [CrossRef]
13. Palma, A.; Ortega, C.; Romero, P.; Garcia-V, A.; Roman, C.; Molina, I.; Santamaria, M. Wiskott-Aldrich syndrome protein (WASp) and relatives: A many-sided family. *Immunologia* **2004**, *23*, 217–230.
14. Chereau, D.; Kerff, F.; Graceffa, P.; Grabarek, Z.; Langsetmo, K.; Dominguez, R. Actin-bound structures of Wiskott-Aldrich syndrome protein (WASP)-homology domain 2 and the implications for filament assembly. *Proc. Natl. Acad. Sci. USA* **2005**, *102*, 16644–16649. [CrossRef] [PubMed]
15. Renault, L.; Deville, C.; van Heijenoort, C. Structural features and interfacial properties of WH2, β-thymosin domains and other intrinsically disordered domains in the regulation of actin cytoskeleton dynamics. *Cytoskeleton* **2013**, *70*, 686–705. [CrossRef]

16. Kiefhaber, T.; Bachmann, A.; Jensen, K.S. Dynamics and mechanisms of coupled protein folding and binding reactions. *Curr. Opin. Struct. Biol.* **2012**, *22*, 21–29. [CrossRef] [PubMed]

17. Liu, X.; Chen, J.; Chen, J. Residual Structure Accelerates Binding of Intrinsically Disordered ACTR by Promoting Efficient Folding upon Encounter. *J. Mol. Biol.* **2019**, *431*, 422–432. [CrossRef] [PubMed]

18. Kozakov, D.; Li, K.; Hall, D.R.; Beglov, D.; Zheng, J.; Vakili, P.; Schueler-Furman, O.; Paschalidis, I.C.; Clore, G.M.; Vajda, S. Encounter complexes and dimensionality reduction in protein–protein association. *eLife* **2014**, *3*, e01370. [CrossRef] [PubMed]

19. Pollard, T.D.; Borisy, G.G. Cellular Motility Driven by Assembly and Disassembly of Actin Filaments. *Cell* **2003**, *112*, 453–465. [CrossRef]

20. Takenawa, T.; Suetsugu, S. The WASP–WAVE protein network: connecting the membrane to the cytoskeleton. *Nat. Rev. Mol. Cell Biol.* **2007**, *8*, 37–48. [CrossRef]

21. Miki, H.; Miura, K.; Takenawa, T. N-WASP, a novel actin-depolymerizing protein, regulates the cortical cytoskeletal rearrangement in a PIP2-dependent manner downstream of tyrosine kinases. *Embo J.* **1996**, *15*, 5326–5335. [CrossRef]

22. Prehoda, K.E.; Scott, J.A.; Mullins, R.D.; Lim, W.A. Integration of Multiple Signals Through Cooperative Regulation of the N-WASP-Arp2/3 Complex. *Science* **2000**, *290*, 801–806. [CrossRef] [PubMed]

23. Fawcett, J.; Pawson, T. N-WASP Regulation—The Sting in the Tail. *Science* **2000**, *290*, 725–726. [CrossRef] [PubMed]

24. Luan, Q.; Zelter, A.; MacCoss, M.J.; Davis, T.N.; Nolen, B.J. Identification of Wiskott-Aldrich syndrome protein (WASP) binding sites on the branched actin filament nucleator Arp2/3 complex. *Proc. Natl. Acad. Sci. USA* **2018**, *115*, E1409–E1418. [CrossRef] [PubMed]

25. Dominguez, R. Actin filament nucleation and elongation factors— Structure–function relationships. *Crit. Rev. Biochem. Mol. Biol.* **2009**, *44*, 351–366. [CrossRef] [PubMed]

26. Rebowski, G.; Namgoong, S.; Boczkowska, M.; Leavis, P.C.; Navaza, J.; Dominguez, R. Structure of a Longitudinal Actin Dimer Assembled by Tandem W Domains: Implications for Actin Filament Nucleation. *J. Mol. Biol.* **2010**, *403*, 11–23. [CrossRef] [PubMed]

27. Gaucher, J.F.; Maugé, C.; Didry, D.; Guichard, B.; Renault, L.; Carlier, M.F. Interactions of Isolated C-terminal Fragments of Neural Wiskott-Aldrich Syndrome Protein (N-WASP) with Actin and Arp2/3 Complex. *J. Biol. Chem.* **2012**, *287*, 34646–34659. [CrossRef] [PubMed]

28. Yamaguchi, H.; Miki, H.; Suetsugu, S.; Ma, L.; Kirschner, M.W.; Takenawa, T. Two tandem verprolin homology domains are necessary for a strong activation of Arp2/3 complex-induced actin polymerization and induction of microspike formation by N-WASP. *Proc. Natl. Acad. Sci. USA* **2000**, *97*, 12631–12636. [CrossRef]

29. Chan-Yao-Chong, M.; Deville, C.; Pinet, L.; van Heijenoort, C.; Durand, D.; Ha-Duong, T. Structural Characterization of N-WASP Domain V Using MD Simulations with NMR and SAXS Data. *Biophys. J.* **2019**, *116*, 1216–1227. [CrossRef]

30. Andreani, J.; Faure, G.; Guerois, R. InterEvScore: a novel coarse-grained interface scoring function using a multi-body statistical potential coupled to evolution. *Bioinformatics* **2013**, *29*, 1742–1749. [CrossRef]

31. Kollmar, M.; Lbik, D.; Enge, S. Evolution of the eukaryotic ARP2/3 activators of the WASP family: WASP, WAVE, WASH, and WHAMM, and the proposed new family members WAWH and WAML. *BMC Res. Notes* **2012**, *5*, 88. [CrossRef]

32. Chen, X.; Ni, F.; Tian, X.; Kondrashkina, E.; Wang, Q.; Ma, J. Structural Basis of Actin Filament Nucleation by Tandem W Domains. *Cell Rep.* **2013**, *3*, 1910–1920. [CrossRef] [PubMed]

33. Merino, F.; Pospich, S.; Funk, J.; Wagner, T.; Küllmer, F.; Arndt, H.D.; Bieling, P.; Raunser, S. Structural transitions of F-actin upon ATP hydrolysis at near-atomic resolution revealed by cryo-EM. *Nat. Struct. Mol. Biol.* **2018**, *25*, 528–537. [CrossRef] [PubMed]

34. Hertzog, M.; van Heijenoort, C.; Didry, D.; Gaudier, M.; Coutant, J.; Gigant, B.; Didelot, G.; Préat, T.; Knossow, M.; Guittet, E.; et al. The β-Thymosin/WH2 Domain: Structural Basis for the Switch from Inhibition to Promotion of Actin Assembly. *Cell* **2004**, *117*, 611–623. [CrossRef]

35. Daura, X.; Gademann, K.; Jaun, B.; Seebach, D.; van Gunsteren, W.F.; Mark, A.E. Peptide Folding: When Simulation Meets Experiment. *Angew. Chem. Int. Ed.* **1999**, *38*, 236–240. [CrossRef]

36. Saladin, A.; Fiorucci, S.; Poulain, P.; Prévost, C.; Zacharias, M. PTools: An opensource molecular docking library. *BMC Struct. Biol.* **2009**, *9*, 27. [CrossRef] [PubMed]

37. Basdevant, N.; Borgis, D.; Ha-Duong, T. A Coarse-Grained Protein–Protein Potential Derived from an All-Atom Force Field. *J. Phys. Chem. B* **2007**, *111*, 9390–9399. [CrossRef] [PubMed]

38. Basdevant, N.; Borgis, D.; Ha-Duong, T. Modeling Protein–Protein Recognition in Solution Using the Coarse-Grained Force Field SCORPION. *J. Chem. Theory Comput.* **2013**, *9*, 803–813. [CrossRef] [PubMed]

39. Abraham, M.J.; Murtola, T.; Schulz, R.; Pall, S.; Smith, J.C.; Hess, B.; Lindahl, E. GROMACS: High performance molecular simulations through multi-level parallelism from laptops to supercomputers. *SoftwareX* **2015**, *1–2*, 19–25. [CrossRef]

40. Essmann, U.; Perera, L.; Berkowitz, M.L.; Darden, T.; Lee, H.; Pedersen, L.G. A smooth particle mesh Ewald method. *J. Chem. Phys.* **1995**, *103*, 8577–8593. [CrossRef]

41. Hess, B. P-LINCS: A Parallel Linear Constraint Solver for Molecular Simulation. *J. Chem. Theory Comput.* **2008**, *4*, 116–122. [CrossRef]

42. Miyamoto, S.; Kollman, P.A. SETTLE: An analytical version of the SHAKE and RATTLE algorithm for rigid water models. *J. Comput. Chem.* **1992**, *13*, 952–962. [CrossRef]

43. Nosé, S. A unified formulation of the constant temperature molecular dynamics methods. *J. Chem. Phys.* **1984**, *81*, 511–519. [CrossRef]

44. Hoover, W.G. Canonical dynamics: Equilibrium phase-space distributions. *Phys. Rev. A* **1985**, *31*, 1695–1697. [CrossRef]

45. Parrinello, M.; Rahman, A. Polymorphic transitions in single crystals: A new molecular dynamics method. *J. Appl. Phys.* **1981**, *52*, 7182–7190. [CrossRef]

46. Best, R.B.; Mittal, J. Protein Simulations with an Optimized Water Model: Cooperative Helix Formation and Temperature-Induced Unfolded State Collapse. *J. Phys. Chem. B* **2010**, *114*, 14916–14923. [CrossRef]

47. Best, R.B.; Zheng, W.; Mittal, J. Balanced Protein–Water Interactions Improve Properties of Disordered Proteins and Non-Specific Protein Association. *J. Chem. Theory Comput.* **2014**, *10*, 5113–5124. [CrossRef] [PubMed]

48. Heinig, M.; Frishman, D. STRIDE: A web server for secondary structure assignment from known atomic coordinates of proteins. *Nucleic Acids Res.* **2004**, *32*, W500–W502. [CrossRef] [PubMed]

10

Analysis of Heterodimeric "Mutual Synergistic Folding"-Complexes

Anikó Mentes [†], Csaba Magyar [†], Erzsébet Fichó and István Simon *

Institute of Enzymology, Research Centre for Natural Sciences, Hungarian Academy of Sciences, Magyar Tudósok krt. 2., H-1117 Budapest, Hungary; mentes.aniko@ttk.mta.hu (A.M.); magyar.csaba@ttk.mta.hu (C.M.); ficho.erzsebet@ttk.mta.hu (E.F.)
* Correspondence: simon.istvan@ttk.mta.hu
† These authors contributed equally to the paper.

Abstract: Several intrinsically disordered proteins (IDPs) are capable to adopt stable structures without interacting with a folded partner. When the folding of all interacting partners happens at the same time, coupled with the interaction in a synergistic manner, the process is called Mutual Synergistic Folding (MSF). These complexes represent a discrete subset of IDPs. Recently, we collected information on their complexes and created the MFIB (Mutual Folding Induced by Binding) database. In a previous study, we compared homodimeric MSF complexes with homodimeric and monomeric globular proteins with similar amino acid sequence lengths. We concluded that MSF homodimers, compared to globular homodimeric proteins, have a greater solvent accessible main-chain surface area on the contact surface of the subunits, which becomes buried during dimerization. The main driving force of the folding is the mutual shielding of the water-accessible backbones, but the formation of further intermolecular interactions can also be relevant. In this paper, we will report analyses of heterodimeric MSF complexes. Our results indicate that the amino acid composition of the heterodimeric MSF monomer subunits slightly diverges from globular monomer proteins, while after dimerization, the amino acid composition of the overall MSF complexes becomes more similar to overall amino acid compositions of globular complexes. We found that inter-subunit interactions are strengthened, and additionally to the shielding of the solvent accessible backbone, other factors might play an important role in the stabilization of the heterodimeric structures, likewise energy gain resulting from the interaction of the two subunits with different amino acid compositions. We suggest that the shielding of the β-sheet backbones and the formation of a buried structural core along with the general strengthening of inter-subunit interactions together could be the driving forces of MSF protein structural ordering upon dimerization.

Keywords: dehydrons; inter-subunit interactions; intrinsically disordered proteins; ion-pairs; mutual synergistic folding; solvent accessible surface area; stabilization centers

1. Introduction

Mutual synergistic folding (MSF) complexes are a unique subset of intrinsically disordered proteins (IDPs). MSF IDPs can adopt a stable structure during the interaction, without a pre-existing folded partner [1–4]. At the time of the mutual synergistic folding process, the participating IDPs of these complexes synergistically fold into a stable, globular complex. Demarest et al. (2002) investigated the first MSF interaction between the p160 transcriptional coactivator protein ACTR and the tumor suppressor CBP proteins. They found that this MSF complex contains many hydrophobic side-chains and highly specific intermolecular hydrogen bonds, as well as buried intermolecular salt bridges, which help to fold the complex [5]. Since IDPs often have a high net charge, and they have a small content of hydrophobic residues, they are usually not able to form a hydrophobic core [6]. However,

MSF complexes contain more hydrophobic residues, presenting an exception to a general view of IDPs [7,8].

While IDPs mostly have low sequence complexity, MSF complexes are rather heterogeneous, like globular proteins. Furthermore, MSF proteinsare also heterogeneous in amino acid composition similar to globular proteins [8]. The residue-based disorder prediction methods, developed for identifying segments bound to folded proteins, cannot be used for detecting of MSF complexes. Systematic analyses are required to understand and predict these MSF interactions. Nevertheless, this is difficult to implement since a severe weakness of the literature is the little information available about these complexes. At present, the most comprehensive and systematic catalog of MSF complexes is the MFIB (Mutual Folding Induced by Binding) database containing 205 entries [9].

A protein in aqueous solution is only stable when it contains a hydrophobic core buried from water by polar residues. Furthermore, these polar residues shield most of the hydrophobic residues from the solvent. For the first criterion, the protein should contain more residues than a required minimum either as a monomer or as an oligomer. The fulfillment of the second criterion depends on the ratio of the polar and hydrophobic residues because the ratio of the surface and buried residues rapidly decreases by increasing the total number of residues. For a given hydrophilic/hydrophobic ratio, either a long polypeptide chain or oligomerization is needed. MSF proteins fulfill both criteria by oligomerization.

Recently, the physical background of homodimeric MSF complexes from MFIB [7] was analyzed. We identified the residues with solvent accessible main-chain patches (RSAMPs) and studied the "under-wrapped" hydrogen bonds (dehydrons), which are not shielded well enough from solvent [10]. Our results suggested that homodimeric MSF complexes contain more RSAMPs and dehydrons than homodimeric complexes where all the interacting chains are globular in their monomeric form. These properties should contribute to their disordered nature in monomeric form and to their folding in the oligomeric state. In this study, the role of this phenomenon for heterodimeric MSF complexes will be discussed. In the case of heterodimers, the interacting polypeptide chains have different amino acid compositions, which discriminates heterodimers from homodimers. The MFIB database contains, unfortunately, a much lower number of heterodimeric structures when compared to homodimeric ones. Furthermore, there are highly similar proteins among them, which makes redundancy filtering necessary.

2. Results and Discussion

2.1. Sequence-based Analysis

In this study, first, we examined the amino acid composition of the MFIB heterodimeric (MFHE) complexes, which were compared with a globular heterodimeric reference dataset (GLHE), which has similar size distribution for the heterodimeric state (see Figure 1). Note that all GLHE subunits are more than 40 residues away from both axes, while the closest distance of an MFHE chain from the x-axis is less than 20 residues. Also, we will show later (see Figure 5) that the smallest identified globular monomer has 35 residues. In some cases, heterodimeric MSF complexes do not have enough amino acids for creating a hydrophobic core, but in most cases, they have as many residues as globular proteins have, thus other factors might also be responsible for the disordered nature of MFHE proteins.

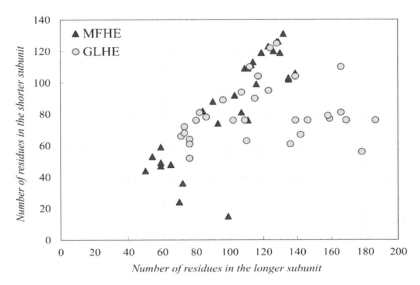

Figure 1. Comparison of the subunit lengths of the Mutual Synergistic Folding (MSF) (MFHE—blue triangles) and globular (globular heterodimeric GLHE—yellow dots) heterodimeric complexes.

Since the beginning of the studies on IDPs, it is known they generally lack hydrophobic residues although alanine has a notably higher content in MFHE complexes compared to GLHE complexes, while the content of other aliphatic residues was similar among the two datasets (see Figure 2A). MFHE complexes have a high net charge, like non-MSF IDPs [11,12].

The amino acid composition of the MFHE and GLHE heterodimers was depicted by a rank-based, indirect gradient analysis method, called Nonmetric MultiDimensional Scaling (NMDS), which creates an ordination based on a distance or dissimilarity matrix, thus it allows decreasing a multidimensional and quantitative, semi-quantitative, qualitative, or mixed variables data set to two dimensions [13]. NMDS demonstrated a separation of MFHE and GLHE complexes and subunits (see Figure 2B,C). The amino acid composition of the subunits, whether globular or MSF complexes are formed, have equal distances from each other as the amino acid compositions of the complexes. Some differences are revealed between the two data sets—the NMDS of the amino acid composition of the MSF heterodimeric complexes showed smaller variation from the globular heterodimeric complexes (see Figure 2C), than the amino acid composition of the MSF subunits from the globular subunits (see Figure 2B). These differences can be explained by the fact that although the amino acid composition of the MSF subunits differs slightly from globular proteins, they are unable to fold into an ordered structure independently. The folding of an MSF subunit requires another partner, which in this case has a different amino acid composition, that could form MSF complexes which have similar amino acid composition than the globular subunits. NMDS also pointed out that MFHE is a diverse group based on their amino acid composition, and these complexes are also clustered according to their structural classes in MFIB [9].

The determination of the amino acids that contribute mainly to the observed difference was revealed by using SIMPER (similarity percentage) analyses. These amino acids were lysine (7.40%; 8.04%), alanine (7.30%; 7.90), leucine (7.14%; 6.64%), glycine (6.86%; 5.83%), arginine (6.39%; 6.70%), and glutamine (6.29%; 6.42%), which values support the similarity of the objects. Mostly aromatic and hydrophobic amino acids cause the amino acid compositions to separate (in slightly different proportions, See Table S1), which case is more common in heterodimeric MSF subunits and complexes if the MSF data were grouped via MFIB for comparison was considered, for the MFIB structural classes (see Figure 2, Table S1), with the exception of glutamine.

Most of the heterodimers from MFIB are histone-type proteins with their high content of lysine and arginine. Acetylated lysine and methylated arginine may interact with proteins containing bromodomains and Tudor domains within the disordered proteins that affect nucleic acid binding and RNA pathways [14].

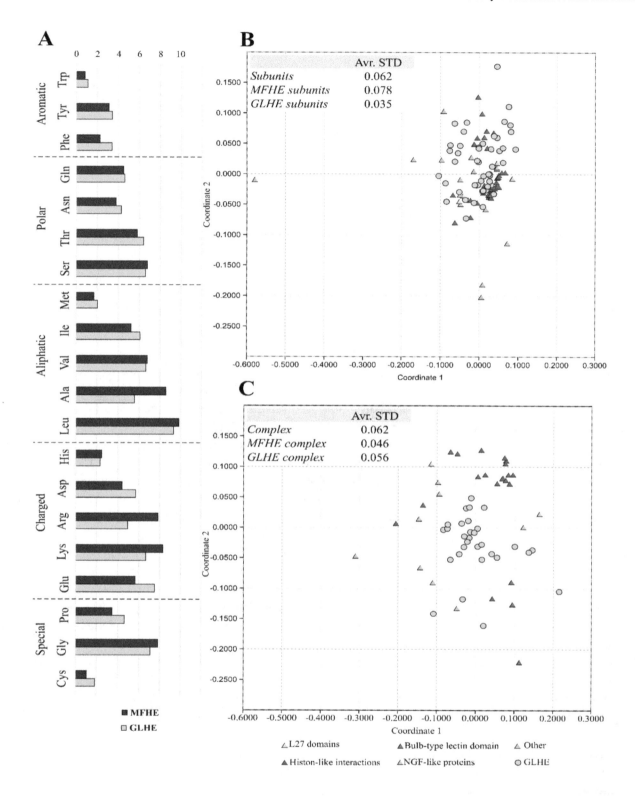

Figure 2. Amino acid composition of the heterodimer datasets, where the types of amino acids were grouped by Mészáros et al. [8] (**A**). The MFHE (triangles) and GLHE (dot) amino acid composition were compared using an indirect gradient analysis method, called Nonmetric Multidimensional Scaling (NMDS), which creates an ordination based on Bray-Curtis distances. In the plot, the objects are protein subunits (**B**) considered separately and complexes (**C**).

The amino acid composition of the homodimeric complexes from MFIB (MFHO), heterodimeric MSF complexes was compared using our small globular protein (SGP) dataset as a standard reference by Kullback-Leibler divergence [15], which measures the extent of the dissimilarity between two probability distributions ($D = \sum_i p_i * \ln \frac{p_i}{q_i}$). MSF heterodimers show about the same similarity to MSF homodimers (D = 1.257) and small globular proteins (D = 1.879), while MSF homodimers are more similar to small globular proteins (D = 0.442). This result is in line with the observation that heterodimeric complexes from MFIB look much more disordered (~20%) than MFIB homodimers (MFHO) (~10%) [7] based on MoRFpred [16] and IUPred [17] results. Some regions of the heterodimeric MFIB complexes are also capable of folding on the surface of a globular protein. Most of these can be found in the DIBS (Disordered Binding Site) database [18]. It is rather rare, but it also shows the elevation of the group inhomogeneity. For example, the cellular tumor antigen p53 protein (UniProt: P04637) is able to establish a coactivator binding domain complex (MFIB: MF2201002, PDB: 2l14) with the CREB-binding MSF protein, although at the same part of the p53 capable to form a transactivator domain complex (DIBS: DI1000009, PDB: 2ly4) with the highly mobile folded B1 protein. We have also found examples of disordered proteins from UniProt (e.g., ID: Q9Y6Q9, Nuclear receptor coactivator 3) which are able to establish an MSF interaction (MF2201001, PDB: 1kbh), and another region is able to form a DIBS interaction (DI1000313, PDB: 3l3x), forming two different types of disordered protein complexes.

It is interesting to note, that a few MFIB homodimers occur in DIBS as ordered interaction partners. For example, the dynein light chain (Tctex-type) protein (UniProt: Q94524), which is disordered in monomeric form based on MFIB (MFIB: MF2110016, PDB: 1ygt), while this homodimeric complex is the ordered part of a DIBS-interaction complex (Cytosolic dynein intermediate chain bound to Tctex-type dynein light chain, DIBS: DI2100002, PDB: 3fm7). An additional example of these multiple structure organizations is the homodimeric S100BEF-hand calcium-binding protein superfamily (MFIB: MF2100013, PDB: 1uwo), which is the ordered component of a DIBS-interaction (RSK1 bound to S100B dimer, DIBS: DI2000012, PDB: 5csf).

Besides the amino acid compositions, other sequential parameters also display differences between GLHE and MFHE. Based on cleverMachine [19] calculations (p-value < 0.0001: 56 scale of all 80) and grouped properties results, membrane proteins (p-value < 0.0001: 7 scale of 10), nucleic acid binding (p-value < 0.0001: 3 scale of 10), disorder propensity (p-value < 0.0001: 8 scale of 10), α-helix (p-value < 0.0001: 9 scale of 10), β-sheet (p-value < 0.0001: 9 scale of 10), aggregation (p-value < 0.0001: 8 scale of 10), burial propensity (p-value < 0.0001: 10 scale of 10), and hydrophobicity (p-value < 0.0001: 2 scale of 10) properties in MFHE are in general stronger than in globular heterodimers (Reference number of the dataset: 196154). While there is no significant difference between the sequences of MFIB homodimers and globular homodimers (GLHO) in most of the properties (exception of some membrane proteins and aggregation scales; p-value < 0.0001: 8 scale of all 80) (Reference number of the dataset: 199533).

We analyzed the Pfam database in conjunction with the intermolecular stabilization centers (SCs, see Chapter 2.2. Structure-based analysis) [20] on MFIB heterodimeric and globular heterodimeric complexes (for detailed results, see Table S2). In the MFHE have found 59 Pfam domains in a total of 19 families, while the GLHE have 64 Pfam domains in a total of 37 families. In the case of globular heterodimers, 3 of the 30 complexes have interactions and SCs between the Pfam domains of the monomers, whereas, for MFIB heterodimers much more, at least 15 of the complexes have Pfam domains in which monomers interactions and intermolecular SCs were found. This result confirms that the folding of the MSF proteins is related to their functional role since, in many cases, the two subunits form the biologically relevant unit.

2.2. Structure-based Analysis

In our recent analysis of MSF homodimeric proteins, we found differences in several structural parameters between our dataset and a globular reference dataset. These structural features were investigated including solvent accessibility, hydrogen bonds, stabilization center content, and ion-pairs with an additional investigation of the buried structural core size.

The inter-subunit interface was identified based on the solvent accessible surface area (SASA) calculations. However, an MSF protein subunit in itself does not have an ordered structure, structural properties were also calculated for their monomeric forms, which were created by deleting a polypeptide chain from the heterodimeric PDB structures. This is referred to as their "monomeric structure" hereafter. The all-atom SASA values were calculated for all residues from the heterodimeric and monomeric structures. If the dimeric SASA value was below 20% of the monomeric value, the residue was identified as an interface residue. In the case of the MFIB heterodimeric dataset, 908 interface residues were identified out of the 4615 residues, that is 19.7% of all residues participate in the formation of the interface. In the globular reference heterodimeric dataset 470 interface residues were identified out of the 5155 total residues, i.e., 9.1% of all residues are forming the interface. As a different measure of the interface region, all-atom SASA values were also compared. In MFHE, 27.3% of the total surface area becomes buried upon dimerization, while in GLHE, only 11.6%. This result is in agreement with the finding of Gunasekaran et al., that the per residue interface area is higher in disordered complexes [3] In MSF proteins, the larger interface contact area underlines the importance of inter-subunit interactions, thus inter-subunit interactions were considered hereafter.

Completely buried residues were identified in the MSF and the globular reference heterodimeric datasets using a stricter definition of burial, defining the core of the protein structure shielded from the solvent. We identified all residues, which have less than 10% relative all-atom solvent accessibility in the heterodimeric and monomeric structures, respectively. In MFHE, 10.8% of all residues are buried in monomeric form, while in GLHE this value is 20.9%. If the dimeric structures were analyzed, the values change to 27.7% and 26.3%, respectively. There are significantly fewer residues buried in the monomeric forms of MSF proteins when compared to globular ones. In the dimeric forms, the ratio of buried residues is similar in both cases. Figure 3 shows the number of buried residues in MSF (see Figure 3A) and globular heterodimeric complexes (see Figure 3B).

It can be seen that in the case of MSF heterodimers, there is a more considerable difference between the number of buried residues in the dimeric and monomeric forms, than in the case of globular heterodimers. In the case of globular heterodimers (see Figure 3B), the sum of the number of buried residues in the two monomeric subunits is close to the number of buried residues in the dimeric form. These subunits are ordered by themselves, and they do not need another subunit to help to order their structures. In the case of MSF heterodimers (see Figure 3A), the sum of the number of buried residues in the monomeric forms is lower than in the case of the globular heterodimers and, more importantly, they are much smaller than the number of buried residues in the dimeric form. These polypeptide chains are disordered by themselves, they need the presence of an interacting partner to help in ordering their structures. These protein chains need each other to form a reasonably sized core, needed for a stable, ordered structure.

The secondary structural element content was determined in the heterodimeric structures using the DSSP program [21]. We found that in the MFHE dataset, 43.6% of the residues have the α-helical conformation and only 16.1% of the residues belonged to β-sheets, in the globular heterodimeric dataset, these values were 21.5% and 27.5%, respectively. In the MSF, heterodimeric dataset β-sheets were less abundant than in globular heterodimeric proteins. This will have some consequences in the interpretation of our later results.

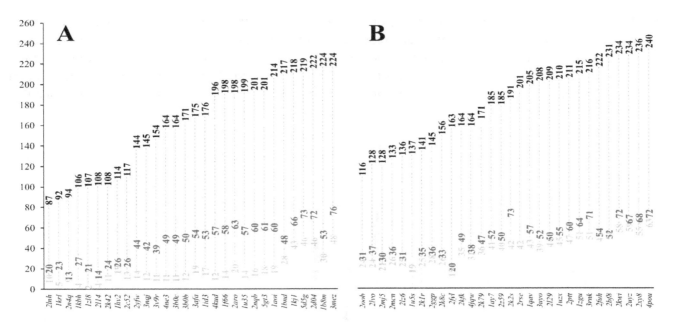

Figure 3. The number of burial residues in MFHE (**A**) and GLHE (**B**) complexes (black: number of all residues in a complex, red: number of buried residues in a heterodimeric complex, blue: sum of numbers of buried residues in the two monomeric subunits. See Figure S1 for the number of buried residues for the homodimeric MFHO and GLHO datasets.

We counted the number of inter-subunit ion-pairs. While there is only a small difference in the number of charged residues between MFHE and GLHE (1224 vs. 1380), the total charge is +320 for all 30 MFHE proteins and −91 for all 30 GLHE proteins. We found only 16 charged residues participating in 8 strong ion-pairs in the MFHE, while 28 residues are participating in 15 ion-pairs in the GLHE dataset. If we also consider weak ion-pairs, these values change to 73 residues participating in 42 ion-pairs for MFHE and 59 residues in 35 ion-pairs for GLHE. This is a 5.25-fold increase for MFHE and only a 2.33-fold increase for GLHE, respectively. Weak ion-pairs, presumably do not contribute to the enthalpic stabilization of the dimers, but probably play a role in the formation of electrostatic complementarity, already observed by Wong et al. in the case of complexes containing IDPs [22] This behavior was unexpected, and further investigation of the role of electrostatic interactions in the stabilization of MSF dimers is planned.

In the case of the MSF homodimers, we found that the main-chain solvent accessibility may play an important role in the stabilization of homodimer structures [8]. We identified residues with solvent accessible main-chain patches (RSAMPs). We have found a total of 161 RSAMPs in the MFHE dataset, and 90 RSAMPs in the GLHE dataset, respectively. There are 2 out of the 30 proteins in the MFHE dataset, which does not contain an RSAMP residue, while there are four such entries in the GLHE dataset. The average RSAMP content was 5.4 per heterodimeric complexes; thus, 17.7% of the interface residues are RSAMPs. In 26 of the 30 globular heterodimeric complexes, the average RSAMP content was 3, thus 19.1% of the interface residues are RSAMPs.

On the one hand, the composition of the RSAMPs of MFIB heterodimers suggested that five types of amino acids (glycine, alanine, isoleucine, leucine, and valine) play a major role in these interactions (see Figure 4). These RSAMP contributing amino acids are mainly hydrophobic, are exposed to the inter-subunit interface. These residues do not contribute to the stabilization of the monomeric form since exposed hydrophobic surfaces are energetically not favorable. However, next to the favorable burial of their main-chain, they might help the formation of the tertiary structure by building sticky hydrophobic patches at the inter-subunit interface. On the other hand, in the case of the globular heterodimer dataset, the two amino acids with the smallest side-chains, glycine and alanine are the most abundant residues under RSAMPs. We investigated the secondary structural distribution of RSAMP, as well. We found that 33.5% of RSAMPs are located in β-sheets and 44.7% in α-helices.

We checked the secondary structural composition of the interface residues, from which RSAMPs are selected. We found that 19.5% of interface residues have β-sheet and 63.9% have α-helical secondary structure. Considering the 3.3-fold higher occurrence of helical secondary structure at the interface, we can conclude that RSAMPs are more abundantly found in β structures, which can be easily broken by disturbing their hydrogen bonding network through interactions with accessible solvent molecules.

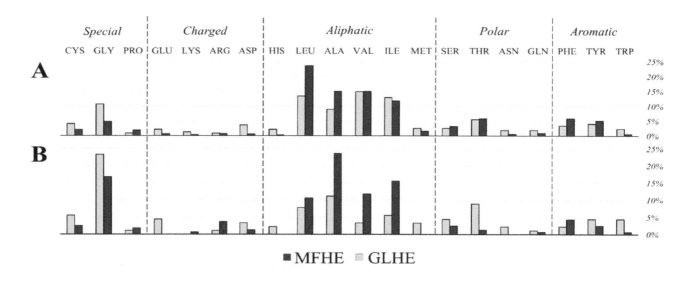

Figure 4. Amino acid composition of the interface (**A**) and the residues with solvent accessible main-chain patches (RSAMPs; (**B**)) of MFHE (blue) and GLHE (yellow) complexes.

We counted the number of inter-subunit hydrogen bonds. We found a total number of 181 H-bonds in the MFHE and only 67 in the GLHE dataset, respectively. This is in agreement with our observation that inter-subunit interactions are of high importance in MSF heterodimers. We calculated the average wrapping of hydrogen bonds [10]. Hydrogen bonds with a low wrapping (dehydrons) are less shielded from the solvent. The average value was 13.8 for the MFHE and 14.6 for the GLHE. Inter-subunit hydrogen bonds are slightly less wrapped in the MSF heterodimers, which also indicates the importance of solvent accessibility.

We also identified inter-subunit stabilization centers in both the MFHE and GLHE datasets. Stabilization centers are special residue pairs, which together with their sequential neighbors, participate in above than average long-range interactions and are believed to contribute to the stabilization of protein structures [23]. The two residues that form a stabilization center are called stabilization center elements (SCEs). In MFHE, the average inter-subunit SCE content was 8.1, and we found at least one inter-subunit SC in 26 of the 30 heterodimers. In GLHE, the average SCE content was 0.5, and we found an inter-subunit SC is only 5 out of the 30 structures.

We investigated if there is a lower size limit for globular proteins, which already bear a buried core structure. Our analysis of monomeric, single-domain globular (SGP) dataset pointed out that proteins with 35 residues are already containing a buried structural core (see Figure 5). Our results, regarding the buried core size of the MFIB heterodimers, indicate that although a couple of polypeptide chains are too small to contain a buried core, this is not a general trend for the MFHE dataset.

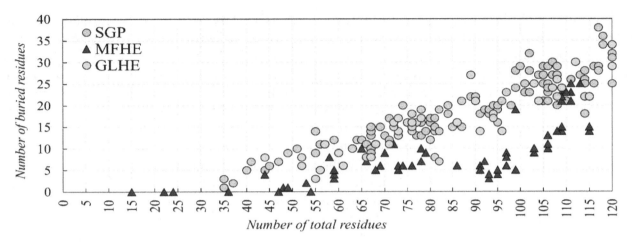

Figure 5. The number of total and buried residues of SGP (grey), GLHE (yellow) and MFHE (blue). For the number of total and buried residues of homodimeric MSF see Figure S2.

3. Conclusions

In our previous article [7], we found that the amino acid composition and sequence properties of MSF homodimers are similar to globular homodimers. However, they have more residues with solvent accessible peptide backbones that make them disordered in monomeric form, but they are ordered in a complex. There are some examples of these interactions in DIBS that prove their ordered nature. According to our results, MFIB heterodimers are less similar to globular proteins than homodimers, based on the calculated sequence and structural features. The MFIB heterodimers like the MFIB homodimers do not lack hydrophobic residues (as non-MSF IDPs), on the contrary, they are enriched in aliphatic residues which would theoretically allow the formation of a hydrophobic core, but in some cases, probably the chain itself is not large enough for the folding.

"Non-MSF" disorder prediction methods identify MFIB protein chains disordered at a short sequence segment which in some cases is confirmed by DIBS. In these DIBS interactions, heterodimeric MFIB subunits could bind to disordered, as well as globular protein regions. Therefore, in the case of heterodimeric MSF complexes, other factors can also affect their disorders than in the case of homodimeric MFIB proteins, because different factors are responsible for order-disordered interactions than for disordered-disordered complexes. This does not exclude that a protein chain can have the capability for both interactions and there has to be another ground why these proteins are unstructured on their own. In most cases of MSF heterodimers, the subunits themselves possibly do not have a low enough "energy" to fold, but the different compositions of the interacting partners may contribute to the stability of the complex. Understanding how sequence and composition and backbone variation affect foldability, will become increasingly crucial in folding protein design methods as more elements are included in the design process [24]. Based on NMDS results, the amino acid composition of the MSF heterodimeric complexes revealed smaller differences from the globular heterodimeric complexes, while the amino acid composition of the subunits showed distant similarity. Aromatic and hydrophobic amino acids are mainly responsible for the separation of the amino acid composition (based on SIMPER analysis) showed on NMDS. The amino acid composition of the MSF subunits is similar to globular proteins, but the MSF subunits together would change the amino acid composition of the complexes for a further reason. The heterodimeric MFIB complexes have a diverse amino acid composition, but they are involved in only a few types of molecular functions, such as DNA or histone binding (based on GO annotations from MFIB), which contributes to functional stability and making improvements in cell interactions [25].

In a recent paper [8], we concluded that MFIB proteins are disordered in monomeric form because they are too small to form a structural core. Our current analyses showed that however there are a couple of MSF protein subunits that do not contain buried residues, this is not a general rule, moreover we found that globular proteins with at least 35 residues already own a buried core (see Figure 5). At the same time, we found that the dimeric structures of MFIB and globular heterodimers contain a similar ratio of buried residues, but in monomeric form the MFIB heterodimers would contain only about half as much buried residues than globular heterodimers. We can conclude that the increased interface area of MFIB heterodimers contributes to the formation of a larger buried core structure. In globular heterodimers, the number of buried residues is increased only by a small margin upon dimerization, while in MFIB heterodimers there is a much larger increase (see Figure 3A). Globular monomers are stable and already own a reasonably sized buried structural core, while MFIB heterodimers are disordered by themselves and they need an interacting partner to form a large enough buried structural core to be stable. According to the structure-based analysis we can deduce that inter-subunit interactions are of high importance in the stabilization of MSF proteins. As in the case of homodimers, shielding of the main-chain from the solvent is an important factor for the stabilization of the heterodimeric structures. Interactions of the main-chain with water molecules might destabilize the secondary structure by breaking the hydrogen bond network, leading to the disruption of the secondary and the tertiary structure. Other interactions, which are identified by our definition of stabilization centers, play an important role in the stabilization of the heterodimeric structures, as well. This is consistent with our results about inter-subunit stabilization centers and Pfam domains, wherein the case of globular heterodimers, a few complexes have interactions and SCs between the Pfam domains of the subunits, whereas for MFIB heterodimers more than half of the complexes have inter-subunit interactions between the Pfam domains of the different chains. This also suggests that the folding of the MSF subunits is related to their functional role.

Though we found that the difference in the number of RSAMPs between MSF and globular proteins is slightly smaller in the case of heterodimers than it was in the case of homodimers, considering the lower β-sheet content of the MFHE dataset, the RSAMP/β-sheet forming residue ratio is correspondingly high in MSF heterodimers and homodimers. We suggest that the shielding of the β-sheet backbones and the formation of a buried structural core together with the general strengthening of inter-subunit interactions together could be the driving forces of MSF protein structural ordering upon dimerization.

Protein folding, the structural organization of proteins in aqueous solution, is realized by monomolecular reactions of intermolecular interactions, even if this is followed later by further macromolecular interactions because of functional or stability reasons. In the case of MSF proteins for the formation of a stable ordered structure intermolecular interactions are needed, therefore it is part of the folding. Opposing the regular folding this is not a monomolecular, but rather a bimolecular reaction, in which the ratio of the participating components and other parameters can be changed. We believe that further experimental and theoretical investigation of the structural organization of MSF proteins can contribute to a more profound understanding of the folding problem.

4. Materials and Methods

There are 49 heterodimeric proteins in the MFIB database. Entries belonging to the "coils and zippers" structural class were excluded, as in the case of homodimers. Since 25 of the 49 heterodimers are histones, filtering of the dataset was necessary to avoid overrepresentation and sequence redundancy of this protein class. Proteins were assigned to the same cluster if their sequence identity was over 90% using the BLASTClust toolkit 2.2.26 [26]. The 2mv7 entry was discarded because it was an outlier due to its fuzzy NMR structure in SASA calculations. One representative structure was kept for the remaining 30 clusters, creating the filtered MFHE dataset (see Table S3). A reference dataset was created of globular heterodimers (GLHE) from the PDBSelect [27] database with a total number of residues less than 240 to match the size distribution of the heterodimer MFIB dataset (see Table S3).

We described the methods in the latest article [7], but briefly: the interface term is used for the contact surface area of the two subunits in the heterodimeric structures. In cases where the term "monomeric structure" is used, calculations were carried out on single polypeptide chains, where the other chain was removed from the PDB files. Residues belonging to the interface region were identified based on solvent accessible surface area (SASA) calculations. All-atom SASA values were calculated using the FreeSASA 2.03 [28] program, residues where the SASA value calculated for the dimeric structure was less than or equal to 20% of the monomeric value, were defined to belong to the interface.

We were looking for residues in the interface that have solvent accessible spots in their main-chain in the monomeric structure, which become buried in the dimeric structures. We identified residues where the main-chain SASA in the dimeric form was less than 20% of the monomeric form value. Only residues with exposed main-chains, with a relative main-chain SASA larger than 0.2 in the monomeric structure, were taken into account. These residues with solvent accessible main-chain patches are called RSAMPs and are believed to be important for structural ordering upon dimerization of the disordered polypeptide chains collected in the MFIB database.

We used an additional Small Globular Protein (SGP) dataset to determine the minimal buried core size of proteins (see Table S3). We collected monomeric single-domain proteins X-ray structures from the PDBSELECT database with less than 120 residues, which do not contain disulfide bonds. Since there was a significant hole in the size distribution of the X-ray structures, monomeric single-domain NMR structures without disulfide bonds were added to the dataset. We excluded rod-like and fuzzy NMR structures using a volume/surface cutoff criterion. Protein volumes were calculated using the ProteinVolume 1.3 program [29].

Secondary structural elements were identified using the DSSP [21] program. Hydrogen bonds were identified using the find_pairs PyMol command using 3.5 Å distance and 45-degree angle criteria [30]. Wrapping of hydrogen bonds was calculated using the dehydron_ter.py program [31]. Stabilization centers (SCs) are special pairs of residues involved in cooperative long-range interactions. The two residues that form a stabilization center are called stabilization center elements (SCEs). SCEs were identified using our SCide server [32]. Ion pairs were defined as pairs of positively and negatively charged residues, with a distance of less than a cutoff value between the charged groups. For strong ion pairs, this value is 4 Å [33], but we introduce additionally, a weak ion-pair definition with a distance cutoff value of 6 Å. Histidine residues were assumed to be neutral in these calculations because of the uncertainty of their protonation states. Ion pairs were identified using our own C++ program. We calculated the total charge of the proteins simply by adding the number of Arg and Lys residues and subtracted the sum of Asp and Glu resides.

Amino acid compositions were determined using MEGA7 software [34]. The amino acid composition of the protein subunits and complexes were visualized in Nonmetric multidimensional scaling (NMDS) in PAST3 [35]. In the plot, one point for each amino acid composition, where close points were more similar in composition (with Bray-Curtis distances). This was followed by a SIMPER analysis (also based on Bray-Curtis distances, in PAST3) to identify those amino acids that contributed most to the observed differences among the type of subunits and complexes. Disorder predictions were revealed by IUPred2A [17] and MoRFpred [16] algorithms.

Supplementary Materials:
Table S1. Contribution of the amino acids for the observed differences in NMDS by using SIMPER analysis in the subunits (A) and the complexes (B) Table S2. Pfam domains and the intermolecular SCs in globular (A) and MFIB (B) heterodimeric complexes. Table S3. List of PDB entries in the MFHE, GLHE and SGP datasets. Figure S1. The number of burial residues in MFHO (A) and GLHO (B) complexes (black: number of all residues in a complex, red: number of buried residues in a heterodimeric complex, blue: sum of number of buried residues in the two monomeric subunits. Figure S2. The number of total and buried residues of SGP, GLHE, GLHO, MFHE and MFHO.

Author Contributions: Conceptualization, I.S., and C.M.; methodology, A.M., E.F.; software, A.M., E.F., C.M.; validation, A.M., C.M; formal analysis, C.M.; investigation, A.M., E.F.; resources, A.M, E.F.; data curation, A.M., E.F., C.M.; writing—original draft preparation, A.M., C.M., I.S.; writing—review and editing, E.F., A.M., C.M; visualization, A.M.; supervision, I.S.; project administration, I.S.; funding acquisition, I.S.

Acknowledgments: The authors acknowledge the support of ELIXIR Hungary.

Abbreviations

IDP	Intrinsically Disordered Protein
MFIB	Mutual Folding Induced by Binding database
DIBS	Disordered Binding Site database
MFHO	MFIB Homodimeric dataset
MFHE	MFIB Heterodimeric dataset
GLHE	Globular Heterodimeric dataset
NMDS	Non Metric Multidimensional Scaling
RSAMPs	Residues with Solvent Accessible Main-chain Patches
SC/SCE	Stabilization centers/Stabilization center elements
SASA	Solvent Accessible Surface Area
SGP	Small Globular Protein dataset

References

1.   Tsai, C.J.; Nussinov, R. Hydrophobic folding units at protein-protein interfaces: Implications to protein folding and to protein-protein association. *Protein Sci.* **1997**, *6*, 1426–1437. [CrossRef] [PubMed]

2.   Xu, D.; Tsai, C.J.; Nussinov, R. Mechanism and evolution of protein dimerization. *Protein Sci.* **1998**, *7*, 533–544.

3.   Gunasekaran, K.; Tsai, C.J.; Nussinov, R. Analysis of ordered and disordered protein complexes reveals structural features discriminating between stable and unstable monomers. *J. Mol. Biol.* **2004**, *341*, 1327–1341. [CrossRef]

4.   Rumfeldt, J.A.; Galvagnion, C.; Vassall, K.A.; Meiering, E.M. Conformational stability and folding mechanisms of dimeric proteins. *Prog. Biophys. Mol. Biol.* **2008**, *98*, 61–84. [CrossRef] [PubMed]

5.   Demarest, S.J.; Martinez-Yamout, M.; Chung, J.; Chen, H.; Xu, W.; Dyson, H.J.; Evans, R.M.; Wright, P.E. Mutual synergistic folding in recruitment of CBP/p300 by p160 nuclear receptor coactivators. *Nature* **2002**, *415*, 549–553. [CrossRef] [PubMed]

6.   Habchi, J.; Tompa, P.; Longhi, S.; Uversky, V.N. Introducing protein intrinsic disorder. *Chem. Rev.* **2014**, *114*, 6561–6588. [CrossRef]

7.   Magyar, C.; Mentes, A.; Fichó, E.; Cserző, M.; Simon, I. Physical Background of the Disordered Nature of "Mutual Synergetic Folding" Proteins. *Int. J. Mol. Sci.* **2018**, *19*, 3340. [CrossRef]

8.   Mészáros, B.; Dobson, L.; Fichó, E.; Tusnády, G.E.; Dosztányi, Z.; Simon, I. Sequential, Structural and Functional Properties of Protein Complexes Are Defined by How Folding and Binding Intertwine. *J. Mol. Biol.* **2019**. [CrossRef]

9.   Fichó, E.; Reményi, I.; Simon, I.; Mészáros, B. MFIB: A repository of protein complexes with mutual folding induced by binding. *Bioinformatics* **2017**, *33*, 3682–3684. [CrossRef]

10. Fernández, A.; Scott, R. Dehydron: A structurally encoded signal for protein interaction. *Biophys. J.* **2003**, *85*, 1914–1928. [CrossRef]

11. Uversky, V.N.; Gillespie, J.R.; Fink, A.L. Why are "natively unfolded" proteins unstructured under physiologic conditions? *Proteins* **2000**, *41*, 415–427. [CrossRef]

12. Campen, A.; Williams, R.M.; Brown, C.J.; Meng, J.; Uversky, V.N.; Dunker, A.K. TOP-IDP-scale: A new amino acid scale measuring propensity for intrinsic disorder. *Protein Pept. Lett.* **2008**, *15*, 956–963. [CrossRef] [PubMed]

13. Taguchi, Y.H.; Oono, Y. Relational patterns of gene expression via non-metric multidimensional scaling analysis. *Bioinformatics* **2005**, *21*, 730–740. [CrossRef] [PubMed]

14. Bah, A.; Forman-Kay, J.D. Modulation of Intrinsically Disordered Protein Function by Post-translational Modifications. *J. Biol. Chem.* **2016**, *291*, 6696–6705. [CrossRef] [PubMed]

15. Kullback, S.; Leibler, R.A. On information and sufficiency. *Ann. Math. Stat.* **1951**, *22*, 79–86. [CrossRef]

16. Disfani, F.M.; Hsu, W.L.; Mizianty, M.J.; Oldfield, C.J.; Xue, B.; Dunker, A.K.; Uversky, V.N.; Kurgan, L. MoRFpred, a computational tool for sequence-based prediction and characterization of short disorder-to-order transitioning binding regions in proteins. *Bioinformatics* **2012**, *28*, i75–i83. [CrossRef] [PubMed]

17. Mészáros, B.; Erdos, G.; Dosztányi, Z. IUPred2A: Context-dependent prediction of protein disorder as a function of redox state and protein binding. *Nucleic Acids. Res.* **2018**, *46*, W329–W337. [CrossRef]

18. Schad, E.; Fichó, E.; Pancsa, R.; Simon, I.; Dosztányi, Z.; Mészáros, B. DIBS: A repository of disordered binding sites mediating interactions with ordered proteins. *Bioinformatics* **2018**, *34*, 535–537. [CrossRef]

19. Klus, P.; Bolognesi, B.; Agostini, F.; Marchese, D.; Zanzoni, A.; Tartaglia, G.G. The cleverSuite approach for protein characterization: Predictions of structural properties, solubility, chaperone requirements and RNA-binding abilities. *Bioinformatics* **2014**, *30*, 1601–1608. [CrossRef]

20. Dosztányi, Z.; Fiser, A.; Simon, I. Stabilization centers in proteins: Identification, characterization and predictions. *J. Mol. Biol.* **1997**, *272*, 597–612. [CrossRef]

21. Kabsch, W.; Sander, C. Dictionary of protein secondary structure: Pattern recognition of hydrogen-bonded and geometrical features. *Biopolymers* **1983**, *22*, 2577–2637. [CrossRef] [PubMed]

22. Wong, E.T.; Na, D.; Gsponer, J. On the importance of polar interactions for complexes containing intrinsically disordered proteins. *PLoS Comput. Biol.* **2013**, *9*, e1003192. [CrossRef] [PubMed]

23. Magyar, C.; Gromiha, M.M.; Sávoly, Z.; Simon, I. The role of stabilization centers in protein thermal stability. *Biochem. Biophys. Res. Commun.* **2016**, *471*, 57–62. [CrossRef] [PubMed]

24. Saven, J.G. Designing protein energy landscapes. *Chem. Rev.* **2001**, *101*, 3113–3130. [CrossRef]

25. Lee, B.M.; Mahadevan, L.C. Stability of histone modifications across mammalian genomes: Implications for 'epigenetic' marking. *J. Cell. Biochem.* **2009**, *108*, 22–34. [CrossRef]

26. Alva, V.; Nam, S.Z.; Söding, J.; Lupas, A.N. The MPI bioinformatics Toolkit as an integrative platform for advanced protein sequence and structure analysis. *Nucleic Acids. Res.* **2016**, *44*, W410–W415. [CrossRef]

27. Griep, S.; Hobohm, U. PDBselect 1992–2009 and PDBfilter-select. *Nucleic Acids. Res.* **2010**, *38*, D318–D319. [CrossRef]

28. Mitternacht, S. FreeSASA: An open source C library for solvent accessible surface area calculations. *F1000Research* **2016**, *5*, 189. [CrossRef]

29. Chen, C.R.; Makhatadze, G.I. ProteinVolume: Calculating molecular van der Waals and void volumes in proteins. *BMC Bioinforma.* **2015**, *16*, 101. [CrossRef]

30. LLC. *The PyMOL Molecular Graphics System*; Schrodinger Version 1.6; LLC: New York, NY, USA, 2011.

31. Martin, O.A. *Wrappy: A Dehydron Calculator Plugin for PyMOL*; IMASL-CONICET: San Louis, Argentina, 2011.

32. Dosztányi, Z.; Magyar, C.; Tusnády, G.; Simon, I. SCide: Identification of stabilization centers in proteins. *Bioinformatics* **2003**, *19*, 899–900. [CrossRef]

33. Barlow, D.J.; Thornton, M.J. Ion-pairs in proteins. *J. Mol. Biol.* **1983**, *168*, 867–885. [CrossRef]

34. Kumar, S.; Stecher, G.; Tamura, K. MEGA7: Molecular Evolutionary Genetics Analysis Version 7.0 for Bigger Datasets. *Mol. Biol. Evol.* **2016**, *33*, 1870–1874. [CrossRef] [PubMed]

35. Hammer, Ø.; Harper, D.A.T.; Ryan, P.D. PAST: Paleontological statistics software package for education and data analysis. *Palaeontol. Electron.* **2002**, *4*, 9.

11

The Significance of the Intrinsically Disordered Regions for the Functions of the bHLH Transcription Factors

Aneta Tarczewska and Beata Greb-Markiewicz *

Department of Biochemistry, Faculty of Chemistry, Wroclaw University of Science and Technology, Wybrzeże Wyspiańskiego 27, 50-370 Wroclaw, Poland; aneta.tarczewska@pwr.edu.pl
* Correspondence: beata.greb-markiewicz@pwr.edu.pl

Abstract: The bHLH proteins are a family of eukaryotic transcription factors regulating expression of a wide range of genes involved in cell differentiation and development. They contain the Helix-Loop-Helix (HLH) domain, preceded by a stretch of basic residues, which are responsible for dimerization and binding to E-box sequences. In addition to the well-preserved DNA-binding bHLH domain, these proteins may contain various additional domains determining the specificity of performed transcriptional regulation. According to this, the family has been divided into distinct classes. Our aim was to emphasize the significance of existing disordered regions within the bHLH transcription factors for their functionality. Flexible, intrinsically disordered regions containing various motives and specific sequences allow for multiple interactions with transcription co-regulators. Also, based on in silico analysis and previous studies, we hypothesize that the bHLH proteins have a general ability to undergo spontaneous phase separation, forming or participating into liquid condensates which constitute functional centers involved in transcription regulation. We shortly introduce recent findings on the crucial role of the thermodynamically liquid-liquid driven phase separation in transcription regulation by disordered regions of regulatory proteins. We believe that further experimental studies should be performed in this field for better understanding of the mechanism of gene expression regulation (among others regarding oncogenes) by important and linked to many diseases the bHLH transcription factors.

Keywords: bHLH; IDP; IDR; LLPS; disorder prediction; LLPS prediction; transcription; phase separation

1. Introduction

The bHLH (basic Helix-Loop-Helix) proteins are the important family of transcription factors (TFs) present in all eukaryotes: from yeasts [1,2] and fungi [3] to plants [4] and metazoans [5–10]. All family members contain the HLH domain responsible for dimerization [11]. This domain is usually preceded by a stretch of basic residues which enable DNA binding [12]. The bHLH TFs recognize tissue-specific enhancers containing E-box sequences which regulate expression of a wide range of genes involved in cell differentiation and development [13].

Currently, a few independent classification systems of the bHLH proteins exists: evolutionary classification based on the phylogenetic studies of the bHLH proteins, which classify the bHLH family members into six A-F classes [7,8,14], and a new one based on the complete amino acid sequence analyses, classifying the bHLH proteins into six clades without assumptions about gene function [15]. Contrary to the previous methods, natural method of classification proposed by Murre [12], which divides the bHLH proteins into seven classes, is based on the presence of additional domains, expression patterns and performed transcriptional function [10]. For purposes of clarity, some attempts to revise and systematize different classification systems were undertaken [16]. In this review we present classification of bHLH proteins according to Murre [12], with some short description of presented classes (Table 1).

Table 1. Classification of bHLH proteins based on [5,7,8,10,12,14,16].

Structural Motif Dimerization	Representative Members	Short Description
class I (E proteins)/ group A		
bHLH, homo- and heterodimerization	Vertebrate: E12, E47 [17], HEB [18,19], TCF4 [20] Invertebrate: Daughterless	transcription activators, ubiquitous expression, neurogenesis, immune cell development, sex development, gonadogenesis
class II/ group A		
bHLH, preferred heterodimerization with class I partners	Vertebrate: MYOD, Myogenin, MYF5-6, Ngn1-3, ATOH, NeuroD, NDRF, MATH, MASH, ASCL1 [21], TAL1/SCL [22], OLIG1-3 [23] Invertebrate: TWIST [24], AS-C	transcription activators, tissue specific expression, muscle development, neuro-genesis, generation of autonomic and olfactory neurons, development of granule neurons and external germinal layer of cerebellum, oligodendrocyte development, specification of blood lineage and maturation of several hematopoietic cells, pancreatic development
class III/ group B		
bHLH-LZ	Vertebrate: MYC [25], USF, TFE3, SREBP1-2 Drosophila: MYC Plants: MYC2	transcription activators/represors, oncogenic transformation, apoptosis, cellular differentiation, proliferation, cholesterol-mediated induction of the low-density lipoprotein receptor, jasmonate signaling (plants)
class IV/ group B		
bHLH, heterodimerisation with each other and MYC proteins	Vertabrate: MAD, MAX [26], MXI1 Drosophila: MNT, MAX	transcription regulators lacking transactivation domain (TAD)
class V/ group D		
HLH (no basic region)	Vertebrate: ID1-4 [27] Invertebrate:EMC	negative transcription regulators of class I and II (group A) proteins, no DNA binding, regulation by sequestration.
class VI/ group B		
bHLH-O, (presence of proline in basic region)	Vertebrate: HES, HEY1-3 [28], STRA13, HERP1-2 [29] Drosophila: HAIRY [30], E(spI)	negative transcription regulators interacting with corepressors (Groucho); neurogenesis, vasculogenesis, mesoderm segmentation, myogenesis, T lymphocyte development, cardiovascular development and homeostasis; effectors of Notch signalling [28]; in Drosophila: regulation of differentiation, anteroposterior segmentation and sex determination
class VII/ group C - subclass I		
bHLH-PAS, heterodimerization with subclass II	Vertebrate: AHR [31], HIF1-3α [32], SIM1-2 [33], CLOCK [34], NPAS1-4 [35–39] Drosophila: MET [40], GCE, SIMA, TRH	transcription regulation in response to physiological and environmental signals: xenobiotics, hypoxia, development, circadian rhytms
class VII/ group C - subclass II		
bHLH-PAS, homo- and heterodimerization with subclass I	Vertebrate: ARNT [41], ARNT2, BMAL1, BMAL2 Drosophila: TANGO, CYCLE	general partners for subclass I bHLH-PAS proteins

Both class I (known as E proteins) and class II of the bHLH TFs do not possess domains additional to the bHLH. Contrary to the class I which is expressed in many tissues, the class II proteins expression is tissue specific. Members of the class II are dimerization partners for the class I transcription factors. Class III comprises proteins possessing Leucine-zipper (LZ) motif in addition to the bHLH. Important members of the class III are proteins belonging to the Myc subfamily, which regulate oncogenic transformation, apoptosis, and cellular differentiation. To class IV belong MAD and MAX which can dimerize with MYC and regulate its activity. Also, MAD/MAX are able to create homo- and heterodimers with each other. Although these TFs do not possess transcription activation domain (TAD), MAD/MAX dimers can influence the transcription in a differentiated way. Class V contains transcriptional inhibitors ID1-3 which are not able to bind DNA and act by the other bHLH proteins sequestration. Interestingly, the fourth member of this class- ID4 function as inhibitor of ID1-3 [42]. Class VI comprise proteins containing additional Orange domain adjacent C-terminally to the bHLH domain (bHLH-O). Transcription factors from the described classes perform regulatory function in various developmental processes including cells differentiation and maintaining pluripotency. For this reason they are often linked to cancer development. Class VII comprise transcription factors which possess PAS (Period-Aryl hydrocarbon receptor nuclear translocator-Single minded) domain located C-terminally to the bHLH domain. PAS domain is crucial for the bHLH-PAS proteins specifity [43]. Structurally, the C-terminal PAS domain is often associated with PAC (C-terminal to PAS) motif [44,45]. bHLH-PAS transcription factors are responsible for sensing environmental signals like the presence of xenobiotics (AHR), hypoxia (HIF) or setting of circadian rhythms of organism (CLOCK, CYCLE, BMAL). The members of subclass II of bHLH-PAS TFs -ARNT proteins are general dimerization partners of the subclass I members.

2. The Role of the bHLH Proteins in Transcription

The regulation of genes expression by multiple transcription factors, cofactors and chromatin regulators establish and maintains a specific state of a cell. Inaccurate regulation of transmitted signals can results in diseases and severe disorders [46]. Therefore, transcription requires balanced orchestration of adjustable complexes of proteins. A key regulator of transcription is Mediator, a multi-subunit Mediator complex which interacts with RNA polymerase II (Pol II), and coordinates the action of numerous co-activators and co-repressors [47–50]. Function of the Mediator is conserved in all eukaryotes, though, the individual subunits have diverged considerably in some organisms [51,52].

Up to date, for some bHLH family representatives, interactions with subunits of the Mediator and/or chromatin remodeling histone acetyltransferases/deacyltransferase, were reported. In plants, the Mediator complex is a core element of transcription regulation important for their immunity [53]. It was shown, that in *Arabidopsis thaliana* important jasmonate signaling and resistance to fungus *Botrytis cinerea*, is dependent on the interaction between MED25 subunit of the Mediator and MYC2 [54–56], and interaction of MED8 subunit of the Mediator with FAMA belonging to the bHLH family [57]. Sterol regulatory element binding proteins (SREBPs) the class II bHLH TFs (Table 1) are transcription activators critical for regulation of cholesterol and fatty acid homeostasis in animals. It was shown that human SREBPs bind CBP/p300 acetyltransferase [58] and MED15 subunit of the Mediator to activate target genes [59]. Also yeast Ino2 was shown to bind MED15 subunit of the Mediator tail [60].

The representative of class II TFs TAL1 (Table 1) is required for the specification of the blood lineage and maturation of several hematopoietic cells. TAL1/SCL is considered as a master TF delineating the cell fate and the identity of progenitor and normal hematopoietic stem cells (HSCs). It regulates other hematopoietic TFs thus has a potential for cell reprogramming [22]. TAL1 also binds CBP/p300 acetyltransferase [61,62]. Similarly MyoD—a myogenic regulatory factor which controls skeletal muscle development binds CBP and recruits histone acetyltransferase to activate myogenic program [63]. Cao et al. showed that of MyoD modify the myoblasts chromatin structure and accessibility [64]. ASCL1 (class II, Table 1) was shown to be a pioneer factor which promotes chromatin accessibility and enables chromatin binding by others TFs [65]. Recently, also AHR (bHLH-PAS, Table 1) was suggested to be a pioneer factor which regulates DNA methylation during embryonic developments in unknown way [66]. In clear cell renal cell carcinoma (ccRCC), the most frequent mutation causes the von Hippel-Lindau (VHL) tumor suppressor inactivation leading to genome-wide enhancer and super-enhancer remodeling. This process is mediated by the interaction of HIF2α and HIF1β (bHLH-PAS, Table 1) with histone acetyltransferase p300 [67]. CLOCK, the other bHLH-PAS subfamily member (Table 1) was shown to mediate histone acetylation in a circadian time-specific manner [68].

Interestingly, the bHLH-O proteins members (class VI, Table 1) HEY proteins can function as transcription repressors as well as transcription activators. They were shown to bind directly DNA and interact with histone deacetylases and other TFs [28,69]. On the other hand, gene activation by HEY is regulated in an indirect way. Multiple HEY binding sites located downstream and close to the transcriptional start site, resulted in a hypothesis that HEY influence the pausing/elongation switch of Pol II [70]. Interestingly, though most of TFs stimulate transcription initiation, MYC (class III, Table 1) was shown to stimulate transcription elongation by recruitment of the elongation factor [71]. The presented studies indicate that the crucial role of the bHLH proteins in maintaining transcriptional regulation of important developmental (e.g., cell differentiation) and oncogenic pathways is dependent on the multiple interactions with basal transcriptional machinery.

3. The bHLH Transcription Factors as IDPs

Intrinsically disordered proteins (IDPs) discovered in 1990s obliterate the paradigm derived from Anfinsen's work, stating that functional proteins must possess a well-defined, ordered, three dimensional structure [72]. Currently it is known, that a large number of proteins is perfectly functional or even multifunctional in a disordered state in which a polypeptide chain undergoes rapid conformational fluctuations [73–76]. Intrinsic disorder can be spread throughout the whole polypeptide chain, or it can be limited to intrinsically disordered regions (IDRs) of various length, which are accompanied by well folded domains [77]. The unique properties of disordered proteins originate from their unusual amino acids composition [78]. IDPs/IDRs are depleted in order promoting amino acid residues (hydrophobic, aromatic, aliphatic side chains). In contrast, they possess unusually high content of charged and hydrophilic amino acid residues [79–81]. As a consequence, disordered polypeptide chains have extremely high net charge and low hydrophobicity [82]. IDPs are pliable and highly dynamic molecules of interconvertible conformations. They may completely or almost completely lack the regular secondary structures. However, the content of secondary structure may also be quite significant and molecules can exist in a molten globule state [83–85]. Various in silico analyses indicated that the proportion of disordered proteins is drastically higher in eukaryotes comparing to prokaryotes [86]. This disproportion reflect the complexity of signaling pathways in which IDPs/IDRs play a crucial role [87]. Due to the flexible and dynamic nature, IDPs/IDRs can form fuzzy complexes, adopting various conformations [88]. According to this, one IDP can form multiple interactions with various partners. Due to a large accessibility of particular residues in a disordered chain, the interaction pattern can be easily modified by posttranslational modifications [89]. For that reason IDPs/IDRs often serve as molecular hubs, modulators and sensors of cellular signals [85].

bHLH TFs are responsible for a control of developmental processes like retinal development, proliferation of progenitors, neurogenesis and gliogenesis. Importantly, this is due to a direct interaction between bHLH TFs and interaction of bHLH TFs with homeodomain factors which create complexes that bind to the specific promoters [90,91]. Transcription of muscle-specific genes during skeletal muscle development is also dependent on the interactions between specific bHLH TFs: MyoD, Myogenin, Myf5 and MRF4 with ubiquitously expressed bHLH E-proteins (E12, E47, TCF4, HEB). Interestingly, it was shown that MyoD interacts with two isoforms of HEB: HEBα and HEBβ. which regulate differentially transcriptional activity of MyoD not only on different, but also on the same promoter [92]. Also interesting is the ability of ID4 to recruit multiple ID proteins to assemble higher order complexes. ID4 restores DNA binding by E47 protein even in the presence of repressing ID1 and ID2. Additionally, the ID proteins can interact with non-bHLH partners expanding regulatory network of ID4 [42]. As a consequence, the ID proteins are proposed as a 'hub' for coordination of multiple cancer events [27]. These examples illustrate the possibility of bHLH TFs to interact with many partners in differentiated way. We suggest that these is related to the disordered character of the bHLH proteins. This hypothesis is substantiated by some experimental studies. Neurogenic bHLH transcriprion factor Neurogenin 2 (Ngn2) was shown to possess long IDR which phosphorylation regulates the activity of the protein [93]. Interestingly, though the bHLH domain was considered as a stable, well ordered structure, partially disordered character of this domain was presented for NeuroD [94], MYC and MAX [95]. We performed in silico analyses to predict the presence of intrinsic disorder and get an insight into the degree of flexibility of bHLH proteins representing all established classes (see Table 1): hHEB (class I), hMYOD (class II), hMYC and atMYC2 (class III) (Figure 1); hMAD1 and hMAX (class IV), hID4 (class V), hHES (class VI) (Figure 2); hAHR, hHIF-1α, hCLOCK and hARNT (class VII) (Figure 3). We used PONDR-VLXT [96,97], http://www.pondr.com/ for the disorder prediction and DynaMine [98,99], http://dynamine.ibsquare.be/submission/ for prediction of the flexibility of proteins backbone.

A representative of the class I, human HEB shows a high content of predicted as disordered and flexible sequences. The only highly ordered/rigid region appears between 577–630 aa which comprise the bHLH domain (Figure 1A). Based on prediction results, we assume HEB as IDP. Also hMyoD, the class II TFs presents a high content of flexible IDRs especially in the C-terminal part of the protein (Figure 1B). As the representatives of the class III we have chosen hMYC (Figure 1C) (for which partial disorder of the bHLH domain was experimentally documented [95]) and *Arabidopsis thaliana* MYC2 (Figure 1D). For both proteins the presence of flexible IDRs was predicted, though they locations were different.

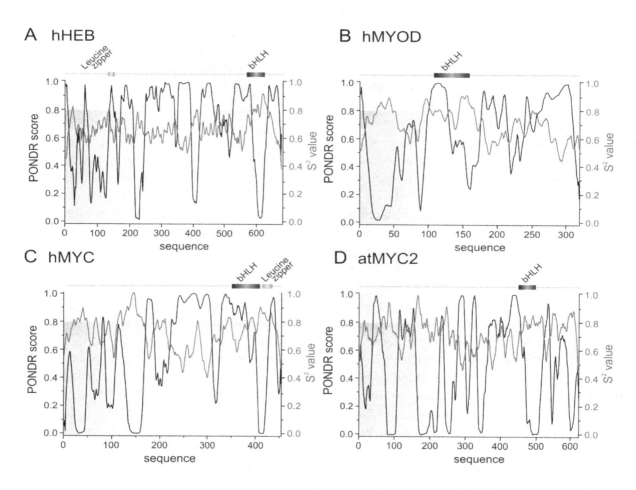

Figure 1. Prediction of intrinsically disordered regions. The top panel presents the domain structure of the analyzed bHLH proteins. Dark grey rectangle indicates the position of bHLH domain, the light grey Leucine zipper. The bottom panel presents a prediction of intrinsically disordered and flexible regions based on the amino acid sequence of proteins. Prediction were performed using PONDR-VLXT (left Y axis) and DynaMine (right Y axis) software. For PONDR prediction, a score above 0.5 indicates disorder. For DynaMine, a S^2 value above 0.8 (blue zone) indicates rigid conformation, 0.69-0.8 (grey zone) is context dependent and a value below 0.69 (green zone) indicates flexible conformation. (**A**) class I human HEB [Q99081], (**B**) class II human MYOD [P15172], (**C**) class III human MYC [P01106-2] and (**D**) *Arabidopsis thaliana* MYC2 [Q39204].

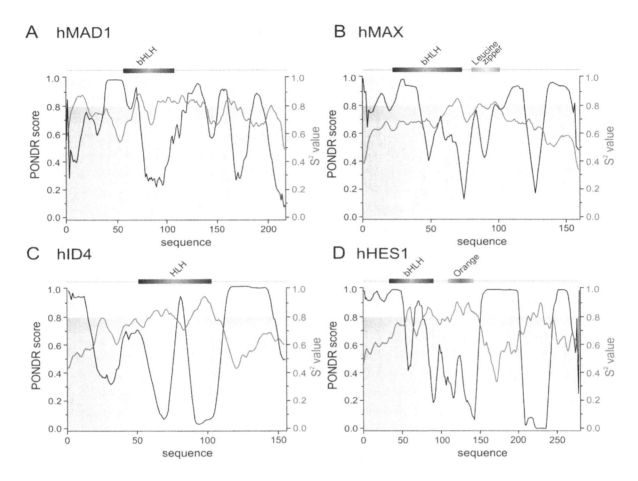

Figure 2. Prediction of intrinsically disordered regions. The top panel presents the domain structure of the analyzed bHLH proteins. Dark grey rectangle indicates the bHLH domain, light grey indicates Leucine zipper or Orange domain. The bottom panel presents a prediction of intrinsically disordered and flexible regions, based on the amino acid sequence of proteins. Predictions were performed using PONDR-VLXT (left Y axis) and DynaMine (right Y axis) software. For PONDR prediction, a score above 0.5 indicates disorder. For Dynamine, a S^2 value above 0.8 (blue zone) indicates rigid conformation, 0.69–0.8 (grey zone) is context dependent and a value below 0.69 (green zone) indicates flexible conformation. (**A**) class IV human MAD [Q9Y6D9] and (**B**) human MAX [P61244], (**C**) class V human ID4 [P47928], (**D**) class VI human HES1 [Q14469].

The representative of the class IV, human MAD1 also shows high content of predicted as disordered and flexible sequences (Figure 2A). Interestingly IDRs of hMAX which belongs to the same class IV are located in the N- and C- protein termini, while the middle part is predicted as possessing more rigid structure (Figure 2B). Also, ID4 belonging to the class V of transcriptional inhibitors presents flexible IDR in the C-terminal part of protein and a shorter one in the N-terminal part (Figure 2C). In addition to similarly located the N- and C-terminal IDRs in the class VI member, human HES1 analysis shows high flexibility/disorder in the central part of protein (Figure 2D).

The class VII proteins comprise the bHLH-PAS subfamily, which additionally to the bHLH domain possess a PAS domain responsible for ligands and co-factors binding. Importantly, their C-termini are usually responsible for the regulation of the protein and created complexes activity [100]. Human AHR, HIF1-α, and CLOCK belong to the subclass I of specialized factors, while human ARNT (the subclass II) is one of the general partners which dimerize with the subclass I proteins and is important for their activity. In contrast to the hAHR, for which relatively short IDRs were predicted within the middle, the N- and the C-terminal part of the protein (Figure 3A), other bHLH-PAS members contain longer IDRs which comprise most of the C-terminal half of proteins and are predicted as highly flexible (hHIF-1α, Figure 3B; hCLOCK, Figure 3C; hARNT, Figure 3D).

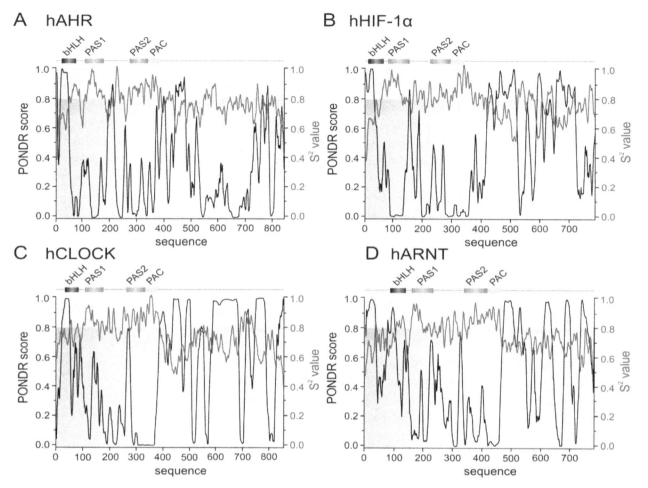

Figure 3. Prediction of intrinsically disordered regions of the class VII bHLH-PAS proteins. The top panel presents the domain structure of the analyzed bHLH–PAS proteins. Dark grey rectangle indicates the bHLH domain, light grey indicates PAS/PAC domains. The bottom panel presents a prediction of intrinsically disordered and flexible regions based on the amino acid sequence of proteins. Prediction were performed using PONDR-VLXT (left Y axis) and DynaMine (right Y axis) software. For PONDR prediction, score above 0.5 indicate disorder. For Dynamine, a S^2 value above 0.8 (blue zone) indicates rigid conformation, 0.69–0.8 (grey zone) is context dependent and a value below 0.69 (green zone) indicates flexible conformation. (**A**) human AHR [P35869], (**B**) human HIF-1α [Q16665], (**C**) human CLOCK [O08785], (**D**) human ARNT [P27540].

To date, the only report, concerning the structure of the full-length bHLH protein is the mentioned study showing Neurogenin as IDP [93]. Based on the presented predictions and our own experience with expression of the selected bHLH proteins (not published), we assume that this is due to the relatively high content of IDRs. This makes overexpression and purification process extremely difficult because of propensity to aggregation and high sensitivity to proteases.

4. The Role of IDPs in Maintaining/Creation of LLPS

Over the last decade, since the pioneering work regarding physical nature of P-bodies was published by Hyman and co-workers [101], many molecular biologists and biophysicists have focused on the significance of spontaneous thermodynamically driven liquid-liquid phase separation (LLPS) in biological systems. LLPS leads to formation of dense, liquid condensates that stably coexist in diluted phase [101,102]. At the molecular level it was shown that LLPS is forced by multiple weak and transient interactions which engage IDPs/IDRs [101,103–106]. Repetitively distributed within IDRs highly charged regions of opposite charges, short motifs such as YG/S-, FG-, RG-, GY-, KSPEA-, SY- and Q/N-rich regions form multivalent interactions between condensate components [107]. A model for the condensate formation and composition proposes that some proteins act as the scaffolds, while others as the clients. The scaffolds are the modular proteins which contain repeated motives that enable heterotypical scaffold-scaffold interaction. As they undergo spontaneous LLPS they are essential for the structural integrity of a condensate [108,109]. Directly interacting sequences called stickers are usually multivalent, whereas the interval sequences which separate stickers, called spacers are responsible for the properties of a condensate [110]. Highly charged and flexible IDRs are in fact frequently identified as scaffolds [108,111]. The clients participate into the condensates by binding to the free, unoccupied scaffold sites [108]. A growing number of evidences indicate that LLPS constitute a fundamental mechanism to compartmentalize the intracellular space. LLPS form the functional centres for biochemical reactions in cytoplasm and membrane-surrounded organelles including nucleus.

The structural and functional organisation of the interior of the nucleus was believed to rely solely on the rigid insoluble nuclear matrix [112]. The rich in A and T DNA sequences known as scaffold/matrix associated regions (S/MARs) attach to nuclear matrix and organise chromatin into higher-order structures which comprise distinct loops and functional units attached to the matrix [113]. That concept is now giving way to a new concept, were dynamic, spontaneously formed condensates, such as nucleolus, splicing speckles, Cajal bodies, PML bodies are the key structural and functional components of the nuclear interior. The barrier-free character of liquid condensates allows for rapid exchange of their components with surrounding so they form an ideal environment for biochemical reactions. On the other hand, nuclear condensates have a stable inert, well-defined structure and can be purified by biochemical methods [114]. It was shown, that the concentration of nucleolar components is close to saturation [115]. It means that small changes in the nucleus can drive spontaneous LLPS. In fact association/dissociation events of nuclear condensates regulate many processes related to gene expression [116] including chromatin structure organisation [117], RNA processing [118], ribosome biogenesis [119]. Importantly, LLPS was shown to be involved in formation of some functional condensates that regulate genes transcription [76,120–122].

5. The Transcription Regulation and LLPS

The genes transcription process require tight regulation to ensure physiological balance of the cell. Knowledge regarding the mechanism of transcription is quite advanced, however some aspects of regulation remains unexplored. Recent findings indicate that regulatory mechanism may tightly depends on the spontaneous LLPS. Transcription of tissue specific gene is initiated at the specific genome regions called super-enhancers (SE). SE first described in embryonic stem cells (ESC) [123] are dense multicomponent assemblies different from typical enhancers [124]. Recently Hnisz [125] performed computational simulation to obtain the probable explanation for typical features of SE. Simulations led to conclusion that formation, activity and unique properties of SE such as sensitivity to concentration of its components, sensitivity to posttranslational modifications, extremely high frequency bursting [126–128] may originate from the fact that SE are liquid condensates assembled/disassembled via spontaneous LLPS [125]. Hnisz and co-workers were the first who point connection and strong dependence between the regulation of transcription initiation at SE and LLPS. Although not experimentally proven, the model serves as the conceptual framework for further research. Recently, Sabari et al. [121] showed that largely disordered BRD4 and MED1 subunit of the Mediator are in close spatial proximity to one another within SE in murine ESC and co-localised puncta show characteristic features of phase separated condensates Moreover, MED1 condensates can incorporate BRD4 and Pol II from nuclear extract [121]. MED1 subunit interacts also with other major pluripotency TFs e.g., OCT-4 [129] and estrogen receptor (ER) [130] forming liquid-like puncta at SE of the key pluripotency genes [121,122]. MED1 condensates depends on the OCT-4 occupancy [122], which are crucial for initiation of tissue specific genes transcription at SE [122,131]. In vitro analyses pointed that formation of MED1-OCT4 liquid condensates occurs via the electrostatic interactions and involves acidic residues enriched in disordered activation domain of the OCT-4 [122]. Interestingly, ER interact with the MED1 subunit by LXXLL motif [132] which is located in the ordered ligand binding domain.

This interaction is regulated by estrogen what means that not only disordered-disordered regions interaction but also disordered-ordered regions interactions play a role in transcription regulation forced by LLPS [122]. Wu et al. [120] showed that largely disordered transcription co-activator TAZ protein forms liquid condensates in vitro and *in vivo*. TAZ condensates compartmentalize DNA binding cofactor TEAD4 and other components of transcription initiation machinery including BRD4, MED1 and CDK9. Importantly, deletion mutant, that is not able to undergo spontaneous LLPS cannot initiate transcription though is able to bind TAZ partners such TEAD4.

Importantly, there are some evidences that not only the initiation, but also the elongation of transcription depends on LLPS. For the transcription elongation essential is hyper-phosphorylation of the YSPTSPS consensus sequence which is repeated multiple times in the disordered C-terminal domain (CTD) of Pol II [133–136]. pTEFb which begins the elongation phase consists of CDK9 kinase associated with cyclin T1 (CycT1). Lu with co-workers [76] concentrated on the function of the lengthy C-terminal IDR of CycT1 in regulation of CDK9 activity. They revealed that a histidine-rich domain (HRD) located in the IDR of CycT1 (residues 480–550) is directly involved in the regulation of the kinase activity [76]. Interestingly, HRD is present also in some other kinases, for example Dyrk1A which phosphorylates CTD of Pol II. Importantly, a homologues kinase Dyrk3 was shown to be responsible for disassembly of stress granules [137] and other cellular condensates during cell division [138]. In vitro studies using a set of recombinant IDRs of the CycT1 and Dyrk1A revealed that the regions can undergo phase separation in a HRD dependent manner. HRD was shown to form condensates which compartmentalize the kinases and the substrate what enables efficient reactions resulting in the hyper-phosphorylation of the CTD of Pol II [76]. Interestingly, the CTD of Pol II can undergo spontaneous LLPS in vitro only in a non-phosphorylated state. The weak CTD-CTD interaction keeps the enzymes molecules in hubs within nucleoplasm. Phosphorylation change the interaction pattern allowing CTD to engage in new multivalent interactions with selected partners [139]. These results indicate that LLPS allows for the condensation of cofactors, that in turn triggers posttranslational modifications leading to the reorganization of the condensate components. Pol II escapes from the promoter site and enables the entry into active elongation stage [76].

Currently not much is known about proteins responsible for formation of the condensates which are important for transcription regulation. The question still remains unanswered which proteins are the scaffolds and which are the clients. Importantly, also not much is known about the involvement of the bHLH TFs in the LLPS process, though they are key players involved in many important cell differentiation and organisms development pathways. As we discussed in previous section, bHLH proteins possess long IDRs which could interact with different partners and be engaged in LLPS. This hypothesis is substantiated by an experimental verification of MyoD possibility to create LLPS [122], and discussed in previous section possibility of some bHLH TFs to interact with the Mediator subunits or other elements of the mechanism which modifies the chromatin accessibility. Interestingly, regulation of circadian clock by BMAL1 comprises binding of CBP, which occurs in discrete nuclear foci. This led to a hypothesis that formation of nuclear bodies containing BMAL1/CBP provides transcriptionally active sites of target genes, like *Per1-2* [34]. Taking the above into consideration, we asked the question if the ability to undergo LLPS is a more general property of the bHLH TFs. As we got positive results for the previously performed prediction of disorder, which was shown to be important for LLPS initiation [76,121,122], we decided to perform in silico analyses to predict if members of the bHLH family comprise putative sequences able to create liquid condensates. We used catGranule program, (http://service.tartaglialab.com/update_submission/216885/dd56e32a89) for computational analyses of the putative propensity to undergo LLPS [140] for the bHLH proteins representing all established classes (see Table 1). Prediction results showed that hHEB (class I), hMyoD (class II), hMYC and 84atMYC2 (class III) (Figure 4) contain sequences with a positive score of propensity to LLPS formation. Interestingly, proteins from the class IV regulators which do not possess TAD: hMAD1 and hMAX, similarly like transcription repressors: hID4 (class V) and hHES (class VI) present very low or even negative score within the whole protein sequence (Figure 5). bHLH-PAS transcription factors representing the class VII, hAHR, hHIF-1α, hCLOCK and hARNT were predicted as containing some sequences with high propensity score (Figure 6). Especially interesting is the observation that the transcription repressors show a very low propensity scoreto undergo LLPS in contrast to the transcription activators such as hHEB or atMYC2. It is possible that the bHLH repressors inhibit transcription by preventing spontaneous phase separation required to form a complete initiation complex. This hypothesis is substantiated by the observation for TAZ mutants [120], discussed in the previous section.

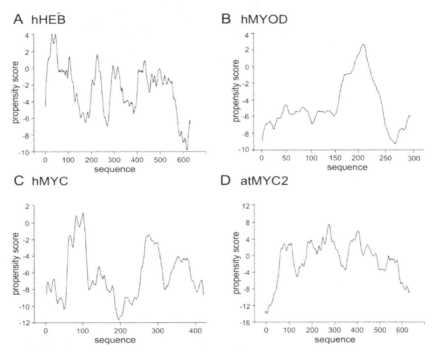

Figure 4. Prediction of propensity of LLPS formation. (**A**) class I human HEB [Q99081], (**B**) class II human MYOD [P15172], (**C**) class III human MYC [P01106-2] and (**D**) *Arabidopsis thaliana* MYC2 [Q39204].

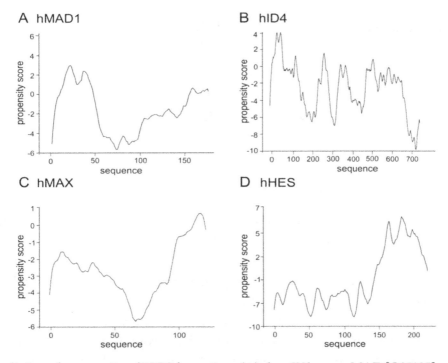

Figure 5. Prediction of propensity of LLPS formation. (**A**) class IV human MAD [Q05195] and (**B**) human MAX [P61244], (**C**) class V human ID4 [P47928], (**D**) class VI human HES1 [Q14469].

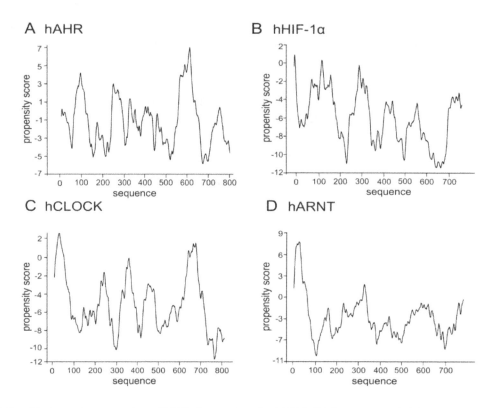

Figure 6. Prediction of propensity of LLPS formation for bHLh-PAS proteins. (**A**) human AHR [P35869], (**B**) human HIF-1α [Q16665], (**C**) human CLOCK [O08785], (**D**) human ARNT [P27540].

As the range of the propensity score is not determined precisely, as a control we performed catGranule prediction for proteins known to create LLPS: nucleophosmin (Figure 7A) and estrogen receptor (Figure 7B) which are deposited in the recently published PhaSePro database (https://phasepro. elte.hu) [141].

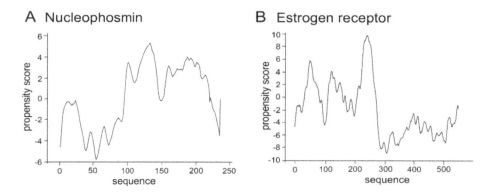

Figure 7. Prediction of propensity of LLPS formation for representative LLPS-enabled proteins. (**A**) nucleophosmin [P06748], (**B**) estrogen receptor [P03372].

Results of performed in silico analyses in comparison to the control show that the selected bHLH proteins have regions that might be involved in multivalent interaction leading to formation of liquid condensates. What would be their role in condensates formation and how would mutations and wrong dimerization/interaction influence formation of the bHLH TFs containing condensate remains a puzzle, however we believe that such an important family of TFs engaged in the crucial pathways and related to many severe disorders like cancer should be the subject of research in this field.

6. Concluding Remarks and Future Perspectives

In eukaryotic cells, regulation of transcription is a dynamic process which requires very precise temporal and spatial coordination of proteins assembling functional complexes. The bHLH family comprises a large group of TFs which utilize conserved DNA binding domain to interact with DNA, but also additional, often disordered domains and motives that allows formation of complex interacting network with various transcription co-factors. It is possible that flexible disordered regions of the bHLH proteins play a role in formation of liquid condensates via LLPS and contribute in this way to regulation of transcription process. Up to date however, there is a lack of experimental evidences. Also recently published PhaSePro database for LLPS does not contain any bHLH TF [141]. We believe that this is due to difficulties with the experimental studies of the bHLH proteins mentioned previously and we expect that some bHLH proteins will be appended in future.

Presented in the previous section predictions may give a hint about the link between LLPS by the bHLH proteins and transcription regulation. This raise a question about functional relevance of this discrepancy between family members. An interesting observation is the predicted low propensity score to form LLPS in the case of transcriptional repressors in contrast to proteins acting as activators. This raise a question about the functional relevance of this discrepancy between family members. Importantly, connection between LLPS and transcription regulation is not limited to the direct interaction between transcription regulators at the active transcription sites. LLPS form nuclear bodies, that maintain, store and modify transcription regulators. Examples include nuclear speckles, polyleukemia bodies, nucleolus, histone locus and others [142]. Within LLPS-formed condensates proteins can undergo acetylation/deacetylation or sumoylation, proteasome-dependent degradation and other posttranslational modifications that influence their functionality [143–145]. Importantly, barrier-free character of these phase separated condensates allows shuttling of its component between the condensates and nucleoplasm, and whenever needed molecules can be recruited from these compartments to the active transcriptionally sites. The discovery that LLPS which is well known in polymer chemistry can play an important role in molecular biology has definitely brought us closer to understanding the cell functionality and regulation of fundamental cellular processes such as transcription. However, our understanding and detailed knowledge is still residual. Many important questions regarding a LLPS concept in transcription regulation remain without answer. We do not know, which components drive association/dissociation events at the active sites. Which molecules serves as a scaffold conditioning formation of liquid condensates and which are just clients. How the type of client molecules influence the function of the phase separated condensates? Also, we do not know which factors and in which way alter LLPS leading to the pathological processes. What would be the role of the bHLH TFs in a condensates formation, and how mutations and incorrect dimerization/interaction of these proteins would impact formation and function of condensates? These questions, as well as many other ones await experimental verification. We believe that such important family of transcription factors which is engaged in crucial pathways and related to many severe diseases like cancer and neurodegenerative disorders, should be the subject of further intensive studies.

Author Contributions: A.T. and B.G.-M. wrote the paper.

Acknowledgments: The authors apologize to investigators whose contributions were not cited more extensively because of space limitations.

Abbreviations

AHR	Aryl hydrocarbon receptor
AS-C	Achaete scute complex
ARNT	Aryl hydrocarbon receptor nuclear translocator
bHLH	Helix–loop–helix
ccRCC	Clear cell renal cell carcinoma
CLOCK	Circadian locomotor output cycles protein kaput
CTD	C-terminal domain
CycT1	Cyclin T1
EMC	Extramacrochaetae
E(spI)	Enhancer of split
ER	Estrogen receptor
ESC	Embryonic stem cells
GCE	Germ cell-expressed protein
GRO	Groucho
HAT	Histone transacetylase
HIF	Hypoxia-inducible factor
HSCa	Hematopoietic stem cells
HRD	histidine reach domain
ID	Inhibitor of DNA binding
IDPs	Intrinsically disordered proteins
IDRs	Intrinsically disordered regions
LLPS	liquid-liquid phase separation
LZ	Leucine zipper motif
MET	Methoprene-tolerant protein
MXI1	Max interacting protein
Ngn2	Neurogenin
NPAS	Neuronal PAS domain-containing protein
PAS	Period-arylhydrocarbon nuclear translocator-single minded domain
Pol II	RNA polymerase II
SE	Super-enhancer
SIM	Single-minded protein
SIMA	Similar protein
S/MARs	Scaffold/matrix associate regions
SREBP	Sterol-responsive element-binding protein
TAD	Transactivation domain
TAZ	Tafazzin
TFs	Transcription factors
TRH	Trachealess protein
USF	Upstream stimulatory factor
VHL	Von Hippel-Lindau tumor suppressor

References

1. Robinson, K.A. A network of yeast basic helix-loop-helix interactions. *Nucleic Acids Res.* **2000**, *28*, 4460–4466. [CrossRef] [PubMed]
2. Robinson, K.A. SURVEY AND SUMMARY: Saccharomyces cerevisiae basic helix-loop-helix proteins regulate diverse biological processes. *Nucleic Acids Res.* **2000**, *28*, 1499–1505. [PubMed]
3. Sailsbery, J.K.; Atchley, W.R.; Dean, R.A. Phylogenetic analysis and classification of the fungal bHLH domain. *Mol. Biol. Evol.* **2012**, *29*, 1301–1318. [CrossRef] [PubMed]
4. Pires, N.; Dolan, L. Origin and diversification of basic-helix-loop-helix proteins in plants. *Mol. Biol. Evol.* **2010**, *27*, 862–874.
5. Massari, M.E.; Murre, C. Helix-loop-helix proteins: Regulators of transcription in eucaryotic organisms. *Mol. Cell. Biol.* **2000**, *20*, 429–440. [CrossRef]

6. Simionato, E.; Ledent, V.; Richards, G.; Thomas-Chollier, M.; Kerner, P.; Coornaert, D.; Degnan, B.M.; Vervoort, M. Origin and diversification of the basic helix-loop-helix gene family in metazoans: Insights from comparative genomics. *BMC Evol. Biol.* **2007**, *7*, 33. [CrossRef]

7. Ledent, V.; Paquet, O.; Vervoort, M. Phylogenetic analysis of the human basic helix-loop-helix proteins. *Genome Biol.* **2002**, *3*, 1–18. [CrossRef]

8. Wang, Y.; Chen, K.; Yao, Q.; Zheng, X.; Yang, Z. Phylogenetic Analysis of Zebrafish Basic Helix-Loop-Helix Transcription Factors. *J. Mol. Evol.* **2009**, *68*, 629–640.

9. Sailsbery, J.K.; Dean, R.A. Accurate discrimination of bHLH domains in plants, animals, and fungi using biologically meaningful sites. *BMC Evol. Biol.* **2012**, *12*, 154.

10. Murre, C. Helix–loop–helix proteins and the advent of cellular diversity: 30 years of discovery. *Genes Dev.* **2019**, *33*, 6–25.

11. Murre, C.; McCaw, P.S.; Baltimore, D. A new DNA binding and dimerization motif in immunoglobulin enhancer binding, daughterless, MyoD, and myc proteins. *Cell* **1989**, *56*, 777–783. [CrossRef]

12. Murre, C.; Bain, G.; van Dijk, M.A.; Engel, I.; Furnari, B.A.; Massari, M.E.; Matthews, J.R.; Quong, M.W.; Rivera, R.R.; Stuiver, M.H. Structure and function of helix-loop-helix proteins. *Biochim. Biophys. Acta (BBA) Gene Struct. Expr.* **1994**, *1281*, 129–135. [CrossRef]

13. Ephrussi, A.; Church, G.; Tonegawa, S.; Gilbert, W. B lineage–specific interactions of an immunoglobulin enhancer with cellular factors in vivo. *Science* **1985**, *227*, 134–140. [CrossRef] [PubMed]

14. Atchley, W.R.; Fitch, W.M. A natural classification of the basic helix-loop-helix class of transcription factors. *Proc. Natl. Acad. Sci. USA* **1997**, *94*, 5172–5176. [CrossRef] [PubMed]

15. Skinner, M.K.; Rawls, A.; Wilson-Rawls, J.; Roalson, E.H. Basic helix-loop-helix transcription factor gene family phylogenetics and nomenclature. *Differ. Res. Biol. Divers.* **2010**, *80*, 1–8. [CrossRef]

16. Jones, S. An overview of the basic helix-loop-helix proteins. *Genome Biol.* **2004**, *5*, 226. [CrossRef]

17. Wöhner, M.; Tagoh, H.; Bilic, I.; Jaritz, M.; Poliakova, D.K.; Fischer, M.; Busslinger, M. Molecular functions of the transcription factors E2A and E2-2 in controlling germinal center B cell and plasma cell development. *J. Exp. Med.* **2016**, *213*, 1201–1221. [CrossRef]

18. Yi, S.; Yu, M.; Yang, S.; Miron, R.J.; Zhang, Y. Tcf12, A Member of Basic Helix-Loop-Helix Transcription Factors, Mediates Bone Marrow Mesenchymal Stem Cell Osteogenic Differentiation In Vitro and In Vivo. *Stem Cells* **2017**, *35*, 386–397. [CrossRef]

19. Li, Y.; Brauer, P.M.; Singh, J.; Xhiku, S.; Yoganathan, K.; Zúñiga-Pflücker, J.C.; Anderson, M.K. Targeted Disruption of TCF12 Reveals HEB as Essential in Human Mesodermal Specification and Hematopoiesis. *Stem Cell Rep.* **2017**, *9*, 779–795. [CrossRef]

20. Quednow, B.B.; Brzózka, M.M.; Rossner, M.J. Transcription factor 4 (TCF4) and schizophrenia: Integrating the animal and the human perspective. *Cell. Mol. Life Sci.* **2014**, *71*, 2815–2835. [CrossRef]

21. Huang, C.; Chan, J.A.; Schuurmans, C. Proneural bHLH Genes in Development and Disease. In *Current Topics in Developmental Biology*; Academic Press: Cambridge, MA, USA, 2014; Volume 110, pp. 75–127.

22. Tan, T.K.; Zhang, C.; Sanda, T. Oncogenic transcriptional program driven by TAL1 in T-cell acute lymphoblastic leukemia. *Int. J. Hematol.* **2019**, *109*, 5–17. [CrossRef] [PubMed]

23. Choudhury, S. Genomics of the OLIG family of a bHLH transcription factor associated with oligo dendrogenesis. *Bioinformation* **2019**, *15*, 430–438. [CrossRef] [PubMed]

24. Bouard, C.; Terreux, R.; Honorat, M.; Manship, B.; Ansieau, S.; Vigneron, A.M.; Puisieux, A.; Payen, L. Deciphering the molecular mechanisms underlying the binding of the TWIST1/E12 complex to regulatory E-box sequences. *Nucleic Acids Res.* **2016**, *44*, 5470–5489. [CrossRef] [PubMed]

25. Whitfield, J.R.; Beaulieu, M.-E.; Soucek, L. Strategies to Inhibit Myc and Their Clinical Applicability. *Front. Cell Dev. Biol.* **2017**, *5*, 10. [CrossRef] [PubMed]

26. Amati, B.; Land, H. Myc-Max-Mad: A transcription factor network controlling cell cycle progression, differentiation and death. *Curr. Opin. Genet. Dev.* **1994**, *4*, 102–108. [CrossRef]

27. Lasorella, A.; Benezra, R.; Iavarone, A. The ID proteins: Master regulators of cancer stem cells and tumour aggressiveness. *Nat. Rev. Cancer* **2014**, *14*, 77–91. [CrossRef]

28. Weber, D.; Wiese, C.; Gessler, M. Hey bHLH transcription factors. In *Current Topics in Developmental Biology*; Academic Press: Cambridge, MA, USA, 2014; Volume 110, pp. 285–315.

29. Iso, T.; Kedes, L.; Hamamori, Y. HES and HERP families: Multiple effectors of the notch signaling pathway. *J. Cell. Physiol.* **2003**, *194*, 237–255. [CrossRef]

30. Saha, T.T.; Shin, S.W.; Dou, W.; Roy, S.; Zhao, B.; Hou, Y.; Wang, X.-L.; Zou, Z.; Girke, T.; Raikhel, A.S. Hairy and Groucho mediate the action of juvenile hormone receptor Methoprene-tolerant in gene repression. *Proc. Natl. Acad. Sci. USA* **2016**, *113*, E735–E743. [CrossRef]

31. Wright, E.J.; Pereira De Castro, K.; Joshi, A.D.; Elferink, C.J. Canonical and non-canonical aryl hydrocarbon receptor signaling pathways. *Curr. Opin. Toxicol.* **2017**, *2*, 87–92. [CrossRef]

32. Semenza, G.L. Hypoxia-inducible factors in physiology and medicine. *Cell* **2012**, *148*, 399–408. [CrossRef]

33. Moffett, P.; Pelletier, J. Different transcriptional properties of mSim-1 and mSim-2. *FEBS Lett.* **2000**, *466*, 80–86. [CrossRef]

34. Lee, Y.; Lee, J.; Kwon, I.; Nakajima, Y.; Ohmiya, Y.; Son, G.H.; Lee, K.H.; Kim, K. Coactivation of the CLOCK-BMAL1 complex by CBP mediates resetting of the circadian clock. *J. Cell Sci.* **2010**, *123*, 3547–3557. [CrossRef] [PubMed]

35. Teh, C.H.L.; Lam, K.K.Y.; Loh, C.C.; Loo, J.M.; Yan, T.; Lim, T.M. Neuronal PAS domain protein 1 is a transcriptional repressor and requires arylhydrocarbon nuclear translocator for its nuclear localization. *J. Biol. Chem.* **2006**, *281*, 34617–34629. [CrossRef] [PubMed]

36. Ohsawa, S.; Hamada, S.; Kakinuma, Y.; Yagi, T.; Miura, M. Novel function of neuronal PAS domain protein 1 in erythropoietin expression in neuronal cells. *J. Neurosci. Res.* **2005**, *79*, 451–458. [CrossRef]

37. Gilles-Gonzalez, M.-A.; Gonzales, G. Signal transduction by heme-containing PAS-domain proteins. *J. Appl. Physiol.* **2004**, *96*, 774–783. [CrossRef] [PubMed]

38. Kamnasaran, D. Disruption of the neuronal PAS3 gene in a family affected with schizophrenia. *J. Med Genet.* **2003**, *40*, 325–332. [CrossRef] [PubMed]

39. Ooe, N.; Saito, K.; Mikami, N.; Nakatuka, I.; Kaneko, H. Identification of a Novel Basic Helix-Loop-Helix-PAS Factor, NXF, Reveals a Sim2 Competitive, Positive Regulatory Role in Dendritic-Cytoskeleton Modulator Drebrin Gene Expression. *Mol. Cell. Biol.* **2003**, *24*, 608–616. [CrossRef]

40. Li, M.; Mead, E.A.; Zhu, J. Heterodimer of two bHLH-PAS proteins mediates juvenile hormone-induced gene expression. *Proc. Natl. Acad. Sci. USA* **2011**, *2011*, 638–643. [CrossRef]

41. Brunnberg, S.; Pettersson, K.; Rydin, E.; Matthews, J.; Hanberg, A.; Pongratz, I. The basic helix-loop-helix-PAS protein ARNT functions as a potent coactivator of estrogen receptor-dependent transcription. *Proc. Natl. Acad. Sci. USA* **2003**, *100*, 6517–6522. [CrossRef]

42. Sharma, P.; Chinaranagari, S.; Chaudhary, J. Inhibitor of differentiation 4 (ID4) acts as an inhibitor of ID-1, -2 and -3 and promotes basic helix loop helix (bHLH) E47 DNA binding and transcriptional activity. *Biochimie* **2015**, *112*, 139–150. [CrossRef]

43. Zelzer, E.; Wappner, P.; Shilo, B.-Z. The PAS domain confers target gene specificity of Drosophila bHLH/PAS proteins. *Genes Dev.* **1997**, *11*, 2079–2089. [CrossRef] [PubMed]

44. Cusanovich, M.A.; Meyer, T.E. Photoactive yellow protein: A prototypic PAS domain sensory protein and development of a common signaling mechanism. *Biochemistry* **2003**, *42*, 4759–4770. [CrossRef] [PubMed]

45. Mö Glich, A.; Ayers, R.A.; Moffat, K. Structure and Signaling Mechanism of Per-ARNT-Sim Domains. *Struct. Fold. Des.* **2009**, *17*, 1282–1294. [CrossRef] [PubMed]

46. Roeder, R.G. Transcriptional regulation and the role of diverse coactivators in animal cells. *FEBS Lett.* **2005**, *579*, 909–915. [CrossRef]

47. Ansari, S.A.; Morse, R.H. Mechanisms of Mediator complex action in transcriptional activation. *Cell. Mol. Life Sci.* **2013**, *70*, 2743–2756. [CrossRef]

48. Conaway, R.C.; Conaway, J.W. Function and regulation of the Mediator complex. *Curr. Opin. Genet. Dev.* **2011**, *21*, 225–230. [CrossRef]

49. Poss, Z.C.; Ebmeier, C.C.; Taatjes, D.J. The Mediator complex and transcription regulation. *Crit. Rev. Biochem. Mol. Biol.* **2013**, *48*, 575–608. [CrossRef]

50. Quevedo, M.; Meert, L.; Dekker, M.R.; Dekkers, D.H.W.; Brandsma, J.H.; van den Berg, D.L.C.; Ozgür, Z.; van IJcken, W.F.J.; Demmers, J.; Fornerod, M.; et al. Mediator complex interaction partners organize the transcriptional network that defines neural stem cells. *Nat. Commun.* **2019**, *10*, 2669. [CrossRef]

51. Malik, S.; Roeder, R.G. The metazoan Mediator co-activator complex as an integrative hub for transcriptional regulation. *Nat. Rev. Genet.* **2010**, *11*, 761–772. [CrossRef]

52. Malik, N.; Agarwal, P.; Tyagi, A. Emerging functions of multi-protein complex Mediator with special emphasis on plants. *Crit. Rev. Biochem. Mol. Biol.* **2017**, *52*, 475–502. [CrossRef]

53. An, C.; Mou, Z. The function of the Mediator complex in plant immunity. *Plant Signal. Behav.* **2013**, *8*, e23182. [CrossRef] [PubMed]

54. Chen, R.; Jiang, H.; Li, L.; Zhai, Q.; Qi, L.; Zhou, W.; Liu, X.; Li, H.; Zheng, W.; Sun, J.; et al. The Arabidopsis Mediator Subunit MED25 Differentially Regulates Jasmonate and Abscisic Acid Signaling through Interacting with the MYC2 and ABI5 Transcription Factors. *Plant Cell* **2012**, *24*, 2898–2916. [CrossRef] [PubMed]

55. An, C.; Li, L.; Zhai, Q.; You, Y.; Deng, L.; Wu, F.; Chen, R.; Jiang, H.; Wang, H.; Chen, Q.; et al. Mediator subunit MED25 links the jasmonate receptor to transcriptionally active chromatin. *Proc. Natl. Acad. Sci. USA* **2017**, *114*, E8930–E8939. [CrossRef] [PubMed]

56. Liu, Y.; Du, M.; Deng, L.; Shen, J.; Fang, M.; Chen, Q.; Lu, Y.; Wang, Q.; Li, C.; Zhai, Q. Myc2 regulates the termination of jasmonate signaling via an autoregulatory negative feedback loop[open]. *Plant Cell* **2019**, *31*, 106–127. [CrossRef]

57. Li, X.; Yang, R.; Chen, H. The arabidopsis thaliana mediator subunit MED8 regulates plant immunity to botrytis cinerea through interacting with the basic helix-loop-helix (bHLH) transcription factor FAMA. *PLoS ONE* **2018**, *13*, e0193458. [CrossRef]

58. Oliner, J.D.; Andresen, J.M.; Hansen, S.K.; Zhou, S.; Tjian, R. SREBP transcriptional activity is mediated through an interaction with the CREB-binding protein. *Genes Dev.* **1996**, *10*, 2903–2911. [CrossRef]

59. Yang, F.; Vought, B.W.; Satterlee, J.S.; Walker, A.K.; Jim Sun, Z.Y.; Watts, J.L.; DeBeaumont, R.; Mako Saito, R.; Hyberts, S.G.; Yang, S.; et al. An ARC/Mediator subunit required for SREBP control of cholesterol and lipid homeostasis. *Nature* **2006**, *442*, 700–704. [CrossRef]

60. Pacheco, D.; Warfield, L.; Brajcich, M.; Robbins, H.; Luo, J.; Ranish, J.; Hahn, S. Transcription Activation Domains of the Yeast Factors Met4 and Ino2: Tandem Activation Domains with Properties Similar to the Yeast Gcn4 Activator. *Mol. Cell. Biol.* **2018**, *38*, e00038-18. [CrossRef]

61. Lazrak, M.; Deleuze, V.; Noel, D.; Haouzi, D.; Chalhoub, E.; Dohet, C.; Robbins, I.; Mathieu, D. The bHLH TAL-1/SCL regulates endothelial cell migration and morphogenesis. *J. Cell Sci.* **2004**, *117*, 1161–1171. [CrossRef]

62. Huang, S.; Qiu, Y.; Stein, R.W.; Brandt, S.J. p300 functions as a transcriptional coactivator for the TAL1/SCL oncoprotein. *Oncogene* **1999**, *18*, 4958–4967. [CrossRef]

63. Puri, P.L.; Avantaggiati, M.L. p300 is required for MyoD-dependent cell cycle arrest and muscle-specific gene transcription arrest in the G 0 phase, irreversible exit from the cell cycle. *EMBO J.* **1997**, *16*, 369–383. [CrossRef] [PubMed]

64. Cao, Y.; Yao, Z.; Sarkar, D.; Lawrence, M.; Sanchez, G.J.; Parker, M.H.; MacQuarrie, K.L.; Davison, J.; Morgan, M.T.; Ruzzo, W.L.; et al. Genome-wide MyoD Binding in Skeletal Muscle Cells: A Potential for Broad Cellular Reprogramming. *Dev. Cell* **2010**, *18*, 662–674. [CrossRef] [PubMed]

65. Raposo, A.A.S.F.; Vasconcelos, F.F.; Drechsel, D.; Marie, C.; Johnston, C.; Dolle, D.; Bithell, A.; Gillotin, S.; van den Berg, D.L.C.; Ettwiller, L.; et al. Ascl1 coordinately regulates gene expression and the chromatin landscape during neurogenesis. *Cell Rep.* **2015**, *10*, 1544–1556. [CrossRef] [PubMed]

66. Ko, C.-I.; Puga, A. Does the aryl hydrocarbon receptor regulate pluripotency? *Curr. Opin. Toxicol.* **2017**, *2*, 1–7. [CrossRef] [PubMed]

67. Yao, X.; Tan, J.; Lim, K.J.; Koh, J.; Ooi, W.F.; Li, Z.; Huang, D.; Xing, M.; Chan, Y.S.; Qu, J.Z.; et al. VHL deficiency drives enhancer activation of oncogenes in clear cell renal cell carcinoma. *Cancer Discov.* **2017**, *7*, 1284–1305. [CrossRef] [PubMed]

68. Doi, M.; Hirayama, J.; Sassone-Corsi, P. Circadian regulator CLOCK is a histone acetyltransferase. *Cell* **2006**, *125*, 497–508. [CrossRef] [PubMed]

69. Harada, A.; Ohkawa, Y.; Imbalzano, A.N. Temporal regulation of chromatin during myoblast differentiation. *Semin. Cell Dev. Biol.* **2017**, *72*, 77–86. [CrossRef]

70. Heisig, J.; Weber, D.; Englberger, E.; Winkler, A.; Kneitz, S.; Sung, W.-K.; Wolf, E.; Eilers, M.; Wei, C.-L.; Gessler, M. Target gene analysis by microarrays and chromatin immunoprecipitation identifies HEY proteins as highly redundant bHLH repressors. *PLoS Genet.* **2012**, *8*, e1002728. [CrossRef]

71. Rahl, P.B.; Lin, C.Y.; Seila, A.C.; Flynn, R.A.; McCuine, S.; Burge, C.B.; Sharp, P.A.; Young, R.A. c-Myc Regulates Transcriptional Pause Release. *Cell* **2010**, *141*, 432–445. [CrossRef]

72. Anfinsen, C.B. Principles that govern the folding of protein chains. *Science* **1973**, *181*, 223–230. [CrossRef]

73. Uversky, V.N. A decade and a half of protein intrinsic disorder: Biology still waits for physics. *Protein Sci. A Publ. Protein Soc.* **2013**, *22*, 693–724. [CrossRef] [PubMed]

74. Dyson, H.J.; Wright, P.E. Intrinsically unstructured proteins and their functions. *Nat. Rev. Mol. Cell Biol.* **2005**, *6*, 197–208. [CrossRef] [PubMed]

75. Dunker, A.K.; Lawson, J.D.; Brown, C.J.; Williams, R.M.; Romero, P.; Oh, J.S.; Oldfield, C.J.; Campen, A.M.; Ratliff, C.M.; Hipps, K.W.; et al. Intrinsically disordered protein. *J. Mol. Graph. Model.* **2001**, *19*, 26–59. [CrossRef]

76. Lu, H.; Yu, D.; Hansen, A.S.; Ganguly, S.; Liu, R.; Heckert, A.; Darzacq, X.; Zhou, Q. Phase-separation mechanism for C-terminal hyperphosphorylation of RNA polymerase II. *Nature* **2018**, *558*, 318. [CrossRef] [PubMed]

77. Uversky, V.N.; Dunker, A.K. Understanding protein non-folding. *Biochim. Biophys. Acta (BBA) Proteins Proteom.* **2010**, *1804*, 1231–1264. [CrossRef]

78. Dosztanyi, Z.; Csizmok, V.; Tompa, P.; Simon, I. The pairwise energy content estimated from amino acid composition discriminates between folded and intrinsically unstructured proteins. *J. Mol. Biol.* **2005**, *347*, 827–839. [CrossRef]

79. Campen, A.; Williams, R.; Brown, C.; Meng, J.; Uversky, V.; Dunker, A. TOP-IDP-Scale: A New Amino Acid Scale Measuring Propensity for Intrinsic Disorder. *Protein Pept. Lett.* **2008**, *15*, 956–963. [CrossRef]

80. Williams, R.M.; Obradovi, Z.; Mathura, V.; Braun, W.; Garner, E.C.; Young, J.; Takayama, S.; Brown, C.J.; Dunker, A.K. The protein non-folding problem: Amino acid determinants of intrinsic order and disorder. *Pac. Symp. Biocomput.* **2001**, *2001*, 89–100.

81. Vacic, V.; Uversky, V.N.; Dunker, A.K.; Lonardi, S. Composition Profiler: A tool for discovery and visualization of amino acid composition differences. *BMC Bioinform.* **2007**, *8*, 211. [CrossRef]

82. Uversky, V.N.; Gillespie, J.R.; Fink, A.L. Why are "natively unfolded" proteins unstructured under physiologic conditions? *Proteins Struct. Funct. Genet.* **2000**, *41*, 415–427. [CrossRef]

83. Dunker, A.K.; Obradovic, Z. The protein trinity—Linking function and disorder. *Nat. Biotechnol.* **2001**, *19*, 805. [CrossRef] [PubMed]

84. Tompa, P. The interplay between structure and function in intrinsically unstructured proteins. *FEBS Lett.* **2005**, *579*, 3346–3354. [CrossRef] [PubMed]

85. Van Der Lee, R.; Buljan, M.; Lang, B.; Weatheritt, R.J.; Daughdrill, G.W.; Dunker, A.K.; Fuxreiter, M.; Gough, J.; Gsponer, J.; Jones, D.T.; et al. Classification of intrinsically disordered regions and proteins. *Chem. Rev.* **2014**, *114*, 6589–6631. [CrossRef] [PubMed]

86. Peng, Z.; Yan, J.; Fan, X.; Mizianty, M.J.; Xue, B.; Wang, K.; Hu, G.; Uversky, V.N.; Kurgan, L. Exceptionally abundant exceptions: Comprehensive characterization of intrinsic disorder in all domains of life. *Cell. Mol. Life Sci.* **2014**, *72*, 137–151. [CrossRef]

87. Wright, P.E.; Dyson, H.J. Intrinsically disordered proteins in cellular signalling and regulation. *Nat. Rev. Mol. Cell Biol.* **2015**, *16*, 18–29. [CrossRef]

88. Tompa, P.; Fuxreiter, M. Fuzzy complexes: Polymorphism and structural disorder in protein-protein interactions. *Trends Biochem. Sci.* **2008**, *33*, 2–8. [CrossRef]

89. Iakoucheva, L.M.; Radivojac, P.; Brown, C.J.; O'Connor, T.R.; Sikes, J.G.; Obradovic, Z.; Dunker, A.K. The importance of intrinsic disorder for protein phosphorylation. *Nucleic Acids Res.* **2004**, *32*, 1037–1049. [CrossRef]

90. Hatakeyama, J.; Kageyama, R. Retinal cell fate determination and bHLH factors. *Semin. Cell Dev. Biol.* **2004**, *15*, 83–89. [CrossRef]

91. Sölter, M.; Locker, M.; Boy, S.; Taelman, V.; Bellefroid, E.J.; Perron, M.; Pieler, T. Characterization and function of the bHLH-O protein XHes2: Insight into the mechanism controlling retinal cell fate decision. *Development* **2006**, *133*, 4097–4108. [CrossRef]

92. Parker, M.H.; Perry, R.L.S.; Fauteux, M.C.; Berkes, C.A.; Rudnicki, M.A. MyoD Synergizes with the E-Protein HEB To Induce Myogenic Differentiation. *Mol. Cell. Biol.* **2006**, *26*, 5771–5783. [CrossRef]

93. McDowell, G.S.; Hindley, C.J.; Lippens, G.; Landrieu, I.; Philpott, A. Phosphorylation in intrinsically disordered regions regulates the activity of Neurogenin2. *BMC Biochem.* **2014**, *15*, 24. [CrossRef] [PubMed]

94. Aguado-Llera, D.; Goormaghtigh, E.; de Geest, N.; Quan, X.-J.; Prieto, A.; Hassan, B.A.; Gómez, J.; Neira, J.L. The basic helix-loop-helix region of human neurogenin 1 is a monomeric natively unfolded protein which forms a "fuzzy" complex upon DNA binding. *Biochemistry* **2010**, *49*, 1577–1589. [CrossRef] [PubMed]

95.   Panova, S.; Cliff, M.J.; Macek, P.; Blackledge, M.; Jensen, M.R.; Nissink, J.W.M.; Embrey, K.J.; Davies, R.; Waltho, J.P. Mapping Hidden Residual Structure within the Myc bHLH-LZ Domain Using Chemical Denaturant Titration. *Structure* **2019**, *27*, 1537–1546. [CrossRef] [PubMed]

96.   Romero, P.; Obradovic, Z.; Dunker, A.K. Sequence Data Analysis for Long Disordered Regions Prediction in the Calcineurin Family. *Genome Inform. Workshop Genome Inform.* **1997**, *8*, 110–124.

97.   Li, X.; Romero, P.; Rani, M.; Dunker, A.; Obradovic, Z. Predicting Protein Disorder for N-, C-, and Internal Regions. *Genome Inform. Workshop Genome Inform.* **1999**, *10*, 30–40.

98.   Cilia, E.; Pancsa, R.; Tompa, P.; Lenaerts, T.; Vranken, W.F. From protein sequence to dynamics and disorder with DynaMine. *Nat. Commun.* **2013**, *4*, 2741. [CrossRef]

99.   Cilia, E.; Pancsa, R.; Tompa, P.; Lenaerts, T.; Vranken, W.F. The DynaMine webserver: Predicting protein dynamics from sequence. *Nucleic Acids Res.* **2014**, *42*, W264–W270. [CrossRef]

100. Furness, S.G.B.; Lees, M.J.; Whitelaw, M.L. The dioxin (aryl hydrocarbon) receptor as a model for adaptive responses of bHLH/PAS transcription factors. *FEBS Lett.* **2007**, *581*, 3616–3625. [CrossRef]

101. Brangwynne, C.P.; Eckmann, C.R.; Courson, D.S.; Rybarska, A.; Hoege, C.; Gharakhani, J.; Jülicher, F.; Hyman, A.A. Germline P granules are liquid droplets that localize by controlled dissolution/condensation. *Science* **2009**, *324*, 1729–1732. [CrossRef]

102. Molliex, A.; Temirov, J.; Lee, J.; Coughlin, M.; Kanagaraj, A.P.; Kim, H.J.; Mittag, T.; Taylor, J.P. Phase Separation by Low Complexity Domains Promotes Stress Granule Assembly and Drives Pathological Fibrillization. *Cell* **2015**, *163*, 123–133. [CrossRef]

103. Nott, T.J.; Petsalaki, E.; Farber, P.; Jervis, D.; Fussner, E.; Plochowietz, A.; Craggs, T.D.; Bazett-Jones, D.P.; Pawson, T.; Forman-Kay, J.D.; et al. Phase Transition of a Disordered Nuage Protein Generates Environmentally Responsive Membraneless Organelles. *Mol. Cell* **2015**, *57*, 936–947. [CrossRef] [PubMed]

104. Elbaum-Garfinkle, S.; Kim, Y.; Szczepaniak, K.; Chen, C.C.H.; Eckmann, C.R.; Myong, S.; Brangwynne, C.P. The disordered P granule protein LAF-1 drives phase separation into droplets with tunable viscosity and dynamics. *Proc. Natl. Acad. Sci. USA* **2015**, *112*, 7189–7194. [CrossRef] [PubMed]

105. Lai, J.; Koh, C.H.; Tjota, M.; Pieuchot, L.; Raman, V.; Chandrababu, K.B.; Yang, D.; Wong, L.; Jedd, G. Intrinsically disordered proteins aggregate at fungal cell-to-cell channels and regulate intercellular connectivity. *Proc. Natl. Acad. Sci. USA* **2012**, *109*, 15781–15786. [CrossRef] [PubMed]

106. Mitrea, D.M.; Cika, J.A.; Guy, C.S.; Ban, D.; Banerjee, P.R.; Stanley, C.B.; Nourse, A.; Deniz, A.A.; Kriwacki, R.W. Nucleophosmin integrates within the nucleolus via multi-modal interactions with proteins displaying R-rich linear motifs and rRNA. *eLife* **2016**, *5*, e13571. [CrossRef] [PubMed]

107. Brangwynne, C.P.; Tompa, P.; Pappu, R.V. Polymer physics of intracellular phase transitions. *Nat. Phys.* **2015**, *11*, 899–904. [CrossRef]

108. Banani, S.F.; Rice, A.M.; Peeples, W.B.; Lin, Y.; Jain, S.; Parker, R.; Rosen, M.K. Compositional Control of Phase-Separated Cellular Bodies. *Cell* **2016**, *166*, 651–663. [CrossRef]

109. Ditlev, J.A.; Case, L.B.; Rosen, M.K. Who's In and Who's Out—Compositional Control of Biomolecular Condensates. *J. Mol. Biol.* **2018**, *430*, 4666–4684. [CrossRef]

110. Harmon, T.S.; Holehouse, A.S.; Rosen, M.K.; Pappu, R.V. Intrinsically disordered linkers determine the interplay between phase separation and gelation in multivalent proteins. *eLife* **2017**, *6*, e30294. [CrossRef]

111. Posey, A.E.; Holehouse, A.S.; Pappu, R.V. Phase Separation of Intrinsically Disordered Proteins. In *Methods in Enzymology*; Elsevier: Amsterdam, The Netherlands, 2018; ISBN 9780128156490.

112. Berezney, R.; Coffey, D.S. Identification of a nuclear protein matrix. *Biochem. Biophys. Res. Commun.* **1974**, *60*, 1410–1417. [CrossRef]

113. Linnemann, A.K.; Platts, A.E.; Krawetz, S.A. Differential nuclear scaffold/matrix attachment marks expressed genes. *Hum. Mol. Genet.* **2009**, *18*, 645–654. [CrossRef]

114. Engelke, R.; Riede, J.; Hegermann, J.; Wuerch, A.; Eimer, S.; Dengjel, J.; Mittler, G. The quantitative nuclear matrix proteome as a biochemical snapshot of nuclear organization. *J. Proteome Res.* **2014**, *13*, 3940–3956. [CrossRef] [PubMed]

115. Weber, S.C.; Brangwynne, C.P. Inverse size scaling of the nucleolus by a concentration-dependent phase transition. *Curr. Biol.* **2015**, *25*, 641–646. [CrossRef] [PubMed]

116. Berry, J.; Weber, S.C.; Vaidya, N.; Haataja, M.; Brangwynne, C.P. RNA transcription modulates phase transition-driven nuclear body assembly. *Proc. Natl. Acad. Sci. USA* **2015**, *112*, E5237–E5245. [CrossRef] [PubMed]

117. Gibson, B.A.; Doolittle, L.K.; Jensen, L.E.; Gamarra, N.; Redding, S.; Rosen, M.K. Organization and Regulation of Chromatin by Liquid-Liquid Phase Separation. *bioRxiv* **2019**. [CrossRef]

118. Duronio, R.J.; Marzluff, W.F. Coordinating cell cycle-regulated histone gene expression through assembly and function of the Histone Locus Body. *RNA Biol.* **2017**, *14*, 726–738. [CrossRef]

119. Mitrea, D.M.; Kriwacki, R.W. Phase separation in biology; Functional organization of a higher order Short linear motifs—The unexplored frontier of the eukaryotic proteome. *Cell Commun. Signal.* **2016**, *14*, 1–20. [CrossRef]

120. Wu, T.; Lu, Y.; Gutman, O.; Lu, H.; Zhou, Q.; Henis, Y.I.; Luo, K. Phase separation of TAZ compartmentalizes the transcription machinery to promote gene expression. *bioRxiv* **2019**. [CrossRef]

121. Sabari, B.R.; Dall'Agnese, A.; Boija, A.; Klein, I.A.; Coffey, E.L.; Shrinivas, K.; Abraham, B.J.; Hannett, N.M.; Zamudio, A.V.; Manteiga, J.C.; et al. Coactivator condensation at super-enhancers links phase separation and gene control. *Science* **2018**, *361*, eaar3958. [CrossRef]

122. Boija, A.; Klein, I.A.; Sabari, B.R.; Dall'Agnese, A.; Coffey, E.L.; Zamudio, A.V.; Li, C.H.; Shrinivas, K.; Manteiga, J.C.; Hannett, N.M.; et al. Transcription Factors Activate Genes through the Phase-Separation Capacity of Their Activation Domains. *Cell* **2018**, *175*, 1842–1855.e16. [CrossRef]

123. Chen, X.; Xu, H.; Yuan, P.; Fang, F.; Huss, M.; Vega, V.B.; Wong, E.; Orlov, Y.L.; Zhang, W.; Jiang, J.; et al. Integration of External Signaling Pathways with the Core Transcriptional Network in Embryonic Stem Cells. *Cell* **2008**, *133*, 1106–1117. [CrossRef]

124. Whyte, W.A.; Orlando, D.A.; Hnisz, D.; Abraham, B.J.; Lin, C.Y.; Kagey, M.H.; Rahl, P.B.; Lee, T.I.; Young, R.A. Master transcription factors and mediator establish super-enhancers at key cell identity genes. *Cell* **2013**, *153*, 307–319. [CrossRef] [PubMed]

125. Hnisz, D.; Shrinivas, K.; Young, R.A.; Chakraborty, A.K.; Sharp, P.A. A Phase Separation Model for Transcriptional Control. *Cell* **2017**, *169*, 13–23. [CrossRef] [PubMed]

126. Mansour, M.R.; Abraham, B.J.; Anders, L.; Berezovskaya, A.; Gutierrez, A.; Durbin, A.D.; Etchin, J.; Lee, L.; Sallan, S.E.; Silverman, L.B.; et al. An oncogenic super-enhancer formed through somatic mutation of a noncoding intergenic element. *Science* **2014**, *346*, 1373–1377. [CrossRef] [PubMed]

127. Brown, J.D.; Lin, C.Y.; Duan, Q.; Griffin, G.; Federation, A.J.; Paranal, R.M.; Bair, S.; Newton, G.; Lichtman, A.H.; Kung, A.L.; et al. Nf-kb directs dynamic super enhancer formation in inflammation and atherogenesis. *Mol. Cell* **2014**, *56*, 219–231. [CrossRef] [PubMed]

128. Lovén, J.; Hoke, H.A.; Lin, C.Y.; Lau, A.; Orlando, D.A.; Vakoc, C.R.; Bradner, J.E.; Lee, T.I.; Young, R.A. Selective inhibition of tumor oncogenes by disruption of super-enhancers. *Cell* **2013**, *153*, 320–334. [CrossRef] [PubMed]

129. Apostolou, E.; Ferrari, F.; Walsh, R.M.; Bar-Nur, O.; Stadtfeld, M.; Cheloufi, S.; Stuart, H.T.; Polo, J.M.; Ohsumi, T.K.; Borowsky, M.L.; et al. Genome-wide Chromatin Interactions of the Nanog Locus in Pluripotency, Differentiation, and Reprogramming. *Cell Stem Cell* **2013**, *12*, 699–712. [CrossRef] [PubMed]

130. Manavathi, B.; Samanthapudi, V.S.K.; Gajulapalli, V.N.R. Estrogen receptor coregulators and pioneer factors: The orchestrators of mammary gland cell fate and development. *Front. Cell Dev. Biol.* **2014**, *12*, 34. [CrossRef]

131. Kagey, M.H.; Newman, J.J.; Bilodeau, S.; Zhan, Y.; Orlando, D.A.; Van Berkum, N.L.; Ebmeier, C.C.; Goossens, J.; Rahl, P.B.; Levine, S.S.; et al. Mediator and cohesin connect gene expression and chromatin architecture. *Nature* **2010**, *467*, 430. [CrossRef]

132. Heery, D.M.; Kalkhoven, E.; Hoare, S.; Parker, M.G. A signature motif in transcriptional co-activators mediates binding to nuclear receptors. *Nature* **1997**, *387*, 733. [CrossRef]

133. Cagas, P.M.; Corden, J.L. Structural studies of a synthetic peptide derived from the carboxyl-terminal domain of RNA polymerase II. *Proteins Struct. Funct. Bioinform.* **1995**, *21*, 149–160. [CrossRef]

134. Portz, B.; Lu, F.; Gibbs, E.B.; Mayfield, J.E.; Rachel Mehaffey, M.; Zhang, Y.J.; Brodbelt, J.S.; Showalter, S.A.; Gilmour, D.S. Structural heterogeneity in the intrinsically disordered RNA polymerase II C-terminal domain. *Nat. Commun.* **2017**, *8*, 15231. [CrossRef] [PubMed]

135. Corden, J.L.; Cadena, D.L.; Ahearn, J.M.; Dahmus, M.E. A unique structure at the carboxyl terminus of the largest subunit of eukaryotic RNA polymerase II. *Proc. Natl. Acad. Sci. USA* **1985**, *82*, 7934–7938. [CrossRef] [PubMed]

136. Dahmus, M.E. Reversible Phosphorylation of the C-terminal Domain of RNA Polymerase II. *J. Biol. Chem.* **1996**, *271*, 19009–19012. [CrossRef]

137. Wippich, F.; Bodenmiller, B.; Trajkovska, M.G.; Wanka, S.; Aebersold, R.; Pelkmans, L. Dual specificity kinase DYRK3 couples stress granule condensation/ dissolution to mTORC1 signaling. *Cell* **2013**, *152*, 791–805. [CrossRef] [PubMed]

138. Rai, A.K.; Chen, J.X.; Selbach, M.; Pelkmans, L. Kinase-controlled phase transition of membraneless organelles in mitosis. *Nature* **2018**, *559*, 211. [CrossRef] [PubMed]

139. Boehning, M.; Dugast-Darzacq, C.; Rankovic, M.; Hansen, A.S.; Yu, T.; Marie-Nelly, H.; McSwiggen, D.T.; Kokic, G.; Dailey, G.M.; Cramer, P.; et al. RNA polymerase II clustering through carboxy-terminal domain phase separation. *Nat. Struct. Mol. Biol.* **2018**, *25*, 833. [CrossRef] [PubMed]

140. Bolognesi, B.; Lorenzo Gotor, N.; Dhar, R.; Cirillo, D.; Baldrighi, M.; Tartaglia, G.G.; Lehner, B. A Concentration-Dependent Liquid Phase Separation Can Cause Toxicity upon Increased Protein Expression. *Cell Rep.* **2016**, *16*, 222–231. [CrossRef] [PubMed]

141. Mészáros, B.; Erdős, G.; Szabó, B.; Schád, É.; Tantos, Á.; Abukhairan, R.; Horváth, T.; Murvai, N.; Kovács, O.P.; Kovács, M.; et al. PhaSePro: The database of proteins driving liquid–liquid phase separation. *Nucleic Acids Res.* **2019**. [CrossRef]

142. Darling, A.L.; Liu, Y.; Oldfield, C.J.; Uversky, V.N. Intrinsically Disordered Proteome of Human Membrane-Less Organelles. *Proteomics* **2018**, *18*, 1700193. [CrossRef]

143. Hyman, A.A.; Weber, C.A.; Jülicher, F. Liquid-Liquid Phase Separation in Biology. *Annu. Rev. Cell Dev. Biol.* **2014**, *30*, 39–58. [CrossRef]

144. Uversky, V.N. Protein intrinsic disorder-based liquid–liquid phase transitions in biological systems: Complex coacervates and membrane-less organelles. *Adv. Colloid Interface Sci.* **2017**, *239*, 97–114. [CrossRef] [PubMed]

145. Shin, Y.; Brangwynne, C.P. Liquid phase condensation in cell physiology and disease. *Science* **2017**, *357*, eaaf4382. [CrossRef] [PubMed]

12

Structural and Functional Properties of the Capsid Protein of Dengue and Related *Flavivirus*

André F. Faustino [1,†], **Ana S. Martins** [1], **Nina Karguth** [1], **Vanessa Artilheiro** [1], **Francisco J. Enguita** [1], **Joana C. Ricardo** [2,‡], **Nuno C. Santos** [1,*] and **Ivo C. Martins** [1,*]

[1]   Instituto de Medicina Molecular, Faculdade de Medicina, Universidade de Lisboa, Av. Prof. Egas Moniz, 1649-028 Lisbon, Portugal

[2]   Centro de Química-Física Molecular, Instituto Superior Técnico, Universidade de Lisboa, 1049-001 Lisbon, Portugal

*   Correspondence: nsantos@fm.ul.pt (N.C.S.); ivomartins@fm.ul.pt (I.C.M.)

†   Present address: Instituto de Biologia Experimental e Tecnológica (iBET), Apartado 12, 2780-901 Oeiras, Portugal.

‡   Present address: Department of Biophysical Chemistry, J. Heyrovský Institute of Physical Chemistry, Czech Academy of Sciences, Dolejškova 3, 182 23 Prague 8, Czech Republic.

Abstract: Dengue, West Nile and Zika, closely related viruses of the Flaviviridae family, are an increasing global threat, due to the expansion of their mosquito vectors. They present a very similar viral particle with an outer lipid bilayer containing two viral proteins and, within it, the nucleocapsid core. This core is composed by the viral RNA complexed with multiple copies of the capsid protein, a crucial structural protein that mediates not only viral assembly, but also encapsidation, by interacting with host lipid systems. The capsid is a homodimeric protein that contains a disordered N-terminal region, an intermediate flexible fold section and a very stable conserved fold region. Since a better understanding of its structure can give light into its biological activity, here, first, we compared and analyzed relevant mosquito-borne *Flavivirus* capsid protein sequences and their predicted structures. Then, we studied the alternative conformations enabled by the N-terminal region. Finally, using dengue virus capsid protein as main model, we correlated the protein size, thermal stability and function with its structure/dynamics features. The findings suggest that the capsid protein interaction with host lipid systems leads to minor allosteric changes that may modulate the specific binding of the protein to the viral RNA. Such mechanism can be targeted in future drug development strategies, namely by using improved versions of pep14-23, a dengue virus capsid protein peptide inhibitor, previously developed by us. Such knowledge can yield promising advances against Zika, dengue and closely related *Flavivirus*.

Keywords: Dengue virus (DENV); capsid protein (C protein); *Flavivirus*; intrinsically disordered protein (IDP); protein–RNA interactions; protein–host lipid systems interaction; circular dichroism; time-resolved fluorescence anisotropy

1. Introduction

Viral hemorrhagic fever is a global problem, with most cases due to dengue virus (DENV), which originates over 390 million infections per year worldwide, being a major socio-economic burden, mainly for tropical and subtropical developing countries [1]. A working vaccine was registered in Mexico in December 2015, approved for official use in some endemic regions of Latin America and Asia and, as of October 2018, also in Europe [2–4]. However, this vaccine is not 100% effective against all DENV serotypes. Thus, research into new prophylactics is still ongoing, with a new vaccine proposed recently being now in phase 3 clinical trials [5]. In spite of these recent developments, fully effective prophylactics approaches are lacking and there are no effective therapies. This is in part, due to a poor understanding of key steps of the viral life cycle.

There are four dengue serotypes occurring: DENV-1, DENV-2, DENV-3 and DENV-4 [6]. Here, if not otherwise indicated, DENV refers to DENV-2. DENV is a member of the *Flavivirus* genus, part of the Flaviviridae family, a genus which comprises 53 viral species [6]. Many of these are important human pathogens as well, such as hepatitis C (HCV), tick-borne encephalitis (TBEV), yellow fever (YFV), West Nile (WNV) and Zika (ZIKV) viruses [6–9]. Flaviviridae are single-stranded positive-sense RNA viruses with approximately 11 kb, containing a single open reading frame [10]. Using the host cell translation machinery, the *Flavivirus* RNA genome is translated into a polyprotein that is co- and post-translationally cleaved by cellular and viral proteases into three structural proteins and seven non-structural proteins [10]. Structural proteins are named as such since they are present in the mature virion structure [11]. Nevertheless, they may also have non-structural roles, such as the capsid (C) protein. This is a structural protein that also mediates viral assembly and encapsidation, crucial steps of the viral life cycle. Given the C protein key roles, it is the focus of this work and will be described in detail below.

DENV C contains 100 amino acid residues, which form an homodimer with an intrinsically disordered protein (IDP) region in the N-terminal followed by four α-helices, α1 to α4, per monomer [12]. Overall, the main structural/dynamics regions consist of the disordered N-terminal, a short flexible intermediate fold and, finally, a large conserved fold region, which greatly stabilizes the protein homodimer structure [12–16]. The C protein has an asymmetric charge distribution: one side of the dimer contains a hydrophobic pocket (α2–α2' interface), responsible for, alongside the disordered N-terminal, the binding to host lipid droplets (LDs) [12–16]. The other is the positively charged C-terminal side (α4–α4' interface), proposed to mediate the C protein binding to the viral RNA [12]. It is noteworthy that several transient conformations for DENV C N-terminal were proposed, which may help modulate DENV C interaction with host lipid systems, via an autoinhibition mechanism [15].

DENV infection affects the host lipid metabolism, increasing host intracellular LDs and unbalancing plasma lipoprotein levels and composition [17–19]. Importantly, DENV C binds LDs, an interaction essential for viral replication [18,20]. DENV C-LDs binding requires potassium ions, the LDs surface protein perilipin 3 (PLIN3) and involves specific amino acid residues of DENV C α2–α2' helical hydrophobic core and of the N-terminal [14,20]. This knowledge led us to design pep14-23, a patented peptide, based on a *Flavivirus* C protein conserved N-terminal motif. We then established that pep14-23 inhibits DENV C-LDs binding [14], acquiring α-helical structure in the presence of anionic phospholipids [15]. Moreover, we also found that DENV C binds specifically to very low-density lipoproteins (VLDL), requiring K^+ ions and a specific VLDL surface protein, apolipoprotein E (APOE), being also inhibited by pep14-23 [21]. This is analogous to DENV C-LDs interaction. The similarities between APOE and PLIN3 further reinforce this, suggesting a common mechanism [22]. The role of LDs in *Flavivirus* infection is well known and has been recently reviewed [14,18,20,23–25]. Given that, pep14-23 is an excellent drug development lead. Further developments require a better understanding of the function of the C protein of dengue and of *Flavivirus* in general.

Therefore, here, we seek to contribute to understand the C proteins biological activity, with a special focus on DENV C. Briefly, we studied DENV C structure-activity relationship in the context of similar and highly homologous mosquito-borne *Flavivirus* C proteins. Our findings shed light into the structure-function relationship behind the C protein biological roles, which may contribute to future therapeutic approaches against DENV and closely related *Flavivirus*.

2. Results

2.1. Analysis of Amino Acid Sequence Conservation Among Flavivirus C proteins

A phylogenetic analysis of the *Flavivirus* C protein and the polyprotein amino acid residue sequences reveals if the C protein is an indicator of phylogenetic similarity (Figure 1). C proteins of Spondweni group viruses, i.e., ZIKV, Spondweni virus (SPOV) and Kedougou virus (KEDV), cluster together, being the most similar to DENV (Figure 1a). Another cluster corresponds to mosquito-borne encephalitis-causing *Flavivirus*: Saint Louis encephalitis (SLEV), WNV, WNV serotype Kunjin (WNV-K), Alfuy (ALFV), Murray Valley encephalitis (MVEV), Usutu (USUV) and Japanese encephalitis (JEV) viruses. The *Flavivirus* polyproteins sequences show similar clusters (Figure 1b). As such, the C protein is a good indicator of viral genetic similarity. Thus, we investigated the C protein amino acid sequences, seeking common patterns relevant to biological activity.

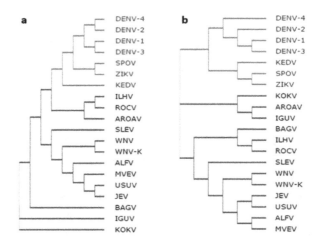

Figure 1. *Flavivirus* phylogenetic trees. Phylogenetic trees of (**a**) *Flavivirus* C proteins, highlighting in red the viruses with the C protein most similar to dengue virus (DENV) C (Spondweni group viruses (ZIKV), Spondweni virus (SPOV) and Kedougou virus (KEDV)) and of the (**b**) entire viral polyproteins of the same *Flavivirus*. Overall, despite some differences, the same general clusters are seen regardless of the clustering being based on the polyprotein or the capsid protein.

The amino acid residues sequences of the *Flavivirus* C proteins identified above were analyzed in the context of the three main regions identified in DENV C sequence, *i.e.*, the conserved fold region, the flexible fold region and the N-terminal IDP region (Figure 2). This was done for all mosquito-borne *Flavivirus* relevant for human diseases (Figure 2a), as well as for the four main DENV C serotypes (Figure 2b). For this, the 16 mosquito-borne *Flavivirus* and the 4 DENV serotypes amino acid sequence of the C protein are jointly aligned. In agreement with previous work [12,14], five conserved motifs are found in the mosquito-borne *Flavivirus* C proteins and deserve attention, namely: the N-terminal conserved ^{13}hNML+R^{18}; ^{40}GXGP^{43} in loop L1-2; ^{44}h+hhLAhhAFF+F^{56} in α2 helix; ^{68}RW^{69} of α3 helix; and, finally, the ^{84}F++−h^{88} motif from α4 (with 'h', '+' and '−' representing hydrophobic, positively charged and negatively charged residues, respectively). Between residues 70–100, other motifs, not previously reported and containing hydrophobic and positively charged residues, are visible. Moreover, amino acid residues G and P, that can break the continuity of α-helices, are conserved in specific positions of the protein, especially in the disordered N-terminal and the flexible fold regions (Figure 2c). Charged residues are also conserved in specific locations. They are mostly in the conserved fold region, especially after position 95 (Figure 2d). Overall, the disordered N-terminal and the flexible fold regions, when compared with the conserved fold region, have an average of, respectively, 10 versus 4 G and P residues (Figure 2c), green, 10 versus 15 K and R residues (Figure 2d), blue, and 1 versus 2 D and E residues (Figure 2d), magenta.

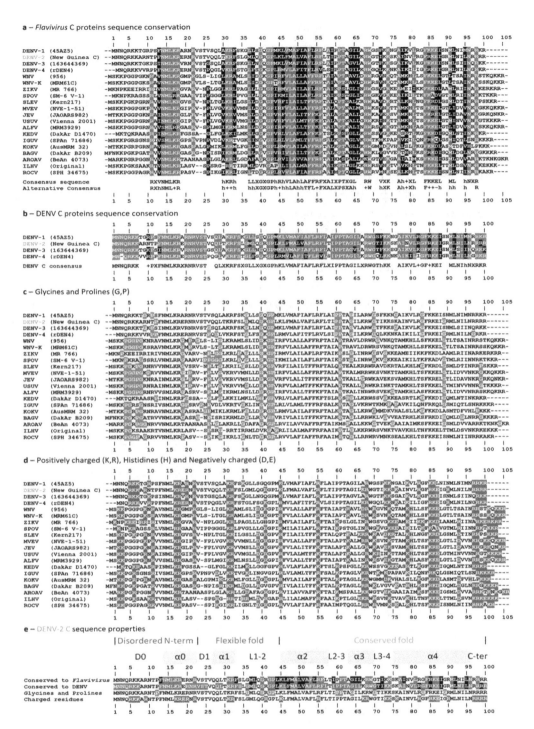

Figure 2. *Flavivirus* C proteins amino acid residues sequence conservation. (**a**) Mosquito-borne *Flavivirus* C protein are 55% conserved, with residues being considered conserved if, in a given position, more than 15 are equal (red) or stereochemically similar (black). (**b**) Conservation between DENV serotypes is 80%, with the same criteria as in (**a**). (**c**) Structure-breaking residues G and P (green). (**d**) Charged residues: dark blue for positively charged residues (K and R), light blue for H, and magenta for negatively charged residues (D or E). (**e**) Overall conserved regions of *Flavivirus* C proteins: the disordered N-terminal and the conserved fold are clearly conserved in terms of charged and G/P amino acids. In contrast, the flexible fold region allows higher variability. Thus, its main role seems to be to connect the disordered N-terminal and the conserved fold regions, and to enable alternative conformations. DENV C serotype 2 is highlighted in blue, with amino acid residues numbered according to its sequence. Amino acid residues are numbered according to the consensus, coinciding with DENV-2 residues numbers. The viruses' full designation is found in the abbreviations section.

Several motifs in the *Flavivirus* C protein sequences can be identified. These represent the main sections of the protein, conserved during evolution as these must be crucial to protein function (Figure 2e). The N-terminal region, although disordered, is highly conserved, in terms of charged amino acid and G/P residues. The flexible fold section allows greater variability, in line with previous reports by us and others, suggesting that it can adopt several conformations [15].

2.2. *Analysis of the Flavirus C Protein Sequences Hydrophobicity and Secondary Structure Propensity*

Hydrophobicity and α-helical propensity predictions were performed as previously reported [15], using the Kite-Doolittle [26] and the Deleage-Roux [27] scales on ProtScale server, respectively, for the 16 mosquito-borne *Flavivirus* C proteins analyzed (Figure 3). The hydrophobicity scale ranges from −4.5, for highly polar amino acids (hydrophilic), to 4.5, for highly hydrophobic amino acid residues [26]. Therefore, when plotting the average values for each amino acid residue of the *Flavivirus* C sequences, negative local minima and positive local maxima indicate, respectively, hydrophilic and hydrophobic regions (Figure 3a,b). All proteins display a similar profile even in the N-terminal and flexible fold regions despite the slightly higher amino acid residues variability (Figure 2). The α0 domain, homologous to pep14-23, is amphipathic, with average values near 0. In the flexible fold region, which is mostly amphipathic too, there is a peak of hydrophobicity between residues 30 and 40, possibly explaining its intermediate structure/dynamics behavior [13,14]. Some peaks of hydrophobicity are observed in the α3 and α4 domains, with the most hydrophobic domain being α2, as expected from the sequence analysis (Figure 2) and from the literature [12,14,18].

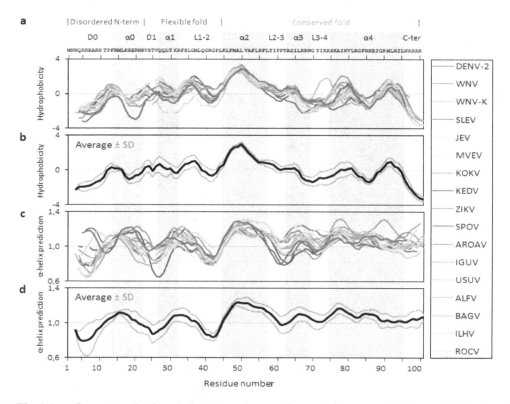

Figure 3. *Flavivirus* C proteins hydrophobicity and secondary structure predictions. (**a**) Hydrophobicity predictions and (**b**) respective average (black line) ± standard deviation, SD (gray lines). (**c**) α-helical secondary structure predictions and (**d**) respective average (black line) ± SD (gray lines). Amino acid residues are numbered according to the consensus, coinciding with DENV 2 residues numbers.

For α-helical predictions secondary structure is highly probable above a threshold of 1.0 [27]. *Flavivirus* C proteins secondary structure predictions correlate well with the known secondary structure of DENV C (Figure 2e) [12]. Such agreement supports the concept of a transient α0 occurring for these proteins, as hypothesized earlier [15]. Roughly, between positions 12 to 20, occurs a disordered region with high tendency to acquire α-helical secondary structure. Importantly, the values of the predictions are similar and the same tendencies are found in all proteins, with peaks and valleys co-localizing (Figure 3). Along with data from the last subsection, these results strengthen the idea that *Flavivirus* C proteins have similar structure and dynamics properties.

2.3. Analysis of the Flavivirus C Protein Tertiary Structure Propensity

Flavivirus C proteins tertiary structure was then investigated, complementing the α-helical predictions, to help understanding the disordered N-terminal region role(s). Following previous work [15], I-TASSER [28–30] was used to predict tertiary structures for the 16 closely related mosquito-borne *Flavivirus* C proteins (Figure 4). Eighty monomer conformations were obtained (several for each sequence) and superimposed with the DENV C homodimer partial structure deposited at the Protein Data Bank (PDB) and obtained via nuclear magnetic resonance (NMR) spectroscopy (PDB ID: 1R6R). Noteworthy, DENV [12,16], WNV [31] and ZIKV [25] C proteins form homodimers, stabilized by hydrophobic and electrostatic interactions involving their conserved fold region [12–14,25,31–33]. Since this is the most conserved region of *Flavivirus* C proteins sequences (Figure 2), a homodimer is thus not only a stable conformational arrangement, but also likely to occur. Thus, as 28 conformers had more than 5 backbone clashes with the other monomer when superimposed in a homodimer structure (not allowing a viable homodimer), those conformers were discarded Table 1. The remaining 52 *Flavivirus* C proteins conformational models were analyzed, while superimposed with DENV C homodimer (PDB ID: 1R6R, model 21 [12]). These were then grouped into four clusters by visual inspection of their similarity (Figure 4).

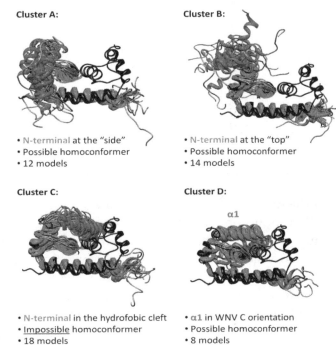

Figure 4. *Flavivirus* C proteins tertiary structure predictions, organized into four conformational clusters. The *Flavivirus* C proteins conformations predicted by I-TASSER are superimposed with DENV C experimental homodimer structure (black). Amino acid residues of the N-terminal region in α-helix conformation are in blue, the other α-helices in red and the loops in gray. From the 80 conformers, 52 can be clustered by similarity of conformations, from cluster A to D. Clusters A, B and C have the α1 helix in the DENV C experimentally determined conformation (Protein Data Bank (PDB) ID: 1R6R [12]). In cluster D the α1 is in West Nile Virus (WNV) C and ZIKV C conformation (PDB IDs: 1SFK [31] and 5YGH [25], respectively). The closed autoinhibitory conformation of cluster C seems the most probable, having the highest number of models. Although unlikely given their transient unstable nature, N-terminal IDP regions may interact with each other. Table 1 specifies each cluster composition.

Most sequences have a conformer in each cluster (Figure 1 and Table 1). In cluster A, some N-terminal amino acid residues are close to α4–α4′ and may interact with RNA, namely the positively charged residues. Cluster B has the most scattered conformers, with the N-terminal region at the "top", not interacting with other protein regions, resembling a transition between more ordered states. In cluster C, the N-terminal region is in an autoinhibitory conformation, blocking the access to the α1–α2–α2′–α1′ region, as previously suggested by us for DENV C [15]. 18 conformer models are predicted in this closed conformation with, at least, one model from most of the C proteins tested (except JEV C and ZIKV C; see Table 1). Therefore, it can occur in most *Flavivirus* C proteins. As for cluster D conformation, the α1 helix is in the conformation of WNV [14,31] and ZIKV [25] C experimental structures, an arrangement not previously reported for DENV C [15]. This closed conformation also involves the N-terminal region and α1 domain, and partially blocks the α2–α2′ hydrophobic cleft (or totally blocks it, when both monomers are in the same conformation). Importantly, both cluster C and D are closed conformations, supporting the autoinhibition hypothesis.

Table 1. Distribution of the I-TASSER predicted models through the four clusters.

Protein	Cluster A	Cluster B	Cluster C	Cluster D	Excluded
ALFV C	1	0	1	0	3
AROAV C	1	1	1	0	2
BAGV C	1	0	2	1	1
DENV C	1	2	1	0	1
IGUV C	1	2	1	0	1
ILHV C	0	2	2	0	1
JEV C	1	1	0	1	2
KEDV C	0	1	1	1	2
KOKV C	1	0	1	1	2
MVEV C	0	2	1	0	2
ROCV C	1	0	1	2	1
SLEV C	1	0	2	0	2
SPOV C	1	2	1	0	1
USUV C	1	0	2	0	2
WNV C	1	0	1	1	2
ZIKV C	0	1	0	1	3
Total	**12**	**14**	**18**	**8**	**28**

Dimers with A or B conformers in one monomer enable the simultaneous co-existence of all other conformers (A to D) on the other monomer. The C conformer neither permits the existence of C-C′ homoconformers (i.e., both monomers in the same conformation) nor the heteroconformers of C-D′ and D-C′. Despite that, D-D′ homoconformers are allowed, similarly to the conformation that WNV C adopts in the crystal form [31]. Moreover, to go from cluster A to cluster C or D, the N-terminal region should pass by cluster B. These constraints suggest a path for transitions between conformations, discussed ahead. Overall, the autoinhibition hypothesis proposed for DENV C [15] is supported and such conformation can occur in other *Flavivirus* C proteins.

2.4. Analysis of Dengue Virus (DENV) C Protein Rotational Correlation Time

Given the close similarities between *Flavivirus* C proteins (Figures 1–4), DENV C can be used as a general model for them. Hence, we proceeded to determine DENV C overall rotational correlation time (τ_c), taking advantage of the tryptophan residue in position 69 (W69) intrinsic fluorescence. Our computational data support three main structure/dynamics regions, including a disordered N-terminal region, which would increase its expected apparent size (as it would not be globular and folded), a property detectable by such an approach. Upon testing molecules in aqueous solution and at room temperature, fluorescence lifetimes are usually in the ns timescale, and the fluorescence decays are sensitive to the anisotropy of the fluorophore, which depends on its τ_c (vd. Equations (1)–(8), describing these relations, in the Methods section [34,35]). Thus, the time-resolved fluorescence decay of DENV C W69 and the corresponding anisotropy decay were determined, both at pH 6.0 and 7.5 (Figure 5).

Table 2. Fitting parameters of DENV C time-resolved fluorescence anisotropy data analysis. Parameters obtained from fitting Equations (5) and (8) to the data of Figure 5. Values are average (±% standard error, SE). * Statistically significant differences ($p < 0.05$) between the values obtained for the two pH values tested.

Parameter	pH 6.0	pH 7.5
τ_1 (ns) *	0.209 (± 3.9%)	0.520 (± 4.0%)
τ_2 (ns)	3.106 (± 0.4%)	3.108 (± 0.9%)
τ_3 (ns) *	6.328 (± 0.4%)	6.506 (± 0.4%)
α_1 *	0.275 (± 0.7%)	0.178 (± 3.4%)
α_2 *	0.315 (± 0.9%)	0.385 (± 0.4%)
α_3 *	0.410 (± 0.4%)	0.437 (± 0.4%)
τ_c (ns) *	16.46 (± 2.9%)	16.41 (± 3.4%)
r_0	0.130 (± 0.8%)	0.131 (± 1.1%)

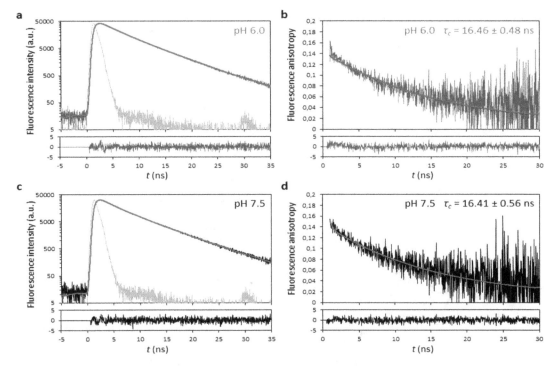

Figure 5. DENV C time-resolved fluorescence anisotropy. Time-resolved fluorescence decay at pH (**a**) 6.0 and (**c**) 7.5, with the corresponding anisotropy decays at pH (**b**) 6.0 and (**d**) 7.5. Fluorescence and anisotropy decays at both pH values are similar (gray and black decays, respectively). Fitting of experimental data (red) took into account the instrument response function (IRF; in green) and the corresponding residuals distribution, displayed below each graph. The equations used for fitting are presented on the Methods Equations (5) and (8). The parameters obtained are shown in Table 2.

Time-resolved fluorescence anisotropy decays at both pH values are similar (Figure 5b,d). Fluorescence lifetime components (τ_1, τ_2 and τ_3) were obtained from the intensity decays Equations (2)–(6) [34,35], with a triple-exponential retrieving the best fit (Figure 5a,c). Fitting the data retrieves similar values Table 2 for τ_1, τ_2 and τ_3, and corresponding weights (α_1, α_2 and α_3 pre-exponential factors, respectively). For accurate calculation of τ_c, the condition $\tau_c < 3 \times \tau_3$ must occur [34,35]. Since τ_3 values were ~6.4 ns (with a significant weight α_3 of ~0.42), this means that, at both pH values, we could measure τ_c values up to a limit of ~19 ns. In both pH conditions, the τ_c measured was

16.4 ± 0.5 ns at 22 °C, within the limit and higher than expected for a purely globular protein of DENV C size, as predicted [13].

Rossi et al. [36] correlated the τ_c of 16 globular proteins at 20 °C with their molecular weight (MW in kDa), based on NMR data, leading to the relation: $\tau_c \approx 0.6$ MW. Assuming DENV C as a 23.5 kDa fully globular homodimer and correcting for the temperature (T) and viscosity (η) [37], the τ_c predicted is 12.0 ns. However, the correlational time must be slightly higher, as the protein will be partially unfolded and disordered (in the N-terminal). Jones et al. [16] measured a τ_c of 13 ns at 27 °C, by NMR, which with the corrections from Equation (10) [37], corresponds to 13.4 ns at 25 °C. Given DENV C size, this implies that the protein is not globular, in line with current knowledge of DENV C structure and dynamics [12–16]. Fluorescence anisotropy supports an even more open and partially disordered DENV C structure, given the τ_c value of 15.2 ± 0.5 ns at 25 °C Table 3, in line with in silico data (Figures 1–4).

Table 3. Comparing DENV C τ_c values (τ_c at 25 °C in H$_2$O were calculated using Equation (10)).

τ_c (ns) at T	T (°C)	τ_c (ns) at 25 °C in H$_2$O	Method	Source
16.4 ± 0.5	22	15.2 ± 0.5	Time-resolved fluorescence anisotropy	This work
13.0	27	13.4	Overall NMR relaxation analysis	Jones et al., 2003 [16]
14.1	20	12.0	τ_c (ns) $\approx 0.6 \times$MW (kDa)	Rossi et al., 2010 [36]

2.5. Analysis of DENV C Conformational Stability

Circular dichroism (CD) spectroscopy was used to study DENV C secondary structure, via its thermal denaturation in solution from 0 to 96 °C, at pH 6.0 and 7.5 (2 °C steps, Figure 6). At both pH values, the α-helical structure is partially lost upon increasing temperature (Figure 6a,b). However, even at 96 °C, the protein does no become completely random coil, as seen from the spectrum shape and its high ellipticity at 222 nm (Figure 6c). Plotting the mean residue molar ellipticity at 222 nm, [θ], as a function of temperature, T, reveals a transition at ~70 °C at both pH (Figure 6c).

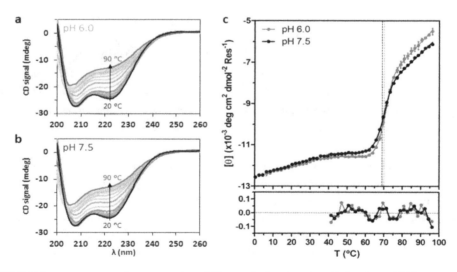

Figure 6. DENV C temperature denaturation followed via circular dichroism (CD) spectroscopy. CD spectra of DENV C, between 20 and 90 °C, at pH (**a**) 6.0 and (**b**) 7.5. For the sake of simplicity, the spectra from 0 to 18 °C and from 92 to 96 °C are not displayed, as they are similar to the 20 °C and the 90 °C spectra, respectively. (**c**) Mean residue molar ellipticity at 222 nm, [θ], as a function of temperature (dots) for pH 6.0 (gray) and 7.5 (black), between 0 and 96 °C. Lines correspond to the fitting of Equation (21) (combined with Equations (20), (22), (24) and (28)). Vertical dashed lines represent experimentally observed T_m, colored according to pH. Error bars represent SD, from three independent experiments. Residuals are shown below the graph, being lower than SD.

DENV C does not display a typical unfolding profile, as the denaturation curves do not reach a flat plateau. Still, ellipticity data were successfully fitted to a denaturation curve (Figure 6c), assuming a homodimer with one-step denaturation [32]. Briefly, Equation (21) was combined with Equations (20), (22) and (24) and fitted to the data. This allows to obtain the thermodynamic parameters of DENV C unfolding Table 4, namely the melting temperature (T_m°), the enthalpy variation at T_m° ($\Delta H^\circ_{T_m^\circ}$) and the entropy variation at T_m° ($\Delta S^\circ_{T_m^\circ}$), with all parameters at standard thermodynamics conditions (symbolized by '°'). Equation (28) was then used to calculate the melting temperature (T_m) at the actual $[P_m]$ (instead of the value at $[P] = 1$ M, details in the Methods). Despite small differences, the parameters obtained are not significantly different between pH values Table 4. A small but consistent variation of the CD spectra between 0 and 40 °C is observable, implying: (i) a conformational equilibrium with temperature and/or (ii) some flexibility of the structure and/or (iii) a transition between alternative conformations. This temperature range covers the physiological conditions of both mosquitoes (20 to 40 °C, depending on the environment) and humans (36 to 40 °C). DENV C can continuously transition between conformations as temperature varies, in line with the previously hypothesized conformational equilibrium [15]. As temperature increases, the disordered conformations become more abundant but only a partial loss of structure is seen. This indicates that the C protein conserved region is thermodynamically stable. Similar observations are expected for other *Flavivirus* C proteins.

Table 4. Fitting parameters of DENV C temperature denaturation CD data. Parameters were estimated by fitting Equation (21) (combined with Equations (20), (22), (24) and (28)) to the data. T_m is the experimentally observed melting temperature (represented by the vertical lines in Figure 6c). Estimations are average ± SE. There were no significant variations between the two pH values tested ($p < 0.05$).

Parameter	pH 6.0	pH 7.5
T_m (°C)	70.02 ± 0.63	69.03 ± 0.65
T_m° (°C)	88.26 ± 0.80	88.80 ± 0.83
$\Delta H^\circ_{T_m^\circ}$ (kJ mol^{-1})	612 ± 26	564 ± 23
$\Delta S^\circ_{T_m^\circ}$ (kJ mol^{-1} K^{-1})	1.693 ± 0.073	1.557 ± 0.065

3. Discussion

Flavivirus C proteins are known to have similar sequences and structure [12–16,25,31]. Here, we go further by examining common features at different structural levels, complemented with data on DENV C size and thermodynamic stability. The phylogenetic analysis of the C proteins and the polyproteins (Figure 1) shows that the former is a marker of *Flavivirus* evolution. There are several conserved motifs, highlighted in previous studies with 16 *Flavivirus* [12,14]. The work is now expanded to include the four DENV serotypes (Figure 2). When these 20 *Flavivirus* C amino acid sequences, with between 96 and 107 amino acid residues each, are jointly analyzed, it is clear that 55% of the residues are conserved or stereochemically similar (Figure 2a). About 80% of amino acid residues are equal or similar and, thus, conserved among the four DENV C serotypes (Figure 2b). From the five major conserved motifs, four are known to be involved in dimer stabilization [14]: the ^{40}GXGP^{43} motif at loop L1-2, that marks the transition from the flexible to the conserved fold region [14]; the ^{68}RW^{69} at α3 forms an hydrophobic pocket that accommodates the W69 side chain involving residues from α2, α3 and α4 [12,32]; and, the ^{44}h+hhLAhhAFF+F^{56} and ^{84}F++–h^{88} motifs, respectively from α2 and α4 helices, maintain the homodimer structure both via the α2–α2' hydrophobic interaction and via the salt bridges of residues [RK]^{45} and [RK]$^{55'}$ with [ED]^{87} [12,14,32]. *Flavivirus* C proteins must have similarly sized secondary structure domains, since G/P are in the same positions and these amino acid residues tend to break the secondary structure (Figure 2c). Charged residues are also conserved (Figure 2d), which makes sense as charges would promote the interaction of the C protein with the negatively charged host lipid systems [12,14,20–22] and the viral RNA [12]. C proteins have a common homodimer conserved fold region (roughly, residues 45–100), as observed for DENV, WNV and ZIKV C structures [12,14,25,31]. Conserved motifs are summarized in (Figure 2e).

The above explains the C proteins similar hydrophobic and α-helix propensities (Figure 3). The conserved motif ^{13}hNML+R^{18}, at the N-terminal region, and the α2–α2′ hydrophobic cleft are of particular importance for DENV C interaction with LDs and VLDL [14,20–22,38]. Mutations in specific residues of DENV C α2–α2′ and α4–α4′ also impair RNA binding. Likewise, ZIKV C also accumulates on LDs surface, with specific mutations on this protein disrupting the association [25]. ZIKV C also binds single-stranded and double-stranded RNAs [25], with, as for DENV C, the high positively charged residues density prompting the binding to LDs and RNA [12,39,40]. Given the match at the level of N-terminal α-helical propensity and α2–α2′ hydrophobicity (Figure 3), the C proteins may all be self-regulated by an autoinhibition mechanism, as proposed for DENV C [15].

The autoinhibition hypothesis is corroborated by the quaternary structure analysis (Figure 4); Table 1. Two clusters, C and D, are autoinhibited conformations. Importantly, cluster D α1 aligns with WNV C [14,31] and ZIKV C [25]. Moreover, if two monomers are in a D conformation (D–D′ homoconformer), the dimer α2–α2′ region is totally inaccessible. Cluster C does not allow a C–C′ homoconformer nor a C–D heteroconformer, imposing restrictions to the simultaneous transitions that are possible between A, B, C and D, as homodimer. The interaction between N-terminal regions within a dimer may be considered. Nonetheless, the disordered nature and high density of positively charged amino acid residues will mostly favor the repulsion between these IDP regions.

It is important to look at the clusters (Figure 4), while considering the number of positively charged residues (Figure 2) in the disordered N-terminal and flexible fold (10 K and R residues) versus those in the conserved fold (15 K and R). The charge distribution in some arrangements implies that the disordered N-terminal is at least in theory able to bind the viral RNA [39,40]. Such binding would be governed by the N-terminal region cationic amino acid residues [41,42]. Here, the structure predictions reveal that, indeed, the first 12 N-terminal residues can locate near α4–α4′ Cluster A (Figure 4), the most likely RNA binding site [12,39,40]. Furthermore, binding to RNA via the C-terminal α4–α4′ interface may be favored by a previous or simultaneous interaction of the protein with host LDs via the N-terminal region and α2–α2′ interface. Access to α2–α2′ (controlled by the N-terminal region) would modulate the interaction (Figure 4) and, thus, viral assembly. In agreement, the binding of the related hepatitis C virus core protein (homologous to DENV C) to host LDs is what enables efficient viral assembly [43]. Thus, the C protein disordered N-terminal would be critical to protein function, enabling crucial structural and functional roles.

To evaluate this, we used DENV C as a model system, measuring its τ_c value by time-resolved fluorescence anisotropy (Figure 5) and its thermal stability by CD spectroscopy (Figure 6), at pH 6.0 and 7.5 (within the usual pH range of its biological microenvironment). A similar τ_c, 15.2 ± 0.5 ns, is obtained at both pH values (Figure 5; Tables 2 and 3), in line with previous work [13]. DENV C maintains its homodimer structure and dynamics behavior between pH 6.0 and 7.5. The τ_c value and respective size are higher than expected, due to the N-terminal disordered nature.

Regarding DENV C thermodynamic stability (Figure 6, Table 4), the protein T_m is ~70 °C at both pH values. These denaturation parameters are in line with other authors, as a chemically synthesized DENV C 21–100 fragment (without most of the disordered N-terminal region) displays a T_m = 71.6 °C [32]. DENV C high thermal stability in physiological conditions is likely due to the large hydrophobic area that is shared by the two monomers [12], but also to the W69 stabilizing interactions and, as experimentally observed [32], the formation of salt bridges (residues K45 and R55′ with E87). As structure/dynamics properties are conserved among *Flavivirus* C proteins (Figures 2–4), these observations can probably be generalized for all these proteins.

These findings must also be considered in light of DENV C biologically relevant interactions with LDs [22] and RNA (Figure 7). DENV C experimental structure [12] contains three distinct structural regions [13]: a disordered N-terminal region (from the N-terminal up to residue R22), a flexible fold (residues V23 to L44, where α-helix 1 is located) and a conserved fold with helices α2, α3 and α4, containing the R68 and W69 amino acid residues, highly conserved among *Flavivirus* [12]. R68 terminates α3 helix, with its side chain pointing to the protein interior [12]. W69 locates at DENV C α4–α4' interface, having a crucial role in the dimer structural stabilization [12]. Along with dimer structural stability, these interactions enable allosteric communication and movements between DENV C more hydrophobic section (α2–α2'dimer interface) and its remaining sections, namely the α4–α4' region. Figure 7 displays this, in the context of the C protein biologically relevant interactions, as they are understood on the basis of recent studies [12–15,18,20–24].

Looking further, it is important to consider that the binding of DENV C to host LDs is mediated by both the N-terminal IDP region and the α2–α2' interface [14]. V51 of α2 is affected by the interaction with LDs and stabilizes the dimer by contacting with α3 (I65). Another interaction via salt bridges, between α2 (K45 and R55') and α4 (E87), stabilizes the homodimer (Figure 7a). The C protein binding to host LDs, which affects the α2–α2', can lead to changes in the α4–α4' structural arrangement (Figure 7b). To investigate this we searched for similar proteins. An RNA-binding protein with a two-helix domain similar to DENV C α4–α4' was identified (Figure 7c), influenza A non-structural protein 1 (NS1, PDB ID: 2ZKO [44]). Influenza NS1 has interesting features: it accumulates in the nuclei of host cells after being translocated by importin α and β and works as a viral immuno-suppressor by weakening the host cell gene expression [45]. DENV C was also reported to have an importin α-like motif in the N-terminal [15,46]. Regarding the targets that may interact with importin α and be transported to the nucleus, they normally contain a nuclear localization sequence (NLS), consisting of a motif of at least 2 consecutive positively charged residues [47–51]. Some of these proteins contain 2 NLS motifs, with at least 8 (up to 40 or even more) residues in between, designated as a bipartite NLS motif [49–51]. Strikingly, *Flavivirus* C proteins have three motifs of two consecutive cationic residues in the N-terminal region and α1 domain, which could form a bipartite NLS. A bipartite NLS formed by the cationic residues before position 10 and at positions 17 and 18, with a spacer of 7 to 13 residues can occur. The other bipartite NLS possibility may be formed by residues at positions 17 and 18, and at positions 31 and 32, with 9 to 12 spacer residues. Possible bipartite NLS are also seen in the conserved fold region but its static nature precludes activity as NLS. If DENV C binds to importin α, it may act as a cargo protein to be transported to the nucleus. This could explain why has DENV C been found in the nucleus of DENV infected cells [46,52,53]. DENV C may directly bind importin β, given the similarities between the N-terminal region of DENV C and importin α [49]. This may allow it to disrupt the normal nuclear import/export system in DENV-infected cells. The conformational plasticity of the N-terminal and flexible fold regions is certainly compatible with interactions with importin(s). As the hypothesized bipartite NLS are conserved among *Flavivirus* C proteins, this may occur in other *Flavivirus*.

The C protein may act as an immuno-suppressor, similarly to influenza NS1, by interacting with importins α and/or importin β. Ivermectin, a specific inhibitor of importin α/β-mediated nuclear import, is able to inhibit HIV-1 and DENV replication [54]. The mechanism of DENV C inhibition might involve the C protein, specifically the intrinsically disordered N-terminal IDP region, which is similar to importin α disordered N-terminal region [15]. Moreover, influenza NS1 can counteract the RNA-activated protein kinase (PKR)-mediated antiviral response through a direct interaction with PKR [55]. Besides, influenza NS1 blocks interferon (IFN) regulatory factor 3 activation, which in turn prevents the induction of IFN-related genes [56]. DENV inhibits the IFN signaling pathway in a similar manner [57]. By its N-terminal region dsRNA-binding ability, influenza NS1 inhibits the nuclear export of mRNAs and modulates pre-mRNA splicing, suppressing antiviral response [44]. Similarities between DENV C and influenza NS1 also extend to the later ability to bind RNA (Figure 7c). Recognition of dsRNA is made by the influenza NS1 RNA-binding domain, which forms a homodimer [44]. Afterwards, a slight change in R38-R38' orientation leads to anchoring the dsRNA to the protein by a hydrogen bond network to the protein [44]. One of the main functions of influenza NS1 binding to RNA is sequestering dsRNA from the 2'–5' oligo(A) synthetase [58]. We propose that, as with influenza NS1, a small conformational change in DENV C α4–α4' interface occurs after the contact of its α2–α2′ interface with LDs, modulated by transitions between alternative N-terminal "open" and "closed" conformations. Binding to LDs requires an open conformation (Figure 7d), decreasing the conformational variability and entropy of the C protein, which trigger the allosteric movements affecting the C-terminal α4–α4′. As with influenza NS1, the *Flavivirus* C protein would remain in the same overall fold, but a small opening of α4–α4′ would facilitate its binding to RNA.

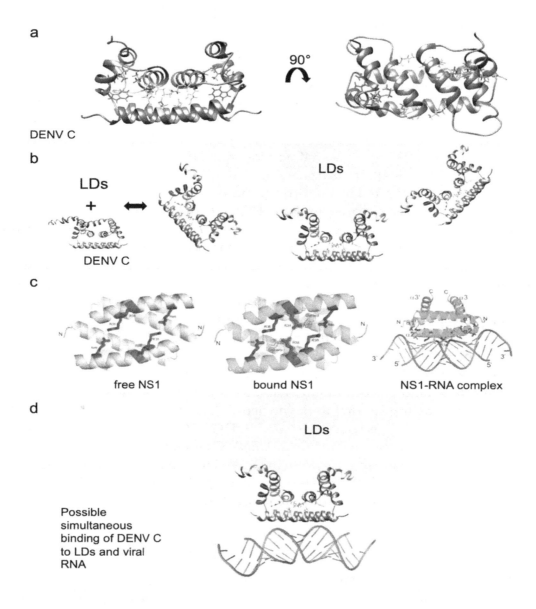

Figure 7. Protein structures of DENV C and influenza NS1. (**a**) DENV C structure from two different angles with the conserved residues R68 and W69 (purple) and the interface stabilizing residues V51 and I65 (green), as well as E87, R55 and K45, forming the salt bridge (cyan). (**b**) DENV C structure in a N-terminal region closed conformation and, next, in an open conformation with schematic binding of lipid droplets (LDs) and the affected amino acid residues (yellow). (**c**) The RNA-binding domain of NS1 protein from influenza A in a RNA-free (left) and RNA-bound state (middle and right), showing an organization similar to DENV C α4–α4′ region (adapted from Cheng et al., 2009 [44]). (**d**) DENV C with schematically bound to a LD and to RNA. DENV C amino acid residues affected by the binding to LDs are colored yellow, while a key internal salt bridge is shown in cyan. DENV C binding to host LDs may enable allosteric rearrangements (eventually involving the salt bridge), allowing a small conformational change in α4 side chains, namely the positively charged residues, prompting stable RNA-C protein binding.

The C-terminal is likely to be the crucial section for RNA binding given its similarity with influenza NS1 (Figure 7). Nevertheless, the N-terminal conformers must also be considered in the context of RNA binding (Figure 4). The A and D conformers allow RNA to be bound to the $\alpha4-\alpha4'$ interface and, simultaneously, to the N-terminal cationic amino acid residues. A–A' and D–D' conformations result in the possible binding of a single continuous portion of RNA to both the C-terminal $\alpha4-\alpha4'$ and the N-terminal IDP region, making the RNA more tightly bound. Moreover, the A–B', B–B' and B–C' conformations would enable the protein to bind two distinct sections of the RNA, one bound to $\alpha4-\alpha4'$ and another to the N-terminal regions. That arrangement may allow to further compact the viral RNA. The N-terminal IDP region putative binding to RNA should not be disregarded given its positive net charge (+7). It compares very well with the C-terminal α-helical region net charge (+8 for a monomer, +16 for $\alpha4-\alpha4'$ dimer interface). Both may thus bind RNA due to, mostly, electrostatic forces. This IDP region can thus provide multi-functionality by several modes of binding and different ligands, enabled by alternative conformations. It must be stressed that this is not unlikely. Viral proteins tend to have IDP regions that increase their biological activity [59–61]. In a proteome as small as that of flaviviruses (10 proteins), IDP regions augment the number of ligands with which it can interact. Less structure often means more function. This is an increasingly hot topic of recent research, leading to design of algorithms to identify these regions [62,63]. Further analysis will help understand the interaction between DENV C and its ligands.

To conclude, the data imply a common structure and functions for mosquito-borne *Flavivirus* C proteins. Moreover, studying DENV C rotational diffusion and thermodynamics reveals a stable protein due to the conserved fold maintaining the homodimer structure. These findings apply to other *Flavivirus* C proteins, supporting a common mechanism for their biological activity. Such understanding of this key protein structure and dynamics properties may contribute to the future development of C protein-targeted drugs to impair dengue virus and other *Flavivirus* infections.

4. Materials and Methods

4.1. Materials

Chromatography columns HiTrap Heparin (1 and 5 mL), Sephadex S200 and the chromatography equipment AKTA-explorer were from GE Healthcare (Little Chalfont, UK). Sodium dodecyl sulphate-polyacrylamide gel electrophoresis (SDS-PAGE) reagents were from BioRad (Hercules, CA, USA). Unless otherwise stated, other chemicals were purchased from Sigma-Aldrich (St. Louis, MO, USA).

4.2. Flavivirus C Proteins Primary, Secondary and Tertiary Structural Predictions

For primary structure alignments we used the 16 non-DENV *Flavivirus* polyprotein sequences identified in reference [14], plus the four DENV reference sequences from NCBI, namely: DENV serotype 1, strain 45AZ5, NCBI ID NP_059433.1; DENV serotype 2, strain New Guinea C, NCBI ID NP_056776.2; DENV serotype 3, strain D3/H/IMTSSA-SRI/2000/1266, NCBI ID YP_001621843.1; and, DENV serotype 4, strain rDEN4, NCBI ID NP_073286.1. For the phylogenetic trees, both the entire polyproteins and the C protein regions were used. For the alignments and subsequent data analysis, the residues next to the NS2B-NS3 protease cleavage site [64,65] were excluded, leaving only the C protein sequences. Alignments and the derived phylogenetic trees were performed via Clustal Omega web tool (http://www.ebi.ac.uk/Tools/msa/clustalo/) [66,67].

Statistical comparison of the disordered N-terminal plus flexible fold regions with the conserved fold region of *Flavivirus* C proteins, for G and P content, as well as charged amino acid residues, was performed via a paired *t*-test, using GraphPad Prism v5 software. *p*-values were always lower than 0.001.

Predictions of hydrophobicity and α-helix propensity were done using ProtScale server (http://web.expasy.org/protscale/) [26,27], tertiary structure predictions were performed via I-TASSER server (http://zhanglab.ccmb.med.umich.edu/I-TASSER/) [28–30], following previous approaches [15]. Briefly, *Flavivirus* C protein sequences from our previous work were employed [14]. DENV and WNV (serotype Kunjin) C structures were excluded, not serving as templates for the tertiary structure prediction. ZIKV C protein structure was also not included, as it was not yet determined when the modeling was conducted. This avoids a bias towards known homologous protein structures. Five I-TASSER models were obtained for each C protein sequence. These were superimposed with DENV C experimental structure (PDB ID 1R6R, model 21) [12] after root-mean-square deviation (RMSD) minimization in UCSF Chimera v1.9 software [68]. Clusters were formed based on the visual similarity between predictions. The number of N-terminal amino acid residues with backbone clashes with the other monomer backbone was calculated for each model. In our previous work [15], a DENV C predicted structure was excluded from further analysis if it had 6 clashes or more, as it would not be viable as an homodimer [15]. Here we excluded models with more than 5 clashes (28 models rejected). These would preclude homodimer formation and, thus, were not considered in the clusters analysis (Table 1 excluded models column).

4.3. Structure Comparison Between DENV C and Influenza NS1

Protein structures coordinates were extracted from the Protein Data Bank (PDB, www.pdb.org). PDB identification codes are specified ahead after each protein name. The protein structures were superimposed through UCSF Chimera 1.13.1 software MatchMaker tool. After that, we carefully analyzed the superposition visually. Then, using the Match-Align tool of UCSF Chimera, which returns a sequence alignment based on the regions and taking into account the structure superimposition, we identified the residues simultaneously similar in structure and sequence. Protein structure figures were obtained using UCSF Chimera 1.13.1 version [68].

4.4. DENV C Recombinant Protein Production and Purification

Recombinant DENV C protein expression and purification was conducted based on previous approaches [13]. We used a pET-21a plasmid containing DENV serotype 2 strain New Guinea C capsid protein gene (encoding amino acid residues 1–100) [69]. The protein was expressed in *Escherichia coli* C41 and C43 bacteria grown in lysogeny broth (LB) medium. The only differences in the purification protocol are the abolition of the ammonium sulfate precipitation step and the addition of a size exclusion chromatography step (with Sephadex S200) after the heparin affinity column chromatography, using an AKTA chromatography equipment. The C protein was purified in a 55 mM KH_2PO_4, pH 6.0, 550 mM KCl. DENV C protein purified fractions were concentrated with Amicon Ultra-4 Centrifugal Filters of 3 or 10 kDa nominal cut-off, from Millipore (Billerica, MA, USA). Concentrated protein samples were stored at −80 °C. Protein samples quality was assessed by SDS-PAGE and matrix-assisted laser desorption/ionization, time-of-flight mass spectrometry (MALDI-TOF MS) analysis. Very low degradation and the highest peak consistent with the expected mass of the protein monomer (11765 Da).

4.5. Time-Resolved Fluorescence Anisotropy

Time-resolved fluorescence spectroscopy measurements were performed in a Life Spec II equipment with an EPLED-280 pulsed excitation light-emitting diode (LED) of 275 nm (Edinburgh Instruments, Livingston, UK), acquiring the emission at 350 nm. DENV C (monomer) concentration was 20 μM in 50 mM KH_2PO_4, 200 mM KCl, pH 6.0 or pH 7.5, with 550 μL total volume, in 0.5 cm × 0.5 cm quartz cuvettes. The instrument response function, $IRF(t)$, was obtained with the same settings, except emission, which was at 280 nm, with a solution of polylatex beads of 60 nm diameter diluted in Mili-Q water. Measurements were performed at 22 °C. Time-resolved fluorescence intensity measurements with picosecond-resolution were obtained by the time-correlated single-photon timing (TCSPT) methodology [35]. Measurements were performed at constant time, with 15 min per decay, acquiring 2048 time points in a 50 ns window. Four intensity decays, $I(t)$, were acquired in each condition, with excitation/emission polarizers, respectively at vertical/vertical positions, $I_{VV}(t)$, vertical/horizontal positions, $I_{VH}(t)$, horizontal/vertical positions, $I_{HV}(t)$, and horizontal/horizontal positions, $I_{HH}(t)$. The instrumental G-factor was calculated as [35]:

$$G = \frac{\int_0^{50} I_{HV}(t)dt}{\int_0^{50} I_{HH}(t)dt} \tag{1}$$

The G-factor value obtained was 1.61. The intensity decay with emission polarizer at the magic angle (~54.7°, with respect to the vertical excitation polarizer), $I_m(t)$, avoids the effects of anisotropy. It can be calculated easily [35]:

$$I_m(t) = I_{VV}(t) + 2GI_{VH}(t) \tag{2}$$

with $I_{VV}(t)$ and $I_{VH}(t)$ depending on the time-resolved fluorescence anisotropy, $r(t)$, as:

$$I_{VV}(t) = \frac{I_m(t)}{3}(1 + 2r(t)) \tag{3}$$

$$I_{VH}(t) = \frac{I_m(t)}{3G}(1 - r(t)) \tag{4}$$

Thus, $I_m(t)$ was used to obtain the fluorescence lifetime components, τ_i, and the respective amplitudes, α_i, for the DENV C W69. $I_m(t)$ was described by a sum of three exponential terms:

$$I_m(t) = \sum_{i=1}^{3} \alpha_i e^{\left(-\frac{t}{\tau_i}\right)} \tag{5}$$

where the index i represents each component of the fluorescence decay. For the fitting to the data, α_i and τ_i values were obtained by iteratively convoluting $I_m(t)$ with the $IRF(t)$:

$$I_m^{calc}(t) = I_m(t) \otimes IRF(t) \tag{6}$$

and fitting $I_m^{calc}(t)$ to the experimental data, $I_m^{exp}(t)$, using a non-linear least squares regression method. The usual statistical criteria, namely a reduced χ^2 value bellow 1.3 and a random distribution of weighted residuals, were used to evaluate the goodness of the fits [35]. Data analysis was performed using the TRFA Data Processing Package v1.4 (Scientific Software Technologies Centre, Belarusian State University, Minsk, Belarus) which allows calculating automatically the standard error (SE) for each fitted parameter [35].

The time-resolved fluorescence anisotropy, $r(t)$, is calculated via $I_{VV}(t)$, $I_{VH}(t)$ and G via_ENREF_52:

$$r(t) = \frac{I_{VV}(t) - GI_{VH}(t)}{I_{VV}(t) + 2GI_{VH}(t)} \tag{7}$$

In this case, the obtained $r(t)$ can be fitted to a single exponential decay [35]:

$$r(t) = r_0 e^{\left(-\frac{t}{\tau_c}\right)} \tag{8}$$

where r_0 is the anisotropy when $t \to 0$ and τ_c is the rotational correlation time. The $r(t)$ decays were globally analyzed in TRFA Data Processing Package v1.4 maintaining the previously obtained α_i and τ_i values constant, and convoluting Equations (3) and (4) with the respective $IRF(t)$, analogously to the analysis of $I_m(t)$, using Equation (8) to fit $r(t)$. Values obtained for both pH conditions were considered statistically different if their 95% confidence intervals (~1.96 × SE) do not overlap (corresponding to $p < 0.05$).

4.6. Rotational Correlation Time Corrections

The τ_c of a molecule in solution is related with the solution viscosity, η, the molecular hydrodynamic volume, V, the Boltzmann constant, k_B, and the absolute temperature, T, as [35,70]:

$$\tau_c = \frac{\eta V}{k_B T} \tag{9}$$

Based on Equation (9), τ_c can be corrected for different temperatures, considering that the molecular volume does not change significantly in a small temperature interval (± 5 °C; i.e., V and k_B are constants), using [70]:

$$\frac{T_a \tau_{c,a}}{\eta_a} = \frac{T_b \tau_{c,b}}{\eta_b} \Leftrightarrow \tau_{c,b} = \tau_{c,a} \frac{\eta_b T_a}{\eta_a T_b} \tag{10}$$

where the indexes 'a' and 'b' represent a different condition of T and η, taking into account the variation of η with T [37]. The η values were assumed to be those of pure H_2O or 10% D_2O in the case of the corrections for the NMR-based values (those from the literature). In this way, Table 5 below shows the values employed on the calculations [37]:

Table 5. Values for η employed in this work, derived from the references and Equations above.

T (°C)	η in H_2O (cP)	η in 10% D_2O (cP)	$\frac{\eta_b T_a}{\eta_a T_b}$ in H_2O	$\frac{\eta_b T_a}{\eta_a T_b}$ in 10% D_2O
20	1.002	1.027	0.8736	0.8523
22	0.955	0.978	0.9231	0.9012
25	0.890	0.911	1	0.9770
27	0.851	0.871	1.0530	1.0293

4.7. Temperature Denaturation Measurements via Circular Dichroism (CD) Spectroscopy

Circular dichroism spectroscopy measurements were carried out in a JASCO J-815 (Tokyo, Japan), using 0.1 cm path length quartz cuvettes, data pitch of 0.5 nm, velocity of 200 nm/min, data integration time (DIT) of 1 s and performing 3 accumulations. Spectra were acquired in the far UV region, between 200 and 260 nm, with 1 nm bandwidth. The temperature was controlled by a JASCO PTC-423S/15 Peltier equipment. It was varied between 0 and 96 °C, in steps of 2 °C, increasing at a rate of 8 °C/min and waiting 100 s after crossing 5 times the target temperature, T. Then, the system was allowed, at least, 120 s to equilibrate (sufficient time for a stable CD signal). Before and after denaturation, spectra were acquired at 25 °C, to determine the reversibility of thermal denaturation. DENV C monomer concentration was 20 µM in 50 mM KH_2PO_4, 200 mM KCl, pH 6.0 or pH 7.5, with 220 µL of total volume. Spectra were smoothed through the means-movement method (using 7 points) and normalized to mean residue molar ellipticity, $[\theta]$ (in deg cm^2 dmol^{-1} Res^{-1}).

For the CD temperature denaturation data treatment, we assumed a dimer to monomer denaturation model [71–73] in which the folded dimer, F_2, separates into unfolded monomers, U, in a single step described by reaction R1:

$$F_2 \Leftrightarrow 2U \tag{R1}$$

In this system, the total protein concentration, $[P_m]$, in monomer equivalents, is described as:

$$[P_m] = 2[F_2] + [U] \tag{11}$$

Hereafter, concentrations are treated as dimensionless, being divided by the standard concentration of 1 M, in order to be at standard thermodynamic conditions. The fractions of monomer in the folded, f_F, and unfolded, f_U, states are calculated by [71,72]:

$$f_F = \frac{2[F_2]}{[P_m]} \tag{12}$$

$$f_U = \frac{[U]}{[P_m]} \tag{13}$$

$$f_F + f_U = 1 \tag{14}$$

and the concentrations of folded dimer and unfolded monomer can be written in terms of f_U:

$$[U] = f_U[P_m] \tag{15}$$

$$[F_2] = \frac{f_F[P_m]}{2} = \frac{(1-f_U)[P_m]}{2} \tag{16}$$

Then, the equilibrium constant, K_{eq}, of R1 is defined in terms of $[U]$ and $[F_2]$, or f_U and $[P_m]$:

$$K_{eq} = \frac{[U]^2}{[F_2]} = \frac{(f_U[P_m])^2}{(1-f_U)[P_m]/2} = \frac{2[P_m] \times f_U^2}{(1-f_U)} \tag{17}$$

which can be solved in order to f_U, with the only solution in which $f_U \in [0;1]$ being:

$$f_U = \frac{\sqrt{8[P_m]K_{eq} + K_{eq}^2} - K_{eq}}{4[P_m]} \tag{18}$$

The $[\theta]$ signal as a function of temperature [71,72,74], $[\theta]_T$, can be described as a linear combination of the signal of the folded, $[\theta]_{T,F}$, and unfolded states, $[\theta]_{T,U}$, weighted by f_U:

$$[\theta]_T = [\theta]_{T,F}(1-f_U) + [\theta]_{T,U}f_U \tag{19}$$

where $[\theta]_{T,F}$ and $[\theta]_{T,U}$ have a variation with T described here by a straight line (*i* can be F or U) [72,74]:

$$[\theta]_{T,i} = m_i \times T + [\theta]_{0,i} \tag{20}$$

Equation (19) can be re-written to evidence f_U and then substitute it by Equation (18) [71,72]:

$$[\theta]_T = [\theta]_{T,F} + \left([\theta]_{T,U} - [\theta]_{T,F}\right)\frac{\sqrt{8[P_m]K_{eq} + K_{eq}^2} - K_{eq}}{4[P_m]} \tag{21}$$

K_{eq} can also be described by the standard Gibbs free-energy, ΔG°, of the reaction R1:

$$K_{eq} = e^{-\frac{\Delta G^\circ}{RT}} \tag{22}$$

where R is the rare gas constant and T is the absolute temperature. The ΔG° function used to fit the data contains both the enthalpic, ΔH°, and entropic, ΔS°, variations with temperature, which take into account $\Delta H^\circ{}_{T_m^\circ}$, the specific heat capacity at constant pressure, ΔC_p°, and the standard conditions' denaturation temperature, T_m°, according to [74]:

$$\Delta G^\circ = \Delta H^\circ{}_{T_m^\circ}\left(1 - \frac{T}{T_m^\circ}\right) - \Delta C_p^\circ\left(T_m^\circ - T + T\ln\left(\frac{T}{T_m^\circ}\right)\right) \tag{23}$$

In our data, ΔC_p° was statistically equal to 0 and, thus, Equation (23) can be simplified to:

$$\Delta G^\circ = \Delta H^\circ{}_{T_m^\circ}\left(1 - \frac{T}{T_m^\circ}\right) \qquad (24)$$

Then, Equation (21) was combined with Equations (20), (22) and (24), and fitted to the data using GraphPad Prism v5 software, via the non-linear least squares method, to extract both the $\Delta H^\circ{}_{T_m^\circ}$ and T_m°, along with the respective SE values. Afterwards, $\Delta S^\circ{}_{T_m^\circ}$ can be obtained, since $\Delta G^\circ = 0$ kJ mol^{-1} at T_m°, via the following Equation:

$$\Delta H^\circ{}_{T_m^\circ} - T_m^\circ \Delta S^\circ{}_{T_m^\circ} = 0 \Rightarrow \Delta S^\circ{}_{T_m^\circ} = \frac{\Delta H^\circ{}_{T_m^\circ}}{T_m^\circ} \qquad (25)$$

The SE of $\Delta S^\circ{}_{T_m^\circ}$ was calculated based on $\Delta H^\circ{}_{T_m^\circ}$, T_m°, and the respective SE values:

$$\mathrm{SE}_{\Delta S^\circ{}_{T_m^\circ}} = \left|\frac{\Delta H^\circ{}_{T_m^\circ}}{T_m^\circ}\right| \times \sqrt{\left(\frac{\mathrm{SE}_{\Delta H^\circ{}_{T_m^\circ}}}{\Delta H^\circ{}_{T_m^\circ}}\right)^2 + \left(\frac{\mathrm{SE}_{T_m^\circ}}{T_m^\circ}\right)^2} \qquad (26)$$

Interestingly, for a dimer to monomer denaturation, K_{eq} depends on $[P_m]$ and, consequently, ΔG° also depends on $[P_m]$. This implies that $\Delta G^\circ = 0$ at T_m° (T_m value estimated if $[P_m] = 1M$), which is considerably higher than the observed T_m (that occurs when $f_U = 0.5$). The dependence of T_m with $[P_m]$ is [72]:

$$\Delta G^\circ{}_{f_U=0.5} = -RT_m \ln([P_m]) \Rightarrow T_m = \frac{\Delta G^\circ{}_{f_U=0.5}}{-R\ln([P_m])} \qquad (27)$$

$$T_m = \frac{\Delta H^\circ{}_{T_m^\circ}}{\Delta S^\circ{}_{T_m^\circ} - R\ln([P_m])} \qquad (28)$$

The SE of T_m was based on the percentual SE value of T_m°.

Values obtained for both pH conditions were statistically evaluated via F-tests to compare two possible fits, one assuming a given parameter as being different for the distinct data sets, and another assuming that parameter to be equal between data sets (while maintaining the other parameters different). No statistically significant difference ($p < 0.05$) was observed.

Author Contributions: Conceptualization, A.F.F., N.C.S. and I.C.M.; In silico studies, A.F.F., V.A., A.S.M., N.K. and I.C.M.; Recombinant protein production, A.F.F., A.S.M., F.J.E. and I.C.M.; Time-resolved fluorescence anisotropy studies, A.F.F. and J.C.R.; Circular dichroism studies, A.F.F. and I.C.M.; Formal analysis, A.F.F., J.C.R. and I.C.M.; Resources, I.C.M., F.J.E., N.C.S.; Writing-original draft preparation, A.F.F., N.K., and I.C.M.; Writing-review and editing, A.F.F., A.S.M., N.K., N.C.S. and I.C.M.; Supervision, N.C.S. and I.C.M.; Project administration, N.C.S. and I.C.M.; Funding acquisition, N.C.S. and I.C.M.

Abbreviations

ALFV	Alfuy virus
APOE	Apolipoprotein E
AROAV	Aroa virus
BAGV	Bagaza virus
C protein	Capsid protein
CD	Circular dichroism
DENV	Dengue virus
ICTV	International Committee on Taxonomy of Viruses
IDP	Intrinsically disordered protein
IFN	Interferon

IGUV	Iguape virus
ILHV	Ilheus virus
JEV	Japanese encephalitis virus
KEDV	Kedougou virus
KOKV	Kokobera virus
LDs	Lipid droplets
MVEV	Murray Valley encephalitis virus
NS1	Non-structural protein 1 from influenza virus A
PDB	Protein Data Bank
pep14-23	Inhibitor peptide pep14-23 (amino acid sequence NMLKRARNRV)
PLIN3	Perilipin 3
ROCV	Rocio virus
SLEV	Saint Louis encephalitis virus
SPOV	Spondweni virus
USUV	Usutu virus
VLDL	Very low-density lipoproteins
WNV	West Nile virus
WNV-K	WNV serotype Kunjin
YFV	Yellow fever virus
ZIKV	Zika virus

References

1. Bhatt, S.; Gething, P.W.; Brady, O.J.; Messina, J.P.; Farlow, A.W.; Moyes, C.L.; Drake, J.M.; Brownstein, J.S.; Hoen, A.G.; Sankoh, O.; et al. The global distribution and burden of dengue. *Nature* **2013**, *496*, 504–507. [CrossRef] [PubMed]

2. Sanofi Pasteur. Available online: https://www.sanofipasteur.com/en/media-room/press-releases/dengvaxia-vaccine-approved-for-prevention-of-dengue-in-europe (accessed on 30 January 2019).

3. Durbin, A.P. A dengue vaccine. *Cell* **2016**, *166*, 1. [CrossRef] [PubMed]

4. Villar, L.; Dayan, G.H.; Arredondo-Garcia, J.L.; Rivera, D.M.; Cunha, R.; Deseda, C.; Reynales, H.; Costa, M.S.; Morales-Ramirez, J.O.; Carrasquilla, G.; et al. Efficacy of a tetravalent dengue vaccine in children in Latin America. *N. Engl. J. Med.* **2015**, *372*, 113–123. [CrossRef] [PubMed]

5. Takeda. Available online: https://www.takeda.com/newsroom/newsreleases/2019/takedas-dengue-vaccine-candidate-meets-primary-endpoint-in-pivotal-phase-3-efficacy-trial/ (accessed on 4 February 2019).

6. ICTV Taxonomy. Available online: https://talk.ictvonline.org/taxonomy/ (accessed on 17 April 2019).

7. Grard, G.; Moureau, G.; Charrel, R.N.; Holmes, E.C.; Gould, E.A.; de Lamballerie, X. Genomics and evolution of Aedes-borne flaviviruses. *J. Gen. Virol.* **2019**, *91*, 87–94. [CrossRef] [PubMed]

8. Schubert, A.M.; Putonti, C. Infection, genetics and evolution of the sequence composition of flaviviruses. *Infect. Genet. Evol.* **2010**, *10*, 129–136. [CrossRef] [PubMed]

9. Calisher, C.H.; Gould, E.A. Taxonomy of the virus family *Flaviviridae*. *Adv. Virus Res.* **2003**, *59*, 1–19. [PubMed]

10. Mukhopadhyay, S.; Kuhn, R.J.; Rossmann, M.G. A structural perspective of the flavivirus life cycle. *Nat. Rev. Microbiol.* **2005**, *3*, 13–22. [CrossRef]

11. Kuhn, R.J.; Zhang, W.; Rossmann, M.G.; Pletnev, S.V.; Corver, J.; Lenches, E.; Jones, C.T.; Mukhopadhyay, S.; Chipman, P.R.; Strauss, E.G.; et al. Structure of dengue virus: Implications for flavivirus organization, maturation, and fusion. *Cell* **2002**, *108*, 717–725. [CrossRef]

12. Ma, L.; Jones, C.T.; Groesch, T.D.; Kuhn, R.J.; Post, C.B. Solution structure of dengue virus capsid protein reveals another fold. *Proc. Natl. Acad. Sci. USA* **2004**, *101*, 3414–3419. [CrossRef]

13. Faustino, A.F.; Barbosa, G.M.; Silva, M.; Castanho, M.A.R.B.; da Poian, A.T.; Cabrita, E.J.; Santos, N.C.; Almeida, F.C.L.; Martins, I.C. Fast NMR method to probe solvent accessibility and disordered regions in proteins. *Sci. Rep.* **2019**, *9*, 1647. [CrossRef]

14. Martins, I.C.; Gomes-Neto, F.; Faustino, A.F.; Carvalho, F.A.; Carneiro, F.A.; Bozza, P.T.; Mohana-Borges, R.; Castanho, M.A.R.B.; Almeida, F.C.L.; Santos, N.C.; et al. The disordered N-terminal region of dengue virus capsid protein contains a lipid-droplet-binding motif. *Biochem. J.* **2012**, *444*, 405–415. [CrossRef]

15. Faustino, A.F.; Guerra, G.M.; Huber, R.G.; Hollmann, A.; Domingues, M.M.; Barbosa, G.M.; Enguita, F.J.; Bond, P.J.; Castanho, M.A.R.B.; da Poian, A.T.; et al. Understanding Dengue virus capsid protein disordered N-terminus and pep14-23-based inhibition. *ACS Chem. Biol.* **2015**, *10*, 517–526. [CrossRef]

16. Jones, C.T.; Ma, L.; Burgner, J.W.; Groesch, T.D.; Post, C.B.; Kuhn, R.J. Flavivirus capsid is a dimeric alpha-helical protein. *J. Virol.* **2003**, *77*, 7143–7149. [CrossRef] [PubMed]

17. Van Gorp, E.C.M.; Suharti, C.; Mairuhu, A.T.A.; Dolmans, W.M.V.; van der Ven, J.; Demacker, P.N.M.; van der Meer, J.W.M. Changes in the plasma lipid profile as a potential predictor of clinical outcome in dengue hemorrhagic fever. *Clin. Infect. Dis.* **2002**, *34*, 1150–1153. [CrossRef]

18. Samsa, M.M.; Mondotte, J.A.; Iglesias, N.G.; Assuncao-Miranda, I.; Barbosa-Lima, G.; da Poian, A.T.; Bozza, P.T.; Gamarnik, A. V Dengue virus capsid protein usurps lipid droplets for viral particle formation. *PLoS Pathog.* **2009**, *5*, e1000632. [CrossRef]

19. Suvarna, J.C.; Rane, P.P. Serum lipid profile: A predictor of clinical outcome in dengue infection. *Trop. Med. Int. Heal.* **2009**, *14*, 576–585. [CrossRef] [PubMed]

20. Carvalho, F.A.; Carneiro, F.A.; Martins, I.C.; Assunção-Miranda, I.; Faustino, A.F.; Pereira, R.M.; Bozza, P.T.; Castanho, M.A.R.B.; Mohana-Borges, R.; da Poian, A.T.; et al. Dengue virus capsid protein binding to hepatic lipid droplets (LD) is potassium ion dependent and is mediated by LD surface proteins. *J. Virol.* **2012**, *86*, 2096–2108. [CrossRef]

21. Faustino, A.F.; Carvalho, F.A.; Martins, I.C.; Castanho, M.A.R.B.; Mohana-Borges, R.; Almeida, F.C.L.; da Poian, A.T.; Santos, N.C. Dengue virus capsid protein interacts specifically with very low-density lipoproteins. *Nanomed. Nanotechnol. Biol. Med.* **2014**, *10*, 247–255. [CrossRef]

22. Faustino, A.F.; Martins, I.C.; Carvalho, F.A.; Castanho, M.A.R.B.; Maurer-Stroh, S.; Santos, N.C. Understanding dengue virus capsid protein interaction with key biological targets. *Sci. Rep.* **2015**, *5*, 10592. [CrossRef]

23. Martins, A.S.; Carvalho, F.A.; Faustino, A.F.; Martins, I.C.; Santos, N.C. West Nile virus capsid protein interacts with biologically relevant host lipid systems. *Front. Cell. Infect. Microbiol.* **2019**, *9*, 8. [CrossRef]

24. Martins, A.S.; Martins, I.C.; Santos, N.C. Methods for lipid droplet biophysical characterization in *Flaviviridae* infections. *Front. Microbiol.* **2018**, *9*, 1951. [CrossRef] [PubMed]

25. Shang, Z.; Song, H.; Shi, Y.; Qi, J.; Gao, G.F. Crystal structure of the capsid protein from Zika virus. *J. Mol. Biol.* **2018**, *430*, 948–962. [CrossRef] [PubMed]

26. Kyte, J.; Doolittle, R.F. A Simple method for displaying the hydropathic character of a protein. *J. Mol. Biol.* **1982**, *157*, 105–132. [CrossRef]

27. Deléage, G.; Roux, B. An algorithm for protein secondary structure prediction based on class prediction. *Protein Eng.* **1987**, *1*, 289–294. [CrossRef] [PubMed]

28. Zhang, Y. I-TASSER server for protein 3D structure prediction. *BMC Bioinform.* **2008**, *8*, 1–8. [CrossRef]

29. Yang, J.; Yan, R.; Roy, A.; Xu, D.; Poisson, J.; Arbor, A.; Arbor, A. The I-TASSER suite: Protein structure and function prediction. *Nat. Methods* **2015**, *12*, 7–8. [CrossRef] [PubMed]

30. Roy, A.; Kucukural, A.; Zhang, Y. I-TASSER: A unified platform for automated protein structure and function prediction. *Nat. Protoc.* **2011**, *5*, 725–738. [CrossRef]

31. Dokland, T.; Walsh, M.; Mackenzie, J.M.; Khromykh, A.A.; Ee, K.-H.; Wang, S. West Nile virus core protein; tetramer structure and ribbon formation. *Structure* **2004**, *12*, 1157–1163. [CrossRef]

32. Zhan, C.; Zhao, L.; Chen, X.; Lu, W.; Lu, W. Total chemical synthesis of dengue 2 virus capsid protein via native chemical ligation: Role of the conserved salt-bridge. *Bioorg. Med. Chem.* **2013**, *21*, 3443–3449. [CrossRef]

33. Morando, M.A.; Barbosa, G.M.; Cruz-Oliveira, C.; da Poian, A.T.; Almeida, F.C.L. Dynamics of Zika virus capsid protein in solution: The properties and exposure of the hydrophobic cleft are controlled by the α-helix 1 sequence. *Biochemistry* **2019**, *58*, 2488–2498. [CrossRef]

34. Kumar, S.; Ravi, V.K.; Swaminathan, R. How do surfactants and DTT affect the size, dynamics, activity and growth of soluble lysozyme aggregates? *Biochem. J.* **2008**, *415*, 275–288. [CrossRef]

35. Lakowicz, J. *Principles of Fluorescence Spectroscopy*, 3rd ed.; Springer Science, LLC: Berlin/Heidelberg, Germany, 2006; ISBN 9780387312781.

36. Rossi, P.; Yuanpeng, G.V.T.S.; James, J.H.; Anklin, C.; Conover, K.; Hamilton, K.; Xiao, R. A microscale protein NMR sample screening pipeline. *J. Biomol. NMR* **2010**, *46*, 11–22. [CrossRef] [PubMed]

37. Cho, C.H.; Urquidi, J.; Singh, S.; Robinson, G.W. Thermal offset viscosities of liquid H_2O, D_2O, and T_2O. *J. Phys. Chem. B* **1999**, *103*, 1991–1994. [CrossRef]

38. Martins, I.C.; Almeida, F.C.L.; Santos, N.C.; da Poian, A.T. DENV–Derived Peptides and Methods for the Inhibition of Flavivirus Replication. International Patent Publication Nr WO/2012/159187, 26 May 2011.

39. Ivanyi-Nagy, R.; Lavergne, J.; Gabus, C.; Ficheux, D.; Darlix, J.; Inserm, L.; Supe, E.N. RNA chaperoning and intrinsic disorder in the core proteins of *Flaviviridae. Nucleic Acids Res.* **2008**, *36*, 712–725. [CrossRef]

40. Ivanyi-Nagy, R.; Darlix, J. Core protein-mediated 5–3 annealing of the West Nile virus genomic RNA in vitro. *Virus Res.* **2012**, *167*, 226–235. [CrossRef]

41. Kumar, M.; Gromiha, M.M.; Raghava, G.P.S. SVM based prediction of RNA-binding proteins using binding residues and evolutionary information. *J. Mol. Recognit.* **2011**, *24*, 303–313. [CrossRef] [PubMed]

42. Järvelin, A.I.; Noerenberg, M.; Davis, I.; Castello, A. The new (dis)order in RNA regulation. *Cell Commun. Signal.* **2016**, *14*, 9. [CrossRef]

43. Shavinskaya, A.; Boulant, S.; Penin, F.; McLauchlan, J.; Bartenschlager, R. The lipid droplet binding domain of hepatitis C virus core protein is a major determinant for efficient virus assembly. *J. Biol. Chem.* **2007**, *282*, 37158–37169. [CrossRef]

44. Cheng, A.; Wong, S.M.; Yuan, Y.A. Structural basis for dsRNA recognition by NS1 protein of influenza A virus. *Cell Res.* **2009**, *19*, 187–195. [CrossRef]

45. Fernandez-Sesma, A.; Marukian, S.; Ebersole, B.J.; Kaminski, D.; Park, M.S.; Yuen, T.; Sealfon, S.C.; Garcia-Sastre, A.; Moran, T.M. Influenza virus evades innate and adaptive immunity via the NS1 protein. *J. Virol.* **2006**, *80*, 6295–6304. [CrossRef]

46. Wang, S.H.; Syu, W.J.; Huang, K.J.; Lei, H.Y.; Yao, C.W.; King, C.C.; Hu, S.T. Intracellular localization and determination of a nuclear localization signal of the core protein of dengue virus. *J. Gen. Virol.* **2002**, *83*, 3093–3102. [CrossRef] [PubMed]

47. Kobe, B. Autoinhibition by an internal nuclear localization signal revealed by the crystal structure of mammalian importin α. *Nat. Struct. Biol.* **1999**, *6*, 388–397. [CrossRef] [PubMed]

48. Catimel, B.; Teh, T.; Fontes, M.R.M.; Jennings, I.G.; Jans, D.A.; Howlett, G.J.; Nice, E.C.; Kobe, B. Biophysical characterization of interactions involving importin-α during nuclear import. *J. Biol. Chem.* **2001**, *276*, 34189–34198. [CrossRef] [PubMed]

49. Marfori, M.; Mynott, A.; Ellis, J.J.; Mehdi, A.M.; Saunders, N.F.W.; Curmi, P.M.; Forwood, J.K.; Boden, M.; Kobe, B. Molecular basis for specificity of nuclear import and prediction of nuclear localization. *Biochim. Biophys. Acta* **2011**, *1813*, 1562–1577. [CrossRef] [PubMed]

50. Fontes, M.R.M.; Teh, T.; Kobe, B. Structural basis of recognition of monopartite and bipartite nuclear localization sequences by mammalian importin-α. *J. Mol. Biol.* **2000**, *297*, 1183–1194. [CrossRef] [PubMed]

51. Marfori, M.; Lonhienne, T.G.; Forwood, J.K.; Kobe, B. Structural basis of high-affinity nuclear localization signal interactions with importin-alpha. *Traffic* **2012**, *13*, 532–548. [CrossRef] [PubMed]

52. Tadano, M.; Makino, Y.; Fukunaga, T.; Okuno, Y.; Fukai, K. Detection of dengue 4 virus core protein in the nucleus I. A monoclonal antibody to dengue 4 virus reacts with the antigen in the nucleus and cytoplasm. *J. Gen. Virol.* **1989**, *70*, 1409–1415. [CrossRef] [PubMed]

53. Makino, Y.; Tadano, M.; Anzai, T.; Ma, S.P.; Yasuda, S.; Žagar, E. Detection of dengue 4 virus core protein in the nucleus II. Antibody against dengue 4 core protein produced by a recombinant baculovirus reacts with the antigen in the nucleus. *J. Gen. Virol.* **1989**, *70*, 1417–1425. [CrossRef] [PubMed]

54. Wagstaff, K.M.; Sivakumaran, H.; Heaton, S.M.; Harrich, D.; Jans, D.A. Ivermectin is a specific inhibitor of importin α/β-mediated nuclear import able to inhibit replication of HIV-1 and dengue virus. *Biochem. J.* **2012**, *443*, 851–856. [CrossRef]

55. Bergmann, M.; Garcia-Sastre, A.; Carnero, E.; Pehamberger, H.; Wolff, K.; Palese, P.; Muster, T. Influenza virus NS1 protein counteracts PKR-mediated inhibition of replication. *J. Virol.* **2000**, *74*, 6203–6206. [CrossRef]

56. Kochs, G.; Garcia-Sastre, A.; Martinez-Sobrido, L. Multiple anti-interferon actions of the influenza A virus NS1 protein. *J. Virol.* **2007**, *81*, 7011–7021. [CrossRef] [PubMed]

57. Rodriguez-Madoz, J.R.; Bernal-Rubio, D.; Kaminski, D.; Boyd, K.; Fernandez-Sesma, A. Dengue virus inhibits the production of type I interferon in primary human dendritic cells. *J. Virol.* **2010**, *84*, 4845–4850. [CrossRef] [PubMed]

58. Min, J.Y.; Krug, R.M. The primary function of RNA binding by the influenza A virus NS1 protein in infected cells: Inhibiting the 2′–5′ oligo (A) synthetase/RNase L pathway. *Proc. Natl. Acad. Sci. USA* **2006**, *103*, 7100–7105. [CrossRef]

59. Uversky, V.N. Intrinsically disordered proteins and their "mysterious" (meta)physics. *Front. Phys.* **2019**, *7*, 10. [CrossRef]

60. Na, J.H.; Lee, W.K.; Yu, Y.G. How do we study the dynamic structure of unstructured proteins: A case study on nopp140 as an example of a large, intrinsically disordered protein. *Int. J. Mol. Sci.* **2018**, *19*, 381. [CrossRef]

61. Uversky, V.N. Introduction to intrinsically disordered proteins (IDPs). *Chem. Rev.* **2014**, *114*, 6557–6560. [CrossRef] [PubMed]

62. Minde, D.P.; Halff, E.F.; Tans, S. Designing disorder: Tales of the unexpected tails. *Intrinsically Disord. Proteins* **2013**, *1*, e26790. [CrossRef]

63. Krystkowiak, I.; Manguy, J.; Davey, N.E. PSSMSearch: A server for modeling, visualization, proteome-wide discovery and annotation of protein motif specificity determinants. *Nucleic Acids Res.* **2018**, *46*, W235–W241. [CrossRef]

64. Bera, A.K.; Kuhn, R.J.; Smith, J.L. Functional characterization of cis and trans activity of the Flavivirus NS2B-NS3 protease. *J. Biol. Chem.* **2007**, *282*, 12883–12892. [CrossRef]

65. Niyomrattanakit, P.; Yahorava, S.; Mutule, I.; Mutulis, F.; Petrovska, R.; Prusis, P.; Katzenmeier, G.; Wikberg, J.E. Probing the substrate specificity of the dengue virus type 2 NS3 serine protease by using internally quenched fluorescent peptides. *Biochem. J.* **2006**, *397*, 203–211. [CrossRef]

66. Sievers, F.; Higgins, D.G. Clustal omega. *Curr. Protoc. Bioinform.* **2014**, *13*, 1–16.

67. Sievers, F.; Wilm, A.; Dineen, D.; Gibson, T.J.; Karplus, K.; Li, W.; Lopez, R.; Thompson, J.D.; Higgins, D.G.; Mcwilliam, H.; et al. Fast, scalable generation of high-quality protein multiple sequence alignments using clustal omega. *Mol. Syst. Biol.* **2011**, *7*, 539. [CrossRef]

68. Pettersen, E.F.; Goddard, T.D.; Huang, C.C.; Couch, G.S.; Greenblatt, D.M.; Meng, E.C.; Ferrin, T.E. UCSF chimera—A visualization system for exploratory research and analysis. *J. Comput. Chem.* **2004**, *25*, 1605–1612. [CrossRef]

69. Irie, K.; Mohan, P.; Sasaguri, Y.; Putnak, R.; Padmanabhan, R. Sequence analysis of cloned dengue virus type 2 genome (New Guinea-C strain). *Gene* **1989**, *75*, 197–211. [CrossRef]

70. Smith, P.; van Gunsteren, W. Translational and rotational diffusion of proteins. *J. Mol. Biol.* **1994**, *236*, 629–636. [CrossRef] [PubMed]

71. Mok, Y.; de Prat Gay, G.; Butler, P.; Bycroft, M. Equilibrium dissociation and unfolding. *Protein Sci.* **1996**, *5*, 310–319. [CrossRef]

72. Rumfeldt, J.; Galvagnion, C.; Vassall, K.; Meiering, E. Conformational stability and folding mechanisms of dimeric proteins. *Prog. Biophys. Mol. Biol.* **2008**, *98*, 61–84. [CrossRef]

73. Neet, K.E.; Timm, D.E. Conformational stability of dimeric proteins: Quantitative studies by equilibrium denaturation. *Protein Sci.* **1994**, *3*, 2167–2174. [CrossRef]

74. Allen, D.L.; Pielak, G.J. Baseline length and automated fitting of denaturation data. *Protein Sci.* **1998**, *7*, 1262–1263. [CrossRef]

13

Sequence and Structure Properties Uncover the Natural Classification of Protein Complexes Formed by Intrinsically Disordered Proteins via Mutual Synergistic Folding

Bálint Mészáros [1,2,3,*], **László Dobson** [4,5], **Erzsébet Fichó** [3] **and István Simon** [3,*]

[1] MTA-ELTE Momentum Bioinformatics Research Group, Department of Biochemistry, Eötvös Loránd University, Pázmány Péter stny 1/c, H-1117 Budapest, Hungary

[2] European Molecular Biology Laboratory, Structural and Computational Biology Unit, Meyerhofstraße 1, 69117 Heidelberg, Germany

[3] Protein Structure Research Group, Institute of Enzymology, RCNS, HAS, Magyar Tudósok krt 2, H-1117 Budapest, Hungary; ficho.erzsebet@ttk.mta.hu

[4] Membrane Protein Bioinformatics Research Group, Institute of Enzymology, RCNS, HAS, Magyar Tudósok krt 2, H-1117 Budapest, Hungary; dobson.laszlo.imre@itk.ppke.hu

[5] Faculty of Information Technology and Bionics, Pázmány Péter Catholic University, Práter u. 50A, H-1083 Budapest, Hungary

* Correspondence: bmeszaros@caesar.elte.hu (B.M.); simon.istvan@ttk.mta.hu (I.S.)

Abstract: Intrinsically disordered proteins mediate crucial biological functions through their interactions with other proteins. Mutual synergistic folding (MSF) occurs when all interacting proteins are disordered, folding into a stable structure in the course of the complex formation. In these cases, the folding and binding processes occur in parallel, lending the resulting structures uniquely heterogeneous features. Currently there are no dedicated classification approaches that take into account the particular biological and biophysical properties of MSF complexes. Here, we present a scalable clustering-based classification scheme, built on redundancy-filtered features that describe the sequence and structure properties of the complexes and the role of the interaction, which is directly responsible for structure formation. Using this approach, we define six major types of MSF complexes, corresponding to biologically meaningful groups. Hence, the presented method also shows that differences in binding strength, subcellular localization, and regulation are encoded in the sequence and structural properties of proteins. While current protein structure classification methods can also handle complex structures, we show that the developed scheme is fundamentally different, and since it takes into account defining features of MSF complexes, it serves as a better representation of structures arising through this specific interaction mode.

Keywords: intrinsically disordered protein; IDP; protein–protein interaction; mutual synergistic folding; coupled folding and binding; structural analysis; structure-based classification; fold recognition

1. Introduction

Intrinsically disordered proteins (IDPs) are crucial elements of the molecular machinery indispensable for complex life [1,2]. IDPs are parts of regulatory pathways [3], control the cell cycle [4,5], function as chaperones [6,7], and regulate protein degradation [8,9], amongst other functions. In accord, IDPs are typically under tight regulation at several levels [3,10]. While some IDPs fulfill their functions directly through their lack of structure, such as spring-like entropic chains, the majority of disordered proteins interact with other macromolecules, most often other proteins [11]. IDP-mediated interactions are essential for many hub proteins [12,13], and several IDPs serve as interaction scaffolds/platforms for macromolecular assembly [14,15]. Mounting evidence also shows that protein disorder plays a crucial role in the assembly of liquid–liquid phase separated non-membrane-bounded organelles [16].

Depending on the partner protein and the specifics of the interaction, IDPs can bind through several mechanisms. Several IDPs recognize and bind to ordered protein domains, usually through a linear sequence motif [17]. While some IDPs retain their inherent flexibility in the bound form as well [18], in most known cases the complex structure lends itself to standard structure determination methods, such as X-ray crystallography or NMR. These cases of coupled folding and binding have been studied intensively [19–21]. However, IDPs can utilize a fundamentally different molecular mechanism for interaction, through which they reach a folded state as well. Complexes that contain only IDPs as constituent protein chains, without the presence of a previously folded domain, are formed via a process called mutual synergistic folding (MSF) [22]—a much less understood way in which protein folding and binding can merge into a single biophysical process.

A major advancement in the field of IDP interactions in recent years was the development of specialized interaction databases for various mechanisms including coupled folding and binding [23,24], fuzzy complexes [25], mutual synergistic folding [26], and proteins driving liquid–liquid phase separation [27]. Out of these aspects, possibly the most understudied one is mutual synergistic folding, owing to the fact that these are the only interactions where none of the partner proteins have a well-defined structure outside of the complex, forcing us to revise our current approaches used for describing protein structures and complexes. The biological and biophysical properties of these interactions are markedly different from those mediated by other types of proteins. While in other interaction types a stable, folded hydrophobic core is already present in at least one partner, here the folding and binding happen at the same time for all partners. Comparative analysis has not only shown that MSF complexes constitute a separate biologically meaningful class, but also highlighted that these complexes are highly heterogeneous in terms of sequence and structure propreties [28–30].

We now have knowledge of over 140,000 protein structures deposited in the Protein Data Bank (PDB) [31], a major part of which contains several proteins in complex. In each of these cases, the proteins achieve stability either before or upon interacting. A major question is how is stability achieved? Can this be a basis of the definition of biologically meaningful classification? In the case of ordered proteins, current hierarchical classification schemes are rooted in the tertiary protein structures, such as in the case of methods/databases as SCOP (Structural Classification of Proteins) [32] and CATH (Class, Architecture, Topology, Homologous superfamily) [33]. While these methods are extended to classify protein complexes as well, they do not explicitly factor in parameters that describe the interactions or the differences in sequence composition between complexes of similar overall structures. However, in the case of MSF complexes, these differences are defining features, as the interaction is the primary reason for the emergence of the structure itself, and this interaction usually requires highly specialized residue compositions [28]. While other classification methods were developed specifically for protein–protein interactions, they only aim to describe the interface, without taking the overall resulting structure into account [34].

Here we present the first classification method designed to identify biologically relevant types of protein complexes formed via mutual synergistic folding. Our work aims to answer specific questions about the types of MSF complexes based on the currently known more than 200 examples. Are there intrinsic classes of MSF complexes or are all known examples basically unique in terms of sequence and structure? If meaningful groups are definable in an objective way, what are the characteristics of each group in terms of sequence composition and adopted structure? In addition, how is the formation of MSF complexes regulated? Are mechanisms known to be important for other molecular interactions relevant to these complexes as well? If so, are there differences between various MSF groups regarding these regulatory mechanisms and other biologically relevant properties, such as binding strength and subcellular localization?

2. Results

2.1. Sequence-Based Properties Define Four Clusters of Complexes

Complexes formed by mutual synergistic folding were taken from the MFIB (Mutual Folding Induced by Binding) database [26], and each complex has been assigned a feature vector describing the sequence composition of its constituent protein chains. To represent the sequence composition, we use the amino acid grouping previously used for investigating protein–protein complexes involving IDPs [28] (see Data and Methods and Figure 1 for definitions, and Supplementary Table S1 for exact values for all complexes). These vectors were used as input for hierarchical clustering (Supplementary Figure S1) to quantify the sequence-based relationship between various complexes. k-means clustering (Supplementary Figure S2) indicates four as a suitable number of clusters, and, therefore, we use four sequence-based clusters in all subsequent analyses. While this choice is not the only acceptable one based on the k-means results, we aim to have a restricted set of clusters to describe the major types of sequential classes. The main features of the four clusters are shown in Figure 1, while cluster numbers for each complex are shown in Supplementary Table S1.

Figure 1 shows the average sequence compositions of each of the four sequence-based clusters. While clusters were defined based on sequence compositions only, Figure 1 also shows the average heterogeneity of the four clusters, meaning the average normalized difference in sequence composition between the interacting proteins of the complexes (see Data and Methods). Complexes in clusters 1 and 2 are both largely devoid of special residues, including Gly (flexible), Pro (rigid), and Cys (cysteine). Members of these two clusters contain an average fraction of hydrophobic residues; however are slightly depleted in aromatic residues, indicating that π–π interactions are not the dominant source of stability. The most characteristic difference between clusters 1 and 2 is that members of cluster 1 typically contain a high fraction of polar residues, while members of cluster 2 are enriched in charged residues. Also, cluster 1 members are typically formed by proteins with highly different compositions (high heterogeneity values), while cluster 2 members are formed by proteins of very similar compositions.

In contrast, members of clusters 3 and 4 are typically enriched in Gly and Pro and contain a higher-than-average fraction of aromatic residues. Again, polar/charged residue balance is a distinguishing feature, with clusters 3 and 4 showing preferences for polar and charged residues, respectively. Also, similarly to clusters 1 and 2, there is a notable difference in heterogeneity values between clusters 3 and 4: members of clusters 3 and 4 are typically composed of proteins with very similar and different residue compositions, respectively.

		Cluster 1	Cluster 2	Cluster 3	Cluster 4	Average	Comparison to average:
Number of members		25	60	38	84	-	
Average amino acid composition	Aromatic (FWY)	0.029	0.049	0.089	0.072	0.064	+30%
	Hydrophobic (AILMV)	0.381	0.324	0.287	0.357	0.336	+20%
	Flexible (G)	0.024	0.021	0.076	0.056	0.046	+10%
	Rigid (P)	0.018	0.005	0.029	0.040	0.026	
	Charged (HKRDE)	0.258	0.392	0.239	0.290	0.308	-10%
	Polar (NQST)	0.281	0.205	0.254	0.179	0.211	-20%
	Cysteine (C)	0.009	0.005	0.027	0.006	0.010	-30%
Heterogeneity (average dissimilarity between subunits)		0.162	0.098	0.057	0.117	0.111	

Figure 1. Average values of sequence features for the four sequence-based clusters. Blue and orange shadings mark values that are over- or under-represented compared with the average of all MSF complexes. Heterogeneity values were not used for cluster definitions.

2.2. Structure-Based Properties Offer A Different Means of Defining Complex Types

The structural properties of the studied complexes were quantified using various features describing secondary structure compositions, various molecular surfaces, and incorporating hydrophobicity measures and atomic contacts (see Supplementary Table S1 and Data and Methods). These structural features were used to describe each complex in the form of a feature vector, and similarly to the analysis of sequence properties, these vectors were input to hierarchical clustering; however, structural features were filtered, and only those that share a modest degree of correlation were kept (see Supplementary Table S2 and Data and Methods for specifics) to avoid bias. The resulting tree is shown in Supplementary Figure S3. In contrast to the sequence-based clustering, k-means within-cluster sum of squares analysis does not indicate any low number of clusters as more optimal than others (Supplementary Figure S4). In order to have a medium number of clusters, we cut the hierarchical tree at a linkage distance that defines five clusters (Supplementary Figure S3), again reflecting our preference to arrive at a moderate number of complex types, to provide a high-level classification scheme. The average values of structural parameters for all five structure classes are shown in Figure 2.

The obtained clusters show distinguishing structural features. Members of cluster 1 incorporate the highest amount of nonhelical secondary structure elements. These complexes heavily rely on a large number of buried hydrophobic residues for stability, and most stabilizing atomic contacts are formed between residues of the same protein, relying less on intermolecular interactions, which tend to be mostly polar in nature.

In contrast, members of cluster 2 adopt mainly helical structures. The stability of these complexes seems to rely more on the interactions formed between the subunits, mostly formed between side chains. The importance of interchain interactions is also reflected in the large relative interface and small relative buried surface areas.

Cluster 3 and 4 complexes exhibit similar features, including a balanced ratio of various secondary structure elements and polar/hydrophobic balance of various molecular surfaces and contacts. For both clusters, interchain contacts rely mostly on side chain–side chain and backbone–backbone contacts. The main difference between the two clusters is the relative role of the interface between the participating proteins. Cluster 3 members have a larger-than-average interface, in terms of both molecular surface and number of contacts, meanwhile cluster 4 complexes have a very restricted interface size, incorporating only a few atomic contacts.

Members of cluster 5 are the most similar to the average in most structural features. There are only weak distinguishing features, including a slightly increased helical content at the expense of extended structural elements, a moderate increase in the role of backbone–side chain interactions in interchain contacts, and the increased ratio of interchain contacts. However, these deviations in average parameter values are modest and—with the exception of the decreased extended structure content—none of them reaches 20% compared to the average values calculated for all complexes.

		Cluster 1	Cluster 2	Cluster 3	Cluster 4	Cluster 5	Average
Number of members		47	58	23	28	52	-
Secondary structures	alpha *	0.153	0.879	0.565	0.691	0.672	0.603
	extended	0.398	0.008	0.208	0.043	0.040	0.131
	coil	0.448	0.113	0.227	0.267	0.288	0.266
Molecular surfaces	SASA – hydrophobic *	0.559	0.570	0.527	0.558	0.547	0.555
	SASA – polar	0.441	0.430	0.473	0.442	0.453	0.445
	Interface – hydrophobic *	0.681	0.783	0.706	0.797	0.768	0.750
	Interface – polar	0.319	0.217	0.294	0.203	0.232	0.251
	Buried surface – hydrophobic *	0.590	0.393	0.515	0.529	0.509	0.498
	Buried surface – polar	0.410	0.607	0.485	0.471	0.491	0.502
	Interface / total *	0.079	0.204	0.217	0.120	0.167	0.157
	Buried / total	0.606	0.400	0.438	0.519	0.474	0.485
Atomic contacts	Interchain hydro:hydro *	0.541	0.656	0.587	0.699	0.650	0.627
	Interchain hydro:polar	0.380	0.290	0.338	0.260	0.300	0.314
	Interchain polar:polar	0.080	0.054	0.074	0.041	0.050	0.059
	Interchain backbone:backbone	0.166	0.012	0.148	0.014	0.063	0.075
	Interchain backbone:side-chain *	0.373	0.309	0.295	0.305	0.379	0.339
	Interchain side-chain:side-chain *	0.462	0.678	0.557	0.682	0.558	0.586
	Intrachain hydro:hydro *	0.459	0.306	0.384	0.400	0.384	0.381
	Intrachain hydro:polar	0.438	0.548	0.491	0.480	0.491	0.494
	Intrachain polar:polar	0.103	0.145	0.125	0.120	0.125	0.125
	Intrachain backbone:backbone	0.284	0.475	0.398	0.361	0.379	0.384
	Intrachain backbone:side-chain *	0.383	0.417	0.379	0.410	0.413	0.403
	Intrachain side-chain:side-chain	0.333	0.107	0.223	0.229	0.209	0.213
	Interchain / total *	0.122	0.222	0.313	0.134	0.234	0.201

Comparison to average:

+30%
+20%
+10%

-10%
-20%
-30%

Figure 2. Average values for structure features for the five structure-based clusters. Blue and orange shadings mark values that are over- or under-represented compared to the average of all MSF complexes. SASA—solvent accessible surface area, hydro:hydro—fraction of contacts that are formed between two hydrophobic atoms. Asterisks mark features that were included in the clustering.

2.3. Defining Interaction Types Based on Sequence and Structure Clusters

Considering together the previously established sequence- and structure-based clusters, in total 20 types of complexes can be defined (Figure 3). The number of known complexes in possible types shows large variations, with some highly favored ones (e.g., type 2[sequence]/2[structure]) and ones with a single known example (e.g., type 2/1), showing that not all sequence compositions are compatible with all types of adopted structures. In order to arrive at a reasonable number of basic complex types, types with 10 or fewer complexes were either merged with the adjacent sequence clusters or were omitted. As structural differences in general are larger between clusters, types corresponding to different structure clusters were never merged. For structure clusters 1 and 2, only two adjacent sequence clusters were merged, as these contain over 95% and 85% of the complexes, respectively. In contrast, for structure classes 3 and 4, all four sequence clusters were merged, as the distribution of complexes is more even across the sequence space. For structure cluster 5, even a single sequence cluster is enough to capture over 85% of complexes, and thus no merging was employed. This approach yielded five main interaction types, each of which has over 20 complexes. In order to include all known MSF complexes, a sixth pseudo-type was introduced, which contains all structures not compatible with any of the previously described five types (see Supplementary Table S1 for an exhaustive list).

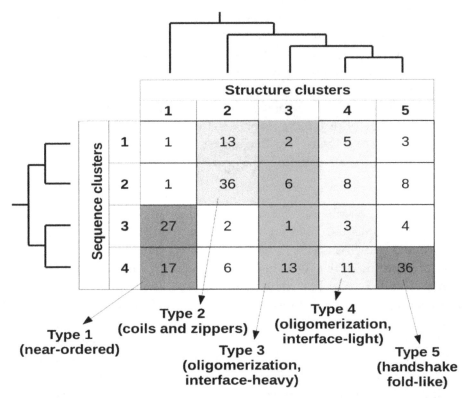

Figure 3.   MSF complex types.   Colored regions mark separate interaction types considering sequence- and structure-based clusters (vertical and horizontal axes, respectively). The relationship of each sequence-and structure-based cluster taken from the hierarchical clustering (Supplementary Figures S1 and S3) is shown on the corresponding side of the table. Each of the six defined types is assigned a randomly selected color (that is of high contrast), and these are used in later figures to denote the corresponding complex types.

The complex types defined so far are based on structure and sequence features. However, if these types represent biologically meaningful classes, there should be other relevant differences between them in terms of the energetics of the interaction, binding strength, subcellular localization, or the biological regulation of the interaction. In the next chapters, we describe each complex type with biologically important characteristics and assess the potential differences between the members of each class.

2.4. Complex Types Show Characteristic Energetic Properties

From a biological perspective, the strength of association between interacting protein chains and the stability of the resulting complex is of utmost importance. Unfortunately, complexes formed exclusively by IDPs via MSF generally lack targeted measurements concerning thermodynamic and stability parameters. However, low-resolution energy calculations and prediction algorithms can give an indication about the characteristic energetics properties of the uncovered complex types in general. While these methods might have fairly large errors in individual cases, they are well equipped for comparative studies between groups of complexes.

In order to assess the energetic properties of complexes, we employed an energy calculation scheme using low-resolution force fields based on statistical potentials (see Data and Methods). As a reference, energetic properties were calculated for complexes formed exclusively by ordered proteins and complexes formed by an IDP binding to an ordered partner via coupled folding and binding (CFB) (see Data and Methods and Supplementary Tables S3 and S4). Figure 4 shows two types of calculated energies for each complex. On one hand, we calculated the total energy per residue in the whole complex, which reflects the overall stability. On the other hand, we also calculated the fraction of this stabilizing energy coming from intermolecular interactions (i.e., how important the interaction is for stability). In accordance with our expectations, complexes formed by ordered proteins feature strongly bound overall structures, with fairly large negative stabilizing energy/residue. In contrast, CFB complexes in general have less favorable per-residue energies, hinting at their comparatively weakly bound overall structures. However, the energetic feature providing the most recognizable difference between ordered and CFB complexes is the energy contribution of interchain contacts to the overall stability. In the case of ordered complexes, this contribution is fairly limited, as individual subunits have a stable structure on their own. In contrast, if the complex features an IDP, the interaction energy becomes a major contributor to stability (Figure 4a).

While ordered and CFB complexes tend to segregate in this energy space, complexes formed by MSF seem to be more heterogeneous, covering the whole available range of energetic values (Figure 4b). In the case of near-ordered proteins (Type 1), the energies resemble that of ordered complexes, hinting at the borderline ordered nature of the constituent IDPs, with the interaction between subunits playing a minor role. In contrast, coiled-coil-like structures (Type 2) on average have a much less stable complex structure, with interaction playing a substantial role in stability. These complexes resemble IDPs bound to ordered domains, and are expected to include several transient interactions. Other types fall largely between these two extreme cases. Energetics properties of the two types of oligomerization modules (Types 3 and 4) reflect the differences in interface surface area and contact numbers, shown in Figure 2 While the overall stability for both types varies in a very wide range, on average, the contribution of the interaction is higher for interface-heavy complexes (Type 3) than for interface-light ones (Type 4). Handshake-like folds (Type 5) show interesting properties: these complexes are quite stable with only limited variation in the per-residue energies. Yet, they achieve this high stability by relying heavily on the interaction between subunits of the dimer. As opposed to the complexes in Figure 4a, MSF complexes show high overlap in the energy space. This shows that very different structures, with potentially very different sequence compositions, can have similar energetic properties. Also, the high variability of energetic properties within complex types (the main reason for high overlap between different groups) shows that depending on the biological function, similar complexes can be required to have very different stabilities. For example, while several dimeric transcription factors can have similar structures that accommodate DNA-binding, the association and dissociation rates of the dimers (regulating their transcriptional activity) have to adapt to the required expression profiles of the genes they regulate.

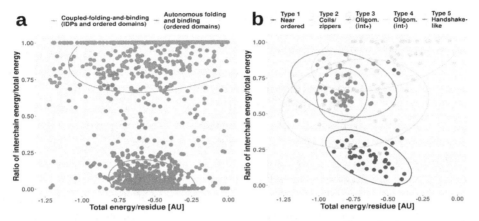

Figure 4. Energetic parameters of various interaction classes. The relative energetic weight of intersubunit interactions in the overall stability (y-axis) as a function of the overall energy per residue (x-axis, measured in arbitrary units, AU) for ordered complexes and complexes formed by coupled folding and binding (**a**), and the five well-defined types of MSF complexes (**b**).

The transient or obligate nature of interactions provides clues about their roles in biological systems. This is at least partially describable through K_d dissociation constants. While there is ample data about K_d values of IDPs binding via CFB to ordered domains [23], these values are largely missing for MSF complexes. In accord, we calculated estimated K_d values for MSF complexes (Supplementary Table S1), with Figure 5 showing the K_d distributions for the six previously defined complex types. In a biological context, actual K_d values can be a nonlinear function of environmental parameters. Unfortunately, this information is largely unknown for most MSF complexes, and such predicted K_d values should be treated with caution and should only be used for comparing group averages, where individual errors can even out. The lowest average K_d values were calculated for complexes with a handshake-like fold (Type 5). The next two types with low K_ds are the near-ordered complexes (Type 1) and interface-heavy oligomerization modules (Type 3). These three types together possibly cover most cases of the interactions where the complex needs to stay stable for an extended period of time, such as histone dimes (Type 5), complexes with enzymatic activity (Type 1) and several transcription factors (Type 3). Coiled-coil-like structures and oligomerization modules with small interfaces in general have a higher K_d, indicating that several transiently bound complexes belong to these types.

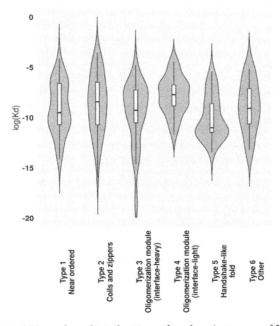

Figure 5. Predicted K_d value distributions for the six types of MSF complexes.

2.5. Interactions Are Heavily Regulated by Several Mechanisms

While the energetics of various interactions can provide clues about their transient/obligatory nature, the regulatory mechanisms can give more direct evidence. For example, while most IDP enzymes (belonging to Type 1) form particularly stable oligomers, indicating an obligate interaction, for example the oligomeric state of superoxide dismutase (SOD1) is known to be controlled by post-translational modification (PTM) serving as an on/off switch [35]; meaning that despite a strong interaction, it is reversible, and the disordered state of the monomers is biologically relevant (Figure 6a). Figure 6a shows additional examples of various regulatory mechanisms of MSF interactions via PTMs. These regulatory steps have already been described in the case of IDPs that bind to ordered domains [36], but have not been studied in the context of IDPs participating in MSF interactions. Apart from the on/off switch exemplified by SOD1, PTMs can control the partner selection of synergistically folding IDPs, such as in the case of another tightly bound complex, formed by H3/H4 histones (Type 5) [37]. PTMs can also tune the affinity of certain interactions, as is the case for the activating p53/CBP interaction (Type 4) [38]. Apart from these mechanisms that directly control the interaction between IDPs, PTMs can have a more indirect effect, modulating the activity of the dimer itself. In the case of the Max dimeric transcription factor, phosphorylation at the N-terminus of the binding region controls the dimer's (Type 4) interaction capacity towards DNA [39]. An even more indirect modulation of function is displayed for the retinoblastoma protein Rb, which in complex with E2F1/DP1 (Type 3) has a strong transcriptional repression activity. Upon methylation, Rb recruits L3MBTL1 [40], which is a direct repressor of transcription via chromatin compaction, augmenting the effect of Rb through a related but separate mechanism extrinsic to the Rb/E2F1/DP1 complex. This way the strength of repression depends on the PTM of the MSF complex, but through an additional protein that is not part of the complex but contributes to the complex function through a parallel mechanism in an indirect way.

To have a more systematic picture of the extent of regulatory mechanisms in MSF interactions, Figure 6b shows the fraction of known MSF complexes with experimentally verified PTM sites (Supplementary Table S5). In total, nearly 30% of studied complexes feature at least one PTM that was experimentally verified in a low-throughput experiment, presenting a regulatory mechanism that is able to directly or indirectly modulate either the interaction itself, or the activity of the resulting complex. The most prevalent PTM is phosphorylation, affecting 22% of complexes, but 10%, 15%, and 5% of MSF complexes contain methylation, acetylation, and ubiquitination sites as well (Figure 6b).

In addition, complex formation can also be regulated through the availability of the subunits participating in the interaction. This availability can depend on the alternative mRNA splicing of the corresponding genes, where certain isoforms lack the binding site (Supplementary Table S6). Also, even if the translated isoform has the binding site, the protein itself can be sequestered by competing interactions with other protein partners (Supplementary Table S7). These mechanisms are present for 11% (alternative splicing) and 16% (competing interactions) of complexes, and together with PTMs, in total 36% of MSF complexes have at least one known regulatory mechanism for modulating the interaction. Furthermore, these regulatory mechanisms often act in cooperation, with seven interactions known to employ PTMs, alternative splicing, and competing interactions as well (Figure 6c).

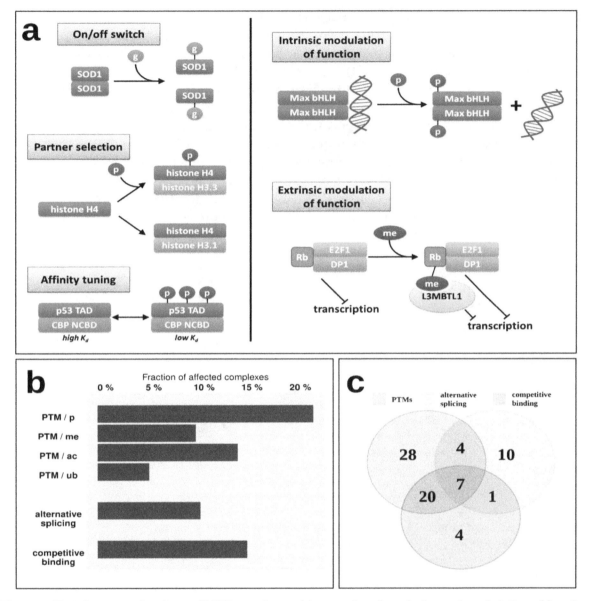

Figure 6. Regulatory mechanisms of MSF complexes. (**a**) examples of regulation and modulation of function through post-translational modifications. p—phosphorylation, g—glutathionylation, me—methylation, SOD1—superoxide dismutase, CBP—CREB-binding protein, Rb—retinoblastoma-associated protein. Colored boxes represent interacting chains forming the MSF complexes. (**b**) The fraction of complexes with verified PTM sites, and the fraction of complexes where at least one interactor is regulated via alternative splicing or by competing interactions. (**c**) Number and overlap of MSF complexes affected by the three types of regulatory mechanisms.

2.6. Various Complex Types Show Differential Subcellular Localization

In addition to regulatory mechanisms detailed in the previous chapter, a crucial element in the spatio-temporal control of protein function is subcellular localization [41]. In order to assess this aspect of MSF complexes, and to understand if the defined interaction types have different properties in terms of cellular localization, we used "cellular component" terms from GeneOntology (GO) [42] (see Data and Methods). Various GO terms were condensed into five categories including "Extracellular", "Intracellular", "Membrane", "Nucleus", and "Other" to enable an overview of the differences in localization between the six complex types (Figure 7) (for exact GO terms for each complex see Supplementary Table S8).

The least amount of information is available for Type 1, near-ordered complexes. Albeit GO terms are lacking for most complexes, even the limited annotations highlight that these complexes are able to efficiently function in the extracellular space, which in general is fairly uncommon for IDPs. Coil- and zipper-type helical complexes (Type 2) are somewhat more often attached to the membrane or function in the intracellular space, or in non-nuclear environments, such as the lysosome. In contrast, oligomerization modules (Types 3 and 4) are most prevalent in the nucleus and the intracellular space, which is in line with the function of the high number of transcription factors in these groups. However, modules with a large interface (Type 3) are relatively often found in other compartments, while modules with smaller interfaces (Type 4) also function in the extracellular space. Complexes adopting a handshake-like fold are enriched in histones, which is reflected in their enrichment in the nucleus and the chromatin (classified as "other" in Figure 7). Type 6 complexes are heterogeneous in terms of localization as well, and hence members can be found in all studied localizations to a comparable degree. These preferences in subcellular localization for different complex types reinforce our notion that even though our classification scheme relies on sequence and structure properties alone, the obtained interaction types also have biological meaning.

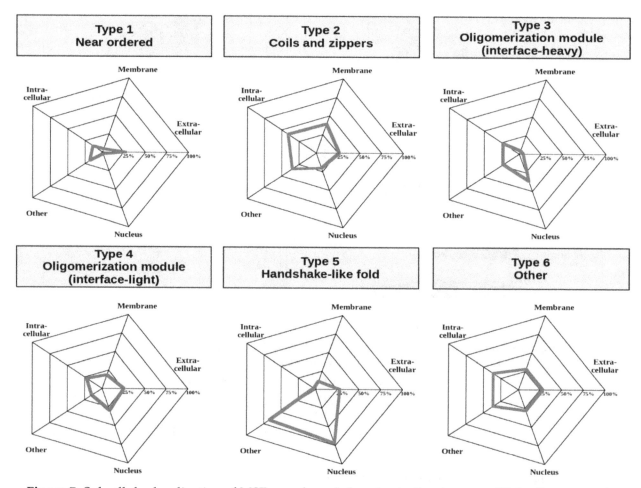

Figure 7. Subcellular localization of MSF complexes belonging to the six types. "Other" contains the "non-membrane-bounded organelle", "secretory granule", "lysosome", "cytoplasmic vesicle lumen", and "transport vesicle" GeneOntology terms.

2.7. The Annotated Catalogue of Complexes Formed via Mutual Synergistic Folding

Considering the previously analyzed features of complexes, averaging the calculated features for the six established interaction types provides the annotated catalogue of MSF interactions (Figure 8). Apart from the main sequential and structural features, Figure 8 also shows example structures, energetic properties, subcellular localization, and the main regulatory mechanisms for each complex type.

Type ID	Type name	# of comp-lexes	Example structure	Sequence: Preferred	Sequence: Depleted	Structure: Secondary structures	Structure: Interface	Structure: Buried surface	Structure: Dominant atomic contacts	Strength: Energy	Strength: Role of inter-action	Regulation	Dominant subcellular localization
1	Near ordered	44	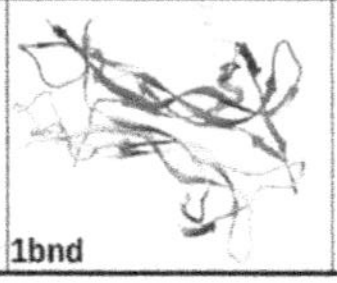 1bnd	Pro/Gly aromatic	charged	extended + coil	small, polar	hydroph.	intrachain	- -	+	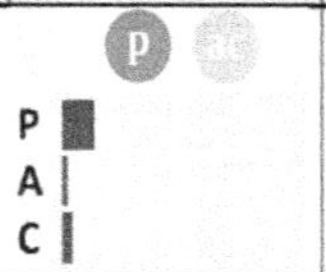 p ac P A C	Extracellular
2	Coils and zippers	49	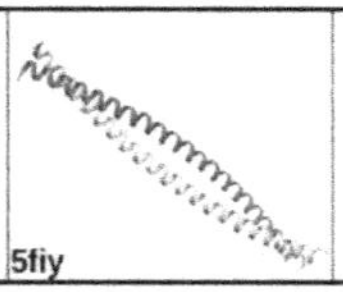 5fiy	hydrophobic charged OR polar	Pro/Gly Cys aromatic	highly helical	large	polar	balanced	- (variable)	+++	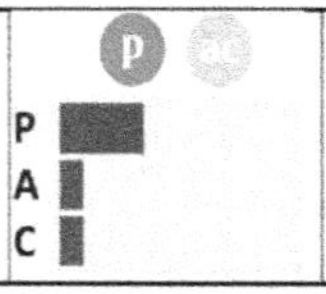 p ac P A C	Membrane Intracellular Other
3	Oligo-merization module (interface heavy)	22	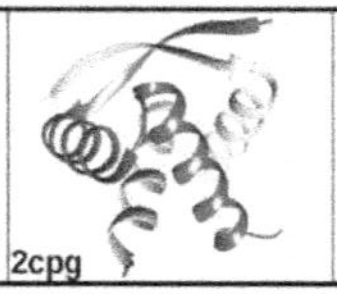 2cpg	highly variable		balanced	large, slightly polar	average	interchain	- - -	+++	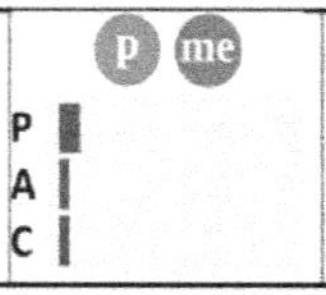 p me P A C	Nucleus Intracellular Other
4	Oligo-merization module (interface light)	27	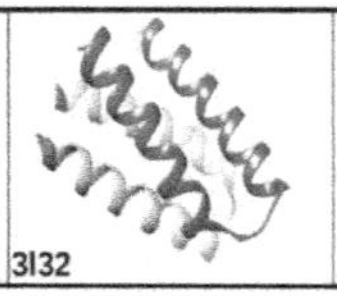 3l32	highly variable		mainly helical	small, hydroph.	average	intrachain	- - (variable)	++	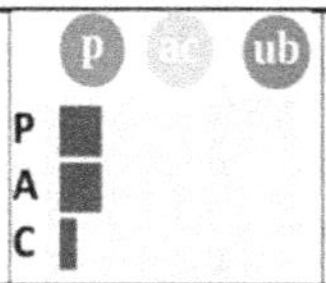 p ac ub P A C	Nucleus Extracellular Intracellular
5	Handshake-like fold	36	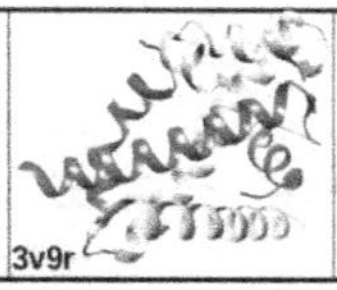 3v9r	Pro/Gly aromatic	Cys polar	mainly helical	average	average	balanced	- - -	++	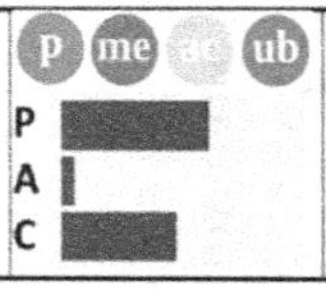 p me ac ub P A C	Nucleus Other
6	Other	25	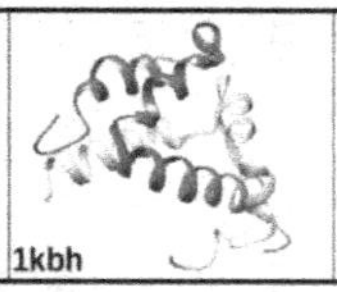 1kbh	highly variable		mixed	-	-	-	- - - (variable)	+ + (variable)	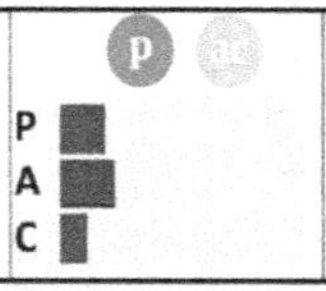 p ac P A C	Heterogeneous

Figure 8. Annotated types of complexes formed by IDPs, based on sequence and structure features. Horizontal bars in the regulation column show the fraction of complexes in a given group involved in various types of regulatory mechanisms (P—post-translational modifications, A—alternative splicing affecting binding regions, C—competing interactions). Color circles mark the dominant post-translational modification(s) for the group (p—phosphorylation, me—methylation, ac—acetylation, ub—ubiquitination).

The first type of complexes bears a high similarity to ordered protein complexes, and hence are named near ordered. The constituent chains are usually similar, in many cases corresponding to homooligomers, with a high Pro/Gly content and typically only a few charges. The main difference compared to protein complexes formed by ordered proteins is that near ordered subunits are depleted in α-helices [28]. For reaching a stable structure through the interaction, they utilize a large number of intrachain contacts, with inter-subunit interactions through a small polar interface playing only a secondary role in the stability of the complex. This group contains a large number of enzymes, transport proteins, and nerve growth factors, where the exact structure is of utmost importance; however, in contrast to monomeric proteins, the presence of this structure relies on the interaction. This interaction type is mostly regulated through phosphorylation and acetylation of binding site residues. These proteins resemble ordered proteins in their localization as well, with extracellular regions being highly representative.

The second type of complexes contains structures with a high overall similarity, mostly consisting of coiled-coils and zippers, structures composed of parallel interacting helical structures, often stabilized by a restricted set of residues, such as leucines, alanines, or tryptophans. In general, constituent proteins are depleted in residues incompatible with α-helix formation, such as Pro and Gly, and also in aromatic residues. In turn, they are abundant in hydrophobic residues and show an enrichment for either polar or charged residues. The constituent helices usually form a fairly weakly bound system, where the interchain interactions via the relatively large interfaces play a major role. Constituent proteins are able to bury only a small fraction of their polar surfaces. Coiled-coil interactions are often regulated, typically via various types of PTMs, most often through phosphorylation or, to a lesser degree, acetylation. Despite their highly similar structures, complexes in this group convey a large variety of functions, mainly pertaining to regulating transcription and performing membrane-associated biological roles, such as organelle and membrane organization.

The third and the fourth type of complexes are both generic oligomerization modules that can be split according to the importance for the interchain interactions, grouping them as either interface-heavy (Type 3) or interface-light (Type 4) complexes. In both cases, the sequences can be highly variable, and the unifying features are mostly structural. Both types typically have an average-sized relative buried area with balanced hydrophobic/polar composition. However, interface-heavy complexes have a large, slightly polar interface that plays a major role in achieving the tightly bound structures. In contrast, interface-light complexes form a more helical structure and have smaller hydrophobic interfaces that play a more diminished role in achieving the stability of a less tightly bound system. This hints at interface-light complexes being more transient, also supported by the fact that these complexes have a higher number of known regulatory PTMs and are also modulated by alternative splicing. Both type 3 and type 4 complexes preferentially occur in nuclear and intracellular processes, as several of them are ribbon–helix–helix (interface-heavy) or basic helix–loop–helix (interface-light) transcription factors, able to shuttle between the nuclear and the intracellular spaces. In addition to the similarities in subcellular localization, type 4 complexes preferentially occur in the extracellular space, and type 3 complexes in other cell compartments, as well.

The fifth type of complexes typically adopts a handshake-like fold, characteristic of histones and homologous proteins. While these structures are usually largely helical, the interacting proteins often contain a relatively high ratio of prolines and glycines, in addition to the enrichment of aromatic residues. While they are depleted in polar residues, both the interface and the buried surface have a fairly balanced hydrophobic/polar makeup. The complexes are relatively tightly bound, and interchain interactions play a fairly large role in stabilizing the interaction. This type of complex has the highest ratio of both PTMs and competitive interactions, providing a large amount of regulation. In addition, PTMs are highly heterogeneous, containing phosphorylations, acetylations, methylations, and ubiquitinations as well. Members of this cluster primarily serve DNA/chromosome-related functions, and hence are usually located in the nucleus.

While types 1–5 represent well-defined groups with members of clear unifying similarities, the final group serves as an umbrella term for complexes that are not members of any previous structural/sequential class. In accord, these complexes cannot be described by simple characteristic features and are the most sequentially and structurally heterogeneous group. This group contains highly specialized interactions that present unique protein complexes, which are regulated through all three control mechanisms and occur in all studied subcellular localizations.

2.8. Interaction Types Present A Novel Classification of Protein Complexes

The described MSF classification method bears similarity to the approach employed in CATH, as both approaches use a hierarchical classification of PDB structures. However, CATH does not consider interactions and simply relies on the secondary structure elements and their connectivity and arrangement, in contrast to the presented analysis taking into account protein chain interactions too, together with sequence composition features.

Figure 9 shows the studied MSF complexes in both our MSF classification system and in CATH, considering the top two levels ("Class" and "Architecture"). The highest-level CATH definitions, corresponding to "Class", reflect the overall secondary structure element distribution of the structures. In this framework, Type 1 near-ordered complexes mostly occupy the "Mainly Beta" CATH class, while complexes from the other five types mostly fall into the "Mainly Alpha" class or the "Other" class. At the next CATH level, "Architecture", certain MSF type complexes (such as type 2 coils and zippers) are segregated into further subclasses.

Considering "Class" and "Architecture" definitions, there is very little correspondence between the CATH and the new MSF classification. If the two schemes showed a high degree of similarity, the matrix in Figure 9 should be close to a diagonal matrix. In reality, however, off-diagonal elements are large, confirming the novelty of the presented MSF classification scheme.

		Mainly Alpha (1)		Mainly Beta (2)		Alpha Beta (3)		Few Secondary Structures (4)	Other
		Orthogonal bundle (1.10)	Up-down bundle (1.20)	Sandwich (2.60)	Orthogonal prism (2.90)	2-layer sandwich (3.30)	3-layer(aba) sandwich (3.40)	Irregular (4.10)	
MSF classification	Near-ordered (Type 1)	2	0	11	9	8	4	0	10
	Coils and Zippers (Type 2)	0	31	0	0	0	0	0	18
	Oligomerization (interface-heavy) (Type 3)	8	1	0	0	1	4	1	7
	Oligomerization (interface-light) (Type 4)	11	3	0	0	3	0	3	7
	Hand-shake fold like (Type 5)	29	2	0	0	0	1	0	4
	Other (Type 6)	8	7	0	0	1	0	2	7

Figure 9. Overlap between CATH and MSF classification.

3. Discussion

Here, we present the first approach aiming at the classification of complex structures formed exclusively by disordered proteins via mutual synergistic folding. We developed and applied a method that can classify these complexes into various types based on sequence- and structure-based properties. The classification scheme takes into account on the one hand, the overall sequence and structure properties of the complex, and on the other hand, the interaction itself, quantifying the role of intra- and intermolecular interactions in relation to the overall contact/surface properties of the structure. As the classification protocol is based on hierarchical clustering, it is freely scalable. Tuning the resolution via changing the number of sequence-based or structure-based clusters, the method can be used to yield any number of types and subtypes. The presented classification is a top-level one highlighting the major types of MSF classes, and this six-way classification scheme will be used to better define MSF complex types in the MFIB [26] database.

While both sequence- and structure-based parameters are taken into account when defining the final complex types, the two sets of descriptors have different roles in the scalability of the method. In our presented approach to defining complex types, the main features are structural properties, while sequence parameters are more descriptive in the sense that they highlight the sequential features needed to be able to fold into a complex of given structural properties (Figure 3). However, sequence features can be used to distinguish subtypes of structure-defined complex types. For example, type 1 near-ordered complexes come in two flavors according to the two sequence clusters they cover (Figures 1 and 3): polar-driven interactions between mostly homodimers, and charge/hydrophobic driven interactions between mostly heterodimers. Also, type 2 complexes (coils and zippers) come in two varieties: relying on polar-driven interactions for heterodimers and charge-driven for homodimers.

In addition to providing a scalable classification scheme, the described method and the defined complex types have biological relevance. The presented complex types have different biological properties; although only information describing the sequence and structure properties were put in, the resulting types show different properties in terms of the energetics and strength of the interactions (Figures 4 and 5), the relevant regulatory processes (Figure 6), and subcellular localization (Figure 7).

The analysis of the energetics properties of the interactions can provide a glimpse into the biophysical details of the binding and folding. The use of low-resolution statistical force fields proved to be a suitable approach to discriminate complexes based on the structural features of constituent chains [28] and to describe the binding of IDPs [43,44]. While complexes of ordered proteins and domain-recognition IDP binding sites have a fairly narrow range in energetics parameters (Figure 4a), complexes formed exclusively by IDPs are more heterogeneous, basically covering the whole range of the energy spectrum (Figure 4b). Furthermore, based on predictions, MSF complexes cover at least 10 orders of magnitude in K_d values (Figure 5). Hence, in terms of binding strength and stability, these complexes have the potential to cover a very wide range of biological functions, overlapping with those of ordered complexes and domain-binding IDPs as well, in agreement with the previous comparative functional analysis of a wide range of interactions [28].

For most known MSF complexes, the resulting structure is instrumental for proper function, such as the coiled-coil structure for the SNAP receptor (SNARE) complex in mediating membrane fusion [45], the dimeric structure for a wide range of transcription factors in precise DNA-binding [46–48], and the proper coordination of catalytic residues for oligomeric enzymes [49,50]. Therefore, for MSF complexes, the interaction de facto switches on the protein function, and hence the precise regulation of the interaction strength is vital in the biological context of these complexes. While structure-based K_d value predictions are informative, in some cases they do not fully describe the interactions. Many MSF complexes are tightly bound, yet they are not necessarily obligate complexes, and their association/dissociation can be under heavy regulation. For example, solely based on K_d values and energetics, type 5 (handshake-like fold) interactions seem to form obligate complexes. However, there are several cases where these interactions do break up in a biological setting, most notably for histones. Histone H4 is able to form dimers with at least eight different H3 variants [51], and it was described that in the case of H3.1 and H3.3, the preference of H4 for these two partners is governed by H4 phosphorylation [37]. The post-translational modifications can enhance complex formation or dissociation in many other cases as well [35]. In addition, competition for the same binding partner and binding site availability as a function of alternative splicing is an additional mechanism for the regulation of the formation of MSF complexes (Figure 6).

Exploring the precise regulatory mechanisms for MSF complexes would be highly informative. Unfortunately, experimental K_d measurements are lacking for the majority of these interactions, and interactions in structural detail have usually been only analyzed in a single PTM state. Therefore, the molecular details and biologically relevant steps of the regulation of these interactions are difficult to assess; but from a biological sense, it is probable that even several low K_d complexes can dissociate rapidly in certain cases. At least some regulatory mechanisms are currently known for about 36% of studied MSF complexes, but the real numbers are bound to be higher. This means that most probably the majority of MSF complexes are not obligate complexes, where the disordered state is physiologically irrelevant, but can exist in both the stable bound state and the disordered unbound state as well, under native conditions. Thus, MSF complexes are integral parts or direct targets of regulatory networks, although the extent of regulation varies with the interaction type considered.

Apart from the studied regulatory mechanisms, additional layers of spatio-temporal regulation can play crucial roles for MSF complexes, similarly to other IDP interactions [41]. An emerging such regulatory mechanism is liquid–liquid phase separation (LLPS). A prime example is the Nck/neuronal Wiskott–Aldrich syndrome protein (N-WASP). N-WASP is known to undergo LLPS when interacting with Nck and nephrin [52], via linear motif-mediated coupled folding and binding. Mutually synergistic folding between the secreted EspFU pathogen protein from enterohaemorrhagic *Escherichia coli* and the autoinhibitory GTPase-binding domain (GBD) in host WASP proteins (MFIB ID:MF2202002, type 5 complex) hijacks the native LLPS-mediated cellular processes [53], showing that competing interactions are not always stoichiometric in nature, and the true extent of MSF regulation is likely to be even more complex than highlighted here.

The difference between complex types in various biological and biophysical properties shows that these type-definitions reflect true biological differences. Apart from being useful for complex classification, the presented method also shows that differences in binding strength, subcellular localization, and regulation are encoded in the sequence and structural properties of proteins. This can be the basis for developing future prediction methods, where these sequence- and structure-based parameters can be used as input for the prediction of biological features of complexes. In addition, the establishment of MSF complex types has direct implications, as knowledge present for a specific complex might be transferable to other complexes of the same type. For example, certain pathological conditions arise through the aggregation of IDPs. A well-known example is transthyretin (TTR) aggregation that can lead to various amyloid diseases, such as senile systemic amyloidosis [54]. Another example from the same near-ordered complex type is the superoxide dismutase SOD1, which is able to form aggregates in amyotrophic lateral sclerosis [55]. While the localization and the biological function of TTR and SOD1 (hormone transport and enzymatic catalysis) are radically different, their potency of malfunctioning (often connected to various mutations) share a high degree of resemblance. On one hand, this marks other type 1 complexes as candidates for toxic aggregation, on the other hand, it indicates that the potential therapeutic techniques for one complex (e.g., CLR01 for TTR) can give clues about potential targeting of other interactions.

Such structural classification approaches can have a high impact on structure research, most importantly in the study of protein structure or evolution, in training and/or benchmarking algorithms, augmenting existing datasets with annotations, and examining the classification of a specific protein or a small set of proteins [56]. Up to date, several structure-based classification approaches have been developed, such as SCOP [32] and CATH [33], which are extended to protein complexes as well. In this sense, previously existing methods are able to classify MSF complexes too. However, the approaches used do not take into account that these structures are only stable in the context of the interaction, and that a certain protein region can adopt fundamentally different structures depending on the interacting partner. The lack of the explicit encoding of parameters describing the properties and importance of the interaction into the classification scheme makes current methods unable to accurately describe the spectrum of MSF complexes, and to date, no such dedicated classification scheme has been proposed. In contrast to previously existing methods that largely encode the same information [57], the presented MSF classification scheme is highly independent (Figure 9), and thus serves as an orthogonal approach capable of properly handling the specific properties of IDP-driven complex formation through mutual synergistic folding.

4. Data and Methods

4.1. Complexes Formed Through Mutual Synergistic Folding (MSF)

MSF complexes were taken from the MFIB database [26]. Two entries, MF2100018 and MF5200001, from the 205 were discarded due to issues with the corresponding PDB structures 1ejp and 1vzj, as constituent chains have an unrealistically low number of interchain contacts. Problems with these two structures are apparent from the high outlier scores and clash scores provided on the PDB server. As the developed classification scheme relies heavily on structural parameters, we opted to leave these two entries out of the calculations. The final list of entries is given in Supplementary Table S1.

4.2. Other Complexes of Ordered and Disordered Proteins

As a reference, two other datasets of protein complexes were used. A set of complexes formed exclusively by ordered single-domain protein interactors was taken from [28]. These 688 complexes (see Supplementary Table S3) are formed via autonomous folding followed by binding, that is, both interacting protein chains adopt a stable structure in their monomeric forms, prior to the interaction. A set of 772 complexes with an IDP interacting with ordered domains was taken from the database of Disordered Binding Sites (DIBS) database [23]. These complexes (see Supplementary Table S4) are formed via coupled folding and binding, where the IDP adopts a stable structure in the context of the interaction.

4.3. Calculating Sequence Features

Similarly to the approach described in [28], the following amino acid groups were used in quantifying sequence composition of proteins: hydrophobic (containing A, I, L, M, V), aromatic (containing F, W, Y), polar (containing N, Q, S, T), charged (containing H, K, R, D, E), rigid (containing only P), flexible (containing only G), and covalently interacting (containing only C). This low-resolution sequence composition at least partially compensates for commonly occurring amino acid substitutions that in most cases do not affect protein structure and function. In all cases, compositions were calculated for the entire complex, including all interacting protein chains. An 8th sequence parameter was used to quantify the compositional difference between subunits. This dissimilarity measure was defined as: $\Delta_{total} = \sum_{i=1}^{7} \Delta_i$, where Δ_i is the largest composition difference of residue group i between any pair of constituent chains. The average dissimilarities for various sequence-based clusters are shown in Figure 1. For exact sequence composition values for all MSF entries, see Supplementary Table S1.

4.4. Calculating Structure Features

Secondary structure assignment was performed by DSSP [58], using a three-state classification distinguishing helical ('H','G','I'), extended ('B','E'), and irregular ('S','T', unassigned) residues.

Molecular surfaces were calculated using Naccess [59]. Solvent accessible surface area (SASA) was defined by the Nacces absolute surface column. Interface is defined as the increase in SASA as a result of removing interaction partners from the structure. Buried surface was calculated by subtracting interface area and SASA from the sum of standard surfaces of residues in the protein chain. Thus, interface and buried surfaces represent the area that is made inaccessible to the solvent by the partner(s) or by the analyzed protein itself. All calculated areas were split into hydrophobic (H) and polar (P) contributions based on the polarity of the corresponding atom. Polar/hydrophobic assignations were taken from Naccess.

Contacts were defined at the atomic level. Two atoms were considered to be in contact if their distance is shorter than the sum of the two atoms' van der Waals radii plus 1 Angstrom. For exact structural feature values for all MSF entries, see Supplementary Table S1.

4.5. Filtering Features for Clustering

Standard Pearson correlation values were calculated between all sequence and structure features (Supplementary Table S2). If two features show a correlation with an absolute value above 0.7, only one was kept. In each case, we discarded the feature that shows a high correlation with a higher number of other features, or the one with the lower standard deviation. In total, none of the seven sequence parameters were discarded, but 13 out of the 24 structure parameters were omitted from subsequent clustering steps.

4.6. Clustering

Both sequence and filtered structure parameters were used as input for clustering separately. First, hierarchical clustering was done using the scaled features as input, using Euclidean distance and Ward's method (Supplementary Figures S1 and S3). Then, k-means clustering was employed, and the within-groups sum of squares were plotted as a function of the number of clusters (Supplementary Figures S2 and S4). k-means clustering analysis did not provide a clear-cut support for the number of clusters to choose, and hence we opted for choosing a low number of clusters in both cases (four and five in the case of sequence- and structure-based clustering, respectively), that are not in contradiction with the k-means analysis. This choice of cluster numbers reflects our preference for providing an overall high-level classification. Clustering was done using R with the Ward.D2 and k-means packages.

4.7. Energetic Features

Interaction energies for residues were calculated using the statistical potentials described in [60]. These interaction potentials were demonstrated to well describe the energetic features of IDP interactions [43], and are the basis for recognizing them from the sequence [44]. These potentials yield dimensionless quantities in arbitrary units, and hence their absolute values bear no direct physical meaning. However, their signs are accurate, and values below 0 correspond to stabilizing interactions. Furthermore, they can be directly compared, and hence more negative values typically correspond to more stable structures. In each analysis, the total energies were calculated from the residue-level interactions from the entire complex. Two residues were considered to be in interaction if there is at least one heavy atom contact between them. Energetic values are given in Supplementary Tables S1 (for MSF complexes), S3 (for ordered complexes), and S4 (for complexes containing both IDPs and ordered domains).

4.8. Prediction of K_d Values

Dissociation constants for MSF complexes were estimated using the method described in [61]. In each case, the modified PDB structures taken from the MFIB database [26] were used as input. For technical reasons, not all structures yield a K_d value prediction, and thus the number of values used in representing the average per-complex type K_ds (Figure 5) is calculated from fewer values than the actual number of complexes per type. K_d values are listed in Supplementary Table S1.

4.9. Post-Translational Modifications (PTMs), Isoforms and Competitive Binding

Post-translational modifications were taken from the 2 October 2017 version of PhosphoSitePlus [62], PhosphoELM [63], and UniProt [64]. Only PTMs that were identified in low-throughput experiments were used. These were mapped to complex structures using BLAST between UniProt and PDB sequences (Supplementary Table S5). Protein isoforms were taken from the 4 October 2017 version of UniProt (Supplementary Table S6). To determine alternative binding partners for IDPs, all oligomer PDB structures containing the same UniProt region were selected. PDB structures listed as related in the corresponding MFIB entry were removed. Structures containing the same interaction partners as the original complex were also removed (Supplementary Table S7).

4.10. GeneOntology Terms for Assessing Subcellular Localization

Subcellular localization was represented using GeneOntology [42] terms from the cellular_component namespace. Terms attached to complexes in MFIB were mapped to a restricted set of terms, called CellLoc GO Slim, used in previous studies [28] to compare localization of protein–protein interactions. Terms in CellLoc GO Slim were split into five categories: extracellular, intracellular, membrane, nucleus, and other, encompassing other membrane-bounded cellular compartments, such as the lysosome, as well as non-membrane-bounded compartments, such as the chromatin. For CellLoc GO terms attached to MSF complexes, see Supplementary Table S8.

Author Contributions: Conceptualization, B.M., I.S.; Methodology, B.M.; Software, B.M., L.D.; Formal Analysis, B.M., L.D., E.F.; Investigation, B.M., I.S.; Resources, B.M., L.D.; Data Curation, B.M.; Writing—Original Draft Preparation, B.M.; Writing—Review & Editing, B.M., L.D., E.F.; Visualization, B.M.; Supervision, B.M., I.S.; Project Administration, B.M., I.S.; Funding Acquisition, B.M., L.D., I.S.

Acknowledgments: The authors express gratitude to Zsuzsanna Dosztányi and Zoltán Gáspári for their comments on the project.

Abbreviations

IDP   Intrinsically Disordered Protein
MSF   Mutual Synergistic Folding
CFB   Coupled Folding and Binding
PTM   Post-Translational Modification
SOD1   Superoxide Dismutase
Rb   Retinoblastoma protein
SCOP   Structural Classification of Proteins
CATH   Class/Architecture/Topology/Homologous superfamily
GO   GeneOntology

References

1. Dyson, H.J.; Wright, P.E. Intrinsically unstructured proteins and their functions. *Nat. Rev. Mol. Cell Biol.* **2005**, *6*, 197–208. [CrossRef] [PubMed]

2. Babu, M.M. The contribution of intrinsically disordered regions to protein function, cellular complexity, and human disease. *Biochem. Soc. Trans.* **2016**, *44*, 1185–1200. [CrossRef] [PubMed]

3. Wright, P.E.; Dyson, H.J. Intrinsically disordered proteins in cellular signalling and regulation. *Nat. Rev. Mol. Cell Biol.* **2015**, *16*, 18–29. [CrossRef] [PubMed]

4. Galea, C.A.; Wang, Y.; Sivakolundu, S.G.; Kriwacki, R.W. Regulation of cell division by intrinsically unstructured proteins: intrinsic flexibility, modularity, and signaling conduits. *Biochemistry* **2008**, *47*, 7598–7609. [CrossRef]

5. Fahmi, M.; Ito, M. Evolutionary Approach of Intrinsically Disordered CIP/KIP Proteins. *Sci. Rep.* **2019**, *9*, 1575. [CrossRef]

6. Tompa, P.; Kovacs, D. Intrinsically disordered chaperones in plants and animals. *Biochem. Cell Biol.* **2010**, *88*, 167–174. [CrossRef]

7. Boczek, E.E.; Alberti, S. One domain fits all: Using disordered regions to sequester misfolded proteins. *J. Cell Biol.* **2018**, *217*, 1173–1175. [CrossRef]

8. He, J.; Chao, W.C.H.; Zhang, Z.; Yang, J.; Cronin, N.; Barford, D. Insights into degron recognition by APC/C coactivators from the structure of an Acm1-Cdh1 complex. *Mol. Cell* **2013**, *50*, 649–660. [CrossRef]

9.   Mészáros, B.; Kumar, M.; Gibson, T.J.; Uyar, B.; Dosztányi, Z. Degrons in cancer. *Sci. Signal.* **2017**, *10*. [CrossRef]

10.   Gsponer, J.; Futschik, M.E.; Teichmann, S.A.; Babu, M.M. Tight regulation of unstructured proteins: from transcript synthesis to protein degradation. *Science* **2008**, *322*, 1365–1368. [CrossRef]

11.   Van der Lee, R.; Buljan, M.; Lang, B.; Weatheritt, R.J.; Daughdrill, G.W.; Dunker, A.K.; Fuxreiter, M.; Gough, J.; Gsponer, J.; Jones, D.T.; et al. Classification of intrinsically disordered regions and proteins. *Chem. Rev.* **2014**, *114*, 6589–6631. [CrossRef] [PubMed]

12.   Dosztányi, Z.; Chen, J.; Dunker, A.K.; Simon, I.; Tompa, P. Disorder and sequence repeats in hub proteins and their implications for network evolution. *J. Proteome Res.* **2006**, *5*, 2985–2995. [CrossRef] [PubMed]

13.   Hu, G.; Wu, Z.; Uversky, V.N.; Kurgan, L. Functional Analysis of Human Hub Proteins and Their Interactors Involved in the Intrinsic Disorder-Enriched Interactions. *Int. J. Mol. Sci.* **2017**, *18*, 2761. [CrossRef] [PubMed]

14.   Cortese, M.S.; Uversky, V.N.; Dunker, A.K. Intrinsic disorder in scaffold proteins: getting more from less. *Prog. Biophys. Mol. Biol.* **2008**, *98*, 85–106. [CrossRef]

15.   Snead, D.; Eliezer, D. Intrinsically disordered proteins in synaptic vesicle trafficking and release. *J. Biol. Chem.* **2019**, *294*, 3325–3342. [CrossRef]

16.   Harmon, T.S.; Holehouse, A.S.; Pappu, R.V. Differential solvation of intrinsically disordered linkers drives the formation of spatially organized droplets in ternary systems of linear multivalent proteins. *New J. Phys.* **2018**, *20*, 045002. [CrossRef]

17.   Davey, N.E.; Van Roey, K.; Weatheritt, R.J.; Toedt, G.; Uyar, B.; Altenberg, B.; Budd, A.; Diella, F.; Dinkel, H.; Gibson, T.J. Attributes of short linear motifs. *Mol. Biosyst.* **2012**, *8*, 268–281. [CrossRef]

18.   Tompa, P.; Fuxreiter, M. Fuzzy complexes: polymorphism and structural disorder in protein-protein interactions. *Trends Biochem. Sci.* **2008**, *33*, 2–8. [CrossRef]

19.   Sugase, K.; Dyson, H.J.; Wright, P.E. Mechanism of coupled folding and binding of an intrinsically disordered protein. *Nature* **2007**, *447*, 1021–1025. [CrossRef]

20.   Wang, Y.; Chu, X.; Longhi, S.; Roche, P.; Han, W.; Wang, E.; Wang, J. Multiscaled exploration of coupled folding and binding of an intrinsically disordered molecular recognition element in measles virus nucleoprotein. *Proc. Natl. Acad. Sci. USA* **2013**, *110*, E3743–E3752. [CrossRef]

21.   Shammas, S.L.; Crabtree, M.D.; Dahal, L.; Wicky, B.I.M.; Clarke, J. Insights into Coupled Folding and Binding Mechanisms from Kinetic Studies. *J. Biol. Chem.* **2016**, *291*, 6689–6695. [CrossRef] [PubMed]

22.   Demarest, S.J.; Martinez-Yamout, M.; Chung, J.; Chen, H.; Xu, W.; Dyson, H.J.; Evans, R.M.; Wright, P.E. Mutual synergistic folding in recruitment of CBP/p300 by p160 nuclear receptor coactivators. *Nature* **2002**, *415*, 549–553. [CrossRef] [PubMed]

23.   Schad, E.; Fichó, E.; Pancsa, R.; Simon, I.; Dosztányi, Z.; Mészáros, B. DIBS: A repository of disordered binding sites mediating interactions with ordered proteins. *Bioinformatics* **2017**. [CrossRef] [PubMed]

24.   Fukuchi, S.; Sakamoto, S.; Nobe, Y.; Murakami, S.D.; Amemiya, T.; Hosoda, K.; Koike, R.; Hiroaki, H.; Ota, M. IDEAL: Intrinsically Disordered proteins with Extensive Annotations and Literature. *Nucleic Acids Res.* **2012**, *40*, D507–D511. [CrossRef] [PubMed]

25.   Miskei, M.; Antal, C.; Fuxreiter, M. FuzDB: database of fuzzy complexes, a tool to develop stochastic structure-function relationships for protein complexes and higher-order assemblies. *Nucleic Acids Res.* **2017**, *45*, D228–D235. [CrossRef]

26.   Fichó, E.; Reményi, I.; Simon, I.; Mészáros, B. MFIB: a repository of protein complexes with mutual folding induced by binding. *Bioinformatics* **2017**, *33*, 3682–3684. [CrossRef]

27.   Mészáros, B.; Erdős, G.; Szabó, B.; Schád, É.; Tantos, Á.; Abukhairan, R.; Horváth, T.; Murvai, N.; Kovács, O.P.; Kovács, M.; et al. PhaSePro: the database of proteins driving liquid-liquid phase separation. *Nucleic Acids Res.* **2019**. [CrossRef]

28.   Mészáros, B.; Dobson, L.; Fichó, E.; Tusnády, G.E.; Dosztányi, Z.; Simon, I. Sequential, Structural and Functional Properties of Protein Complexes Are Defined by How Folding and Binding Intertwine. *J. Mol. Biol.* **2019**. [CrossRef]

29.   Mentes, A.; Magyar, C.; Fichó, E.; Simon, I. Analysis of Heterodimeric "Mutual Synergistic Folding"-Complexes. *Int. J. Mol. Sci.* **2019**, *20*, E5136. [CrossRef]

30.   Magyar, C.; Mentes, A.; Fichó, E.; Cserző, M.; Simon, I. Physical Background of the Disordered Nature of "Mutual Synergetic Folding" Proteins. *Int. J. Mol. Sci.* **2018**, *19*, E3340. [CrossRef]

31.   wwPDB consortium. Protein Data Bank: the single global archive for 3D macromolecular structure data. *Nucleic Acids Res.* **2019**, *47*, D520–D528. [CrossRef] [PubMed]

32. Chandonia, J.-M.; Fox, N.K.; Brenner, S.E. SCOPe: classification of large macromolecular structures in the structural classification of proteins-extended database. *Nucleic Acids Res.* **2019**, *47*, D475–D481. [CrossRef] [PubMed]

33. Sillitoe, I.; Dawson, N.; Lewis, T.E.; Das, S.; Lees, J.G.; Ashford, P.; Tolulope, A.; Scholes, H.M.; Senatorov, I.; Bujan, A.; et al. CATH: expanding the horizons of structure-based functional annotations for genome sequences. *Nucleic Acids Res.* **2019**, *47*, D280–D284. [CrossRef] [PubMed]

34. Zhao, N.; Pang, B.; Shyu, C.-R.; Korkin, D. Structural similarity and classification of protein interaction interfaces. *PLoS One* **2011**, *6*, e19554. [CrossRef] [PubMed]

35. Redler, R.L.; Wilcox, K.C.; Proctor, E.A.; Fee, L.; Caplow, M.; Dokholyan, N.V. Glutathionylation at Cys-111 induces dissociation of wild type and FALS mutant SOD1 dimers. *Biochemistry* **2011**, *50*, 7057–7066. [CrossRef] [PubMed]

36. Van Roey, K.; Gibson, T.J.; Davey, N.E. Motif switches: decision-making in cell regulation. *Curr. Opin. Struct. Biol.* **2012**, *22*, 378–385. [CrossRef]

37. Kang, B.; Pu, M.; Hu, G.; Wen, W.; Dong, Z.; Zhao, K.; Stillman, B.; Zhang, Z. Phosphorylation of H4 Ser 47 promotes HIRA-mediated nucleosome assembly. *Genes Dev.* **2011**, *25*, 1359–1364. [CrossRef]

38. Lee, C.W.; Ferreon, J.C.; Ferreon, A.C.M.; Arai, M.; Wright, P.E. Graded enhancement of p53 binding to CREB-binding protein (CBP) by multisite phosphorylation. *Proc. Natl. Acad. Sci. USA* **2010**, *107*, 19290–19295. [CrossRef]

39. Bousset, K.; Oelgeschläger, M.H.; Henriksson, M.; Schreek, S.; Burkhardt, H.; Litchfield, D.W.; Lüscher-Firzlaff, J.M.; Lüscher, B. Regulation of transcription factors c-Myc, Max, and c-Myb by casein kinase II. *Cell. Mol. Biol. Res.* **1994**, *40*, 501–511.

40. Saddic, L.A.; West, L.E.; Aslanian, A.; Yates, J.R., 3rd; Rubin, S.M.; Gozani, O.; Sage, J. Methylation of the retinoblastoma tumor suppressor by SMYD2. *J. Biol. Chem.* **2010**, *285*, 37733–37740. [CrossRef]

41. Gibson, T.J. Cell regulation: determined to signal discrete cooperation. *Trends Biochem. Sci.* **2009**, *34*, 471–482. [CrossRef] [PubMed]

42. The Gene Ontology Consortium. The Gene Ontology Resource: 20 years and still GOing strong. *Nucleic Acids Res.* **2019**, *47*, D330–D338. [CrossRef] [PubMed]

43. Mészáros, B.; Tompa, P.; Simon, I.; Dosztányi, Z. Molecular principles of the interactions of disordered proteins. *J. Mol. Biol.* **2007**, *372*, 549–561. [CrossRef] [PubMed]

44. Mészáros, B.; Erdos, G.; Dosztányi, Z. IUPred2A: context-dependent prediction of protein disorder as a function of redox state and protein binding. *Nucleic Acids Res.* **2018**, *46*, W329–W337. [CrossRef] [PubMed]

45. Strop, P.; Kaiser, S.E.; Vrljic, M.; Brunger, A.T. The structure of the yeast plasma membrane SNARE complex reveals destabilizing water-filled cavities. *J. Biol. Chem.* **2008**, *283*, 1113–1119. [CrossRef] [PubMed]

46. Bonvin, A.M.; Vis, H.; Breg, J.N.; Burgering, M.J.; Boelens, R.; Kaptein, R. Nuclear magnetic resonance solution structure of the Arc repressor using relaxation matrix calculations. *J. Mol. Biol.* **1994**, *236*, 328–341. [CrossRef]

47. Madl, T.; Van Melderen, L.; Mine, N.; Respondek, M.; Oberer, M.; Keller, W.; Khatai, L.; Zangger, K. Structural basis for nucleic acid and toxin recognition of the bacterial antitoxin CcdA. *J. Mol. Biol.* **2006**, *364*, 170–185. [CrossRef]

48. Sauvé, S.; Tremblay, L.; Lavigne, P. The NMR solution structure of a mutant of the Max b/HLH/LZ free of DNA: insights into the specific and reversible DNA binding mechanism of dimeric transcription factors. *J. Mol. Biol.* **2004**, *342*, 813–832. [CrossRef]

49. Le Trong, I.; Stenkamp, R.E.; Ibarra, C.; Atkins, W.M.; Adman, E.T. 1.3-A resolution structure of human glutathione S-transferase with S-hexyl glutathione bound reveals possible extended ligandin binding site. *Proteins* **2002**, *48*, 618–627. [CrossRef]

50. Dams, T.; Auerbach, G.; Bader, G.; Jacob, U.; Ploom, T.; Huber, R.; Jaenicke, R. The crystal structure of dihydrofolate reductase from Thermotoga maritima: molecular features of thermostability. *J. Mol. Biol.* **2000**, *297*, 659–672. [CrossRef]

51. Tachiwana, H.; Osakabe, A.; Shiga, T.; Miya, Y.; Kimura, H.; Kagawa, W.; Kurumizaka, H. Structures of human nucleosomes containing major histone H3 variants. *Acta Crystallogr. D Biol. Crystallogr.* **2011**, *67*, 578–583. [CrossRef] [PubMed]

52. Banjade, S.; Wu, Q.; Mittal, A.; Peeples, W.B.; Pappu, R.V.; Rosen, M.K. Conserved interdomain linker

promotes phase separation of the multivalent adaptor protein Nck. *Proc. Natl. Acad. Sci. USA* **2015**, *112*, E6426–E6435. [CrossRef] [PubMed]

53. Cheng, H.-C.; Skehan, B.M.; Campellone, K.G.; Leong, J.M.; Rosen, M.K. Structural mechanism of WASP activation by the enterohaemorrhagic E. coli effector EspF(U). *Nature* **2008**, *454*, 1009–1013. [CrossRef] [PubMed]

54. Westermark, P.; Sletten, K.; Johansson, B.; Cornwell, G.G., 3rd. Fibril in senile systemic amyloidosis is derived from normal transthyretin. *Proc. Natl. Acad. Sci. USA* **1990**, *87*, 2843–2845. [CrossRef]

55. Pansarasa, O.; Bordoni, M.; Diamanti, L.; Sproviero, D.; Gagliardi, S.; Cereda, C. SOD1 in Amyotrophic Lateral Sclerosis: "Ambivalent" Behavior Connected to the Disease. *Int. J. Mol. Sci.* **2018**, *19*, E1375. [CrossRef]

56. Fox, N.K.; Brenner, S.E.; Chandonia, J.-M. The value of protein structure classification information-Surveying the scientific literature. *Proteins* **2015**, *83*, 2025–2038. [CrossRef]

57. Hadley, C.; Jones, D.T. A systematic comparison of protein structure classifications: SCOP, CATH and FSSP. *Structure* **1999**, *7*, 1099–1112. [CrossRef]

58. Touw, W.G.; Baakman, C.; Black, J.; te Beek, T.A.H.; Krieger, E.; Joosten, R.P.; Vriend, G. A series of PDB-related databanks for everyday needs. *Nucleic Acids Res.* **2015**, *43*, D364–D368. [CrossRef]

59. Hubbard, S.; Thornton, J. NACCESS Computer Program. 1992. Available online: http://wolf.bms.umist.ac.uk/naccess/ (accessed on 31 October 2019).

60. Dosztányi, Z.; Csizmók, V.; Tompa, P.; Simon, I. The pairwise energy content estimated from amino acid composition discriminates between folded and intrinsically unstructured proteins. *J. Mol. Biol.* **2005**, *347*, 827–839. [CrossRef]

61. Vangone, A.; Bonvin, A.M. Contacts-based prediction of binding affinity in protein-protein complexes. *Elife* **2015**, *4*, e07454. [CrossRef]

62. Hornbeck, P.V.; Zhang, B.; Murray, B.; Kornhauser, J.M.; Latham, V.; Skrzypek, E. PhosphoSitePlus, 2014: mutations, PTMs and recalibrations. *Nucleic Acids Res.* **2015**, *43*, D512–D520. [CrossRef] [PubMed]

63. Dinkel, H.; Chica, C.; Via, A.; Gould, C.M.; Jensen, L.J.; Gibson, T.J.; Diella, F. Phospho.ELM: A database of phosphorylation sites–update 2011. *Nucleic Acids Res.* **2011**, *39*, D261–D267. [CrossRef] [PubMed]

64. UniProt Consortium. UniProt: A worldwide hub of protein knowledge. *Nucleic Acids Res.* **2019**, *47*, D506–D515. [CrossRef] [PubMed]

14

Structural Determinants of the Prion Protein N-Terminus and its Adducts with Copper Ions

Carolina Sánchez-López [1,†], **Giulia Rossetti** [2,3,4], **Liliana Quintanar** [1,*] and **Paolo Carloni** [2,5,6,*]

[1] Department of Chemistry, Center for Research and Advanced Studies (Cinvestav), 07360 Mexico City, Mexico; magdacarolina29@hotmail.com

[2] Institute of Neuroscience and Medicine (INM-9) and Institute for Advanced Simulation (IAS-5), Forschungszentrum Jülich, Wilhelm-Johnen-Strasse, 52425 Jülich, Germany; g.rossetti@fz-juelich.de

[3] Jülich Supercomputing Center (JSC), Forschungszentrum Jülich, 52428 Jülich, Germany

[4] Department of Oncology, Hematology and Stem Cell Transplantation, Faculty of Medicine, RWTH Aachen University, Pauwelsstraße 30, 52074 Aachen, Germany

[5] Department of Physics and Department of Neurobiology, RWTH Aachen University, 52078 Aachen, Germany

[6] Institute for Neuroscience and Medicine (INM)-11, Forschungszentrum Jülich, 52428 Jülich, Germany

* Correspondence: lilianaq@cinvestav.mx (L.Q.); p.carloni@fz-juelich.de (P.C.)

† Current Address: Instituto de Biología Molecular y Celular de Rosario (IBR-CONICET), Ocampo y Esmeralda, 2000 Rosario, Argentina.

Abstract: The N-terminus of the prion protein is a large intrinsically disordered region encompassing approximately 125 amino acids. In this paper, we review its structural and functional properties, with a particular emphasis on its binding to copper ions. The latter is exploited by the region's conformational flexibility to yield a variety of biological functions. Disease-linked mutations and proteolytic processing of the protein can impact its copper-binding properties, with important structural and functional implications, both in health and disease progression.

Keywords: N-terminal prion protein; copper binding; prion disease mutations

1. Introduction

Prion diseases or transmissible spongiform encephalopathies (TSEs) are rare neurodegenerative diseases exhibiting symptoms of both cognitive and motor dysfunction, vacuolation of the grey matter in the human central nervous system, neuronal loss, and astrogliosis [1]. A crucial event for the diseases' development is the misfolding of the extracellular, membrane-anchored human prion protein (HuPrP^C) into the fibril-forming isoform called "scrapie" (HuPrP^{Sc}), the major or only component of the infectious particle [2]. This eventually leads to protofibril and fibrillar structures. Accordingly, with the "Protein only hypothesis" by Nobel Laureate Prusiner [3], the feature to undergo induced or spontaneous misfolding depends basically on intrinsic features of the protein. These include the amino acid sequence [4,5] as well as secondary structure elements [6–8], the highly flexible amino terminal region of the protein [9], and posttranslational modification elements [10]. The propensity to form the scrapie form is modulated by a variety of external factors. These include pH [11–13], cofactors like metal ions [14,15], or the presence of proteins [16,17]. Pathogenic mutations (PM) in HuPrP^C are linked to the spontaneous generation of prion diseases [18–21].

HuPrP^C is ubiquitously expressed throughout the body. It is mostly found in the central nervous system. After being synthesized in the rough endoplasmic reticulum, it transits through the Golgi compartment, and it is released to the cell surface where it resides in lipid membrane domains [22]. Though its physiological role is still not clear, HuPrP^C might be involved in neuronal development, cell adhesion, apoptotic events, and cell signaling in the central nervous system. Moreover, HuPrP[C] can interact with different neuronal proteins or proteins of the extracellular matrix, as well as with other binders including glycosaminoglycans, nucleic acids, and copper ions [23]. Hence, HuPrP[C] has been also proposed as a copper sensing or transport protein [24].

The protein features two signal peptides (1–22 and 232–235, Figure 1), a folded globular domain (GD, residues 125–231), and a naturally unfolded N-terminal tail (N-term_HuPrP[C], hereafter, residues 23–124), which is the focus of this review. The GD consists of two β-sheets (S1 and S2), three α-helices (H1, H2, and H3), one disulfide bond (SS) between cysteine residues 179 and 214, and two potential sites for N-linked glycosylation (green forks in Figure 1) at residues 181 and 197 [25]. H2 and H3 helices linked by the SS-bond constitute the H2 + H3 domain. A glycosylphosphatidylinositol anchor (GPI, in blue in Figure 1) is attached to the C-terminus, which is located on the outside cellular membrane.

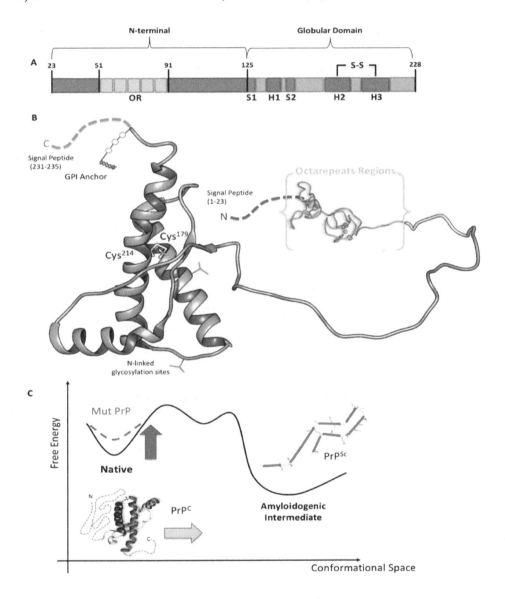

Figure 1. (**A**) Schematic and (**B**) tridimensional view of HuPrP[C]. (**C**) Qualitative scheme illustrating the Gibbs free energy change in the conversion from HuPrP[C] (left) to HuPrP[Sc] (right) [26]. The depicted amyloidogenic intermediate is the parallel, in-register β-structure model for the core of recombinant PrP90–231 amyloid fibrils formed in vitro [27], one of the models among others [28–30], whereas the native globular domain (GD) of the HuPrP[C] is the nuclear magnetic resonance (NMR) structure by Zahn et al. [25]. Adapted from [31,32].

The HuPrP^C→HuPrP^{Sc} interconversion involves mostly the GD. It may entail increasingly β-stranded intermediate structures [33] (Figure 1C), leading to small aggregates, protofibrils, and finally ordered rigid fibrils [34–38]. Experimental structural information for these is lacking [34–38].

While the structure of the GD of HuPrP^C has been resolved experimentally, the intrinsically disordered nature of the N-term_HuPrP^C has represented a challenge for structural studies. In this paper, the structural properties of the N-term_HuPrP^C are discussed, with a focus on recent insights obtained from computational approaches and on the functional and disease-related implications of copper–N-term_HuPrP^C interactions.

2. The N-Term: Function and Structural Determinants

This naturally unfolded domain contains the major part of the so-called transmembrane domain (termed TM1, comprising roughly residues 112–135) and the preceding "stop transfer effector" (STE, a hydrophilic region containing roughly residues 104–111) [39,40] (Figure 1B). STE and TM1 act in concert to control the co-translational translocation at the endoplasmic reticulum (ER) during the biosynthesis of the protein [41,42].

N-term_HuPrP^C functions as a broad-spectrum molecular sensor [43]. Along with the highly homologous protein from mouse (N-term_MoPrP^C, 93% sequence identity), it interacts with copper ions (see below) and sulphated glycosaminoglycans [44]. In addition, N-term_MoPrP^C interacts with vitronectin [45], the stress-inducible protein 1 (STI1) [46], amyloid-β (Aβ) multimers [47–49], lipoprotein receptor-related protein 1 (LRP1) [50], and the neural cell adhesion molecule (NCAM) [51].

Because experimental structural information on the full-length N-term_HuPrP^C is currently lacking, one has to resort to biocomputing-based predictions. Recently, some of us have used a combination of bioinformatics along with replica-exchange-based Monte Carlo simulation at room temperature, based on a simplified force field, to predict the conformational ensemble on the full-length N-term_MoPrP^C [31,52].

This is expected to be quite similar to that from *Homo sapiens*, given the extremely high sequence identity (93%) with N-term_HuPrP^C [31,52]. Monte Carlo simulations suggest that the N-term_MoPrP^C consists of several regions characterized by different secondary structure elements, consistently with biophysical data [53–57]. Specifically, it contains 19 ± 8% α-helix, 8 ± 5% β-sheet, 7 ± 3% β-bridge, 27 ± 5% β-turn, 12 ± 4% bend, 4 ± 3% 3$_{10}$-helix, and 1 ± 1% π-helix. The secondary structure elements are distributed among the N-term in a highly heterogeneous manner (Figure 2A): residues 23-30 are mainly coil/β-turn/bend; residues 31-50 are mainly β-turn/coil/bend/β-bridge; and residues 59-90 form four sequential octarepeat (OR) peptides, with sequence PHGGGWGQ, and are mainly β-turn/coil/bend/β-sheet conformations. In particular, the loop/β-turn conformations in the OR region resemble (backbone RMSD < 2.5 Å) those identified by NMR [57]; residues 89-98 are mainly coil/β-turn/bend/β-sheet; residues 99-117 feature the highest content α-helix of N-term_MoPrP^C regions; and residues 118-125 display a comparable percentage of α-helix and β-turn. Residues 105-125, the "amyloidogenic region", feature transient helical structures (the last eight residues also have a comparable content of beta turn). This is consistent with circular dichroism (CD), nuclear magnetic resonance (NMR), and Fourier transform infrared (FTIR) studies on HuPrP^C fragments [54–56] (Figure 2). The same simulation procedure can be carried out for the known disease-linked mutations (Figure 2).

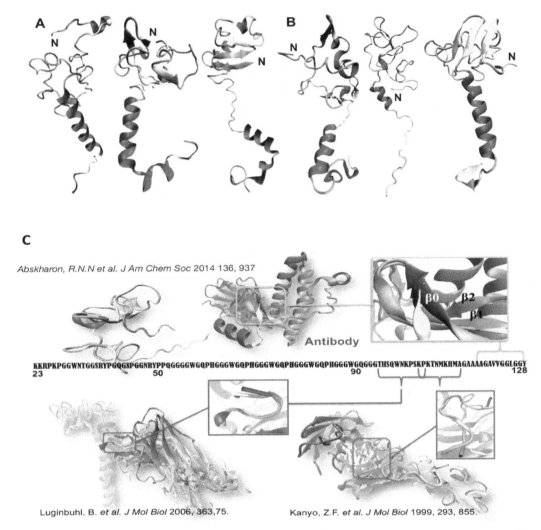

Figure 2. Selected conformations of (**A**) WT N-term_MoPrP^C and (**B**) one PM (N-term_MoPrP^C_Q52P) emerging from molecular simulation [31,52]. These contain transient α-helix (in violet), β-sheet (yellow), β-bridge (orange), β-turn (cyan), 3_{10}-helix (blue), and p-helix (red) elements. (**C**) Superimposition of our conformational ensemble (orange) with available fragments of N-term deposited structures. Readapted from [31,52].

While many PMs in the GD are known to modify significantly the folded structure and to increase its flexibility [58–61], our Monte Carlo calculations suggest that those in the N-term do not impact significantly the global structural properties of the N-term. This finding is consistent with experimental findings showing that PMs in N-term_HuPrP^C do not affect the thermostability or misfolding kinetics of the protein [58,62–64]. On the contrary, our Monte Carlo simulations show that the PMs at the N-term modify local features at the binding sites for known cellular partners, as well as of interdomain interactions. This points to an interference of the PMs with the related physiological functions.

The major differences in the presence of the PMs were observed in the residues binding Cu^{2+} and sulphated GAG (i.e., the OR region and the H110 Cu^{2+}-binding site mouse sequence, H111 in the human sequence). In addition, the PMs affect the SS and the flexibility and increase the hydrophobicity of STE/TM1. The latter contains the putative binding sites for in vivo binding partner proteins such as vitronectin [45] and STI1 [46]. This might affect the biological function of these interactions, which involves the signaling for axonal growth [45] and that for neuroprotection [46], respectively.

The PM Q52P in the OR region, interestingly, affects the flexibility of STE/TM1, while the other six PMs in STE/TM1 also alter the intra-molecular contacts in the OR region suggesting a role played by PMs in altering transient interdomain interactions between the OR region and STE/TM1. Recent studies suggest that N-Term and GD interactions might also serve to regulate the activity and/or toxicity of the PrP^C N-term [65]. Unfortunately, in the reported Monte Carlo study [52], the GD was not taken into account.

The altered local features in STE/TM1 might also impact the interactions of the protein with trans-acting factors in the cytosol and in the ER membrane [66]. This result is consistent with the in vitro data that PMs P101L, P104L, and A116V increase the interactions between MoPrP^C STE/TM1 and a membrane mimetic at pH 7 [67].

3. Copper Binding

Copper ions bind to the N-term of HuPrP^C in vivo [24]. Since the protein is anchored to the neuronal membrane, facing the extracellular space, it is exposed to fluctuations in Cu^{2+} ion concentrations, that can reach 100 μM during synaptic transmission [68]. This represents orders of magnitude larger than the experimentally measured range of binding affinities for Cu(II) ions at the N-term (nM to μM) [69]. Thus, it is plausible that the protein responds to Cu(II) ion concentration changes at the synapse. On the other hand, upon endocytosis, HuPrP^C is exposed to the intracellular reducing environment, where the interaction of its N-term with Cu$^+$ ions would also be relevant. Two main functions of copper binding to the N-term have been proposed so far: (i) stimulation of HuPrP^C endocytosis [70–73] and (ii) copper sensing associated to cell signaling. Copper-induced endocytosis of HuPrP^C requires its N-term terminus, specifically the octarepeat region, and it might involve conformational alterations of the N-term with the subsequent delivery of copper ions to endosomes. This has led to a proposed role for HuPrP^C in copper transport. However, it is unlikely that HuPrP^C delivers copper efficiently into the cytosol, since high concentrations are needed for copper-induced endocytosis (150–300 μM) [71,73]. On the other hand, HuPrP^C can interact with the human N-methyl-D-aspartate receptors (HuNMDAR) and alpha-amino-3-hydroxy-5-methyl-4-isoxazolepropionic acid receptors (HuAMPAR) involved in synaptic transmission, while both receptors are regulated by Cu ions [74–76]. In the case of HuNMDAR activity, copper binding to HuPrP^C is necessary to regulate the activity of this receptor [75]. Indeed, HuPrP^C and Cu^{2+} are required to inhibit HuNMDAR activity through a mechanism that involves post-translational S-nitrosylation of cysteine residues in HuNMDAR [77]. Overall, these important findings underscore a role for Cu-HuPrP^C interactions in neuroprotective mechanisms, which could be disrupted by other Cu-binding proteins or peptides at the synapse. For instance, human amyloid-β (Aβ) neurotoxicity has recently been linked to its ability to compete for Cu^{2+} ions with HuPrP^C, thereby interfering with the modulation of HuNMDAR activity [75].

The N-term region of HuPrP^C contains six His residues that may serve as anchoring sites for Cu^{2+} ions [78]. The ion binds to different sites within the protein [79–82], which are conserved in mammalian species [83], a fact that underscores its physiological relevance. Four of them are located in the OR region, spanning residues 60-91 with four repeats of the highly conserved octapeptide PHGGGWGQ (Figure 3). Beyond the OR region, two additional His residues, 96 and 111, also act as copper-binding sites in the 92-115 region. Studies on synthetic peptide fragments have suggested that metal coordination modes depend on copper concentration [69], as well as the relative copper:protein ratio and proton concentration [79,80]. At physiological pH, three distinct Cu^{2+} coordination modes have been identified by electron paramagnetic resonance (EPR) [84]. At low Cu:protein ratios, three or four His residues can chelate one metal ion, leading to a multiple histidine Cu-binding mode, named Component 3 (Figure 3). At higher Cu:protein ratios, a species with two His ligands forming a 2N2O equatorial coordination mode is observed (Component 2 in Figure 3). When enough Cu^{2+} is provided to reach a 1:1 ratio for each octapeptide fragment, a species with a 3N1O equatorial coordination mode is formed, named Component 1, where the coordinating residues are as follows: one His imidazole ligand, two deprotonated backbone amide groups, and a carbonyl group from the glycine residues that follow the anchoring His in the sequence (Figure 3) [78,85]. X-ray crystallographic studies of the Cu^{2+} complex with one octapeptide PHGGGWGQ fragment also revealed the participation of a water molecule as an axial ligand, stabilized by hydrogen bonding to the Trp residue [86]. This is the only Cu-binding site fragment that has been characterized so far by X-ray crystallography.

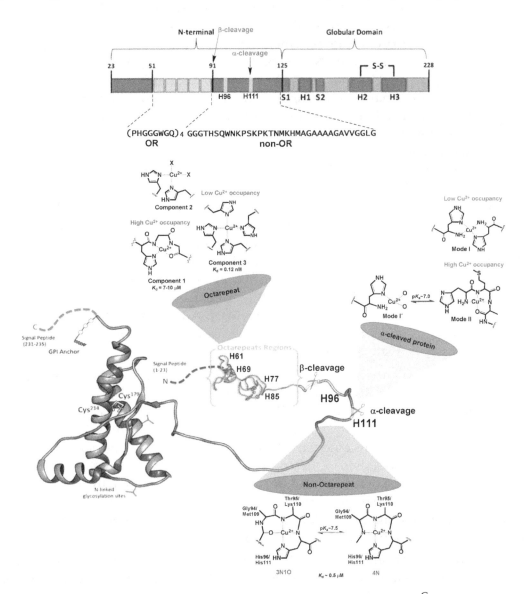

Figure 3. Cu coordination properties of the N-terminal region of human HuPrP^C. The six His residues that act as anchoring sites for Cu ions are highlighted: His61, His69, His77, and His85 in the OR region, and His96 and His111 in the non-OR region. The models for the different Cu^{2+} coordination modes identified at each site at physiological pH are drawn. The impact of α-cleavage processing on the His111 binding site is also shown.

The K_d for the low occupancy multiple-His coordination mode (Component 3) is 0.12 nM, whereas it ranges from 7 to 10 µM for the high-occupancy 3N1O mode (Component 1) [69]. Hence, Cu binding to the OR region displays a negative cooperativity. This is consistent with the formation of intermediate species such as Component 2. Electrochemical studies have determined that the high-occupancy Component 1 species is capable of reducing dioxygen to produce low levels of hydrogen peroxide that may be relevant for cell signaling, whereas the low-occupancy multiple-His mode cannot activate dioxygen at all [87,88].

Outside the OR region, His 96 and 111 act as anchoring sites for Cu^{2+} ions, constituting the closest Cu-binding sites to the amyloidogenic region of HuPrP^C [81,89,90]. An X-ray absorption spectroscopy (XAS) study of a HuPrP (90–231) construct of the protein showed that at low pH Cu ions can coordinate to both His residues in the non-OR region, while at physiological pH only the His111 site is populated [91]. The peptide fragments mostly used to characterize the individual non-OR Cu-binding sites are HuPrP (92-96) and HuPrP (106-115) with sequences GGGTH and KTNMKHMAGA, respectively [85,92]. EPR and other spectroscopic studies have determined that Cu^{2+} coordination is highly pH-dependent in these sites, yielding two different equatorial coordination modes at physiological pH, 3N1O and 4N, related by a pKa value near 7.5 (7.8 for the His 96 site and 7.5 for the His 111 site) [93–96]. In both sites, Cu^{2+} coordination in the 3N1O mode involves the His imidazole group, two deprotonated amide nitrogens, and a carbonyl group from the backbone amide bonds that precede the His residue in the sequence, while a third deprotonated amide group replaces the carbonyl moiety in the 4N equatorial mode (Figure 3). Although the equatorial coordination modes of Cu^{2+} bound to these sites are identical, the presence of two Met residues in the His 111 site provides it with distinct properties, particularly in terms of relative binding affinity, and redox properties. The thioether groups of Met 109 and Met 112 can participate in Cu^{1+} coordination to the PrP(106-115) fragment, as demonstrated by XAS studies, yielding coordination modes where the His111 imidazole ring, the two Met residues, and a backbone carbonyl group stabilize a tetra-coordinated Cu^{1+} species at physiological pH, where Met109 plays a more important role in metal coordination, as compared to Met112 [97]. Anchoring of Cu^{1+} ions by Met residues persists even at low pH values (<5), as those found in endosomes. Thus, the MKHM motif and Cu coordination features of the His111 site assure that the metal ion would still be bound to the protein, even under decreased pH and reducing conditions, such as those encountered upon endocytosis. Additionally, the capability of the His111 site to stabilize both Cu^{2+} and Cu^{1+} makes it a unique site in the N-term region of HuPrP^C that may support redox activity to activate dioxygen.

The relative binding affinity of Cu^{2+} for the non-OR sites has also been studied [98–102]. While a slight preference for Cu^{2+} binding to His111 over His96 has been observed spectroscopically and ascribed to the Met residues nearby, the two sites get loaded simultaneously with K_d values in the range of 0.4 to 0.7 μM at physiological pH [102]. Given the Cu-binding affinity features of the OR region, this implies that upon increasing Cu^{2+} levels, the multiple His species (Component 3) would first form, followed by the population of the non-octarepeat His96/His111 sites, before the OR region is fully loaded to yield the high occupancy mode (Component 1). Overall, the Cu(II) coordination and binding features of the N-term region provide HuPrP^C with the ability to respond to a wide range of Cu concentrations, adopting different metal coordination modes, which in turn may impose different conformations to this unstructured region of the protein.

The conformational flexibility of the N-term_HuPrP^C and the presence of several His residues as Cu anchoring sites provide a platform to accommodate different Cu^{2+} coordination modes as a function of relative metal:protein concentrations. Unlike the static (entatic) Cu active sites of cuproenzymes, where the protein structure imposes restrictions on the metal coordination and geometry, the preferred coordination modes at each Cu-binding site in the flexible N-term domain of HuPrP^C are dictated by the geometric and electronic preferences of the metal ion, which can actually impose metal-induced conformations with potentially different functional implications or a propensity to aggregate. For example, the participation of deprotonated backbone amides in Cu^{2+} coordination, as in the high-occupancy (Component 1) mode of the OR region, inevitably imposes a certain turn in the backbone chain, which is known to yield more compact conformations [81]. Indeed, full loading of Cu^{2+} ions into the OR region yields a conformation where the average Cu–Cu distance is 4 to 7 Å, as determined by EPR [84]. Given the relatively high Cu concentrations needed for endocytosis of HuPrP^C and the Cu-binding affinity features of the OR region, this high-occupancy conformation of the OR region is likely involved in the mechanism of endocytosis.

Recent studies have revealed metal-induced interactions between the N-term and C-term regions of MoPrP^C [103–105]. An NMR study suggested interactions of the region 90-120, containing His96 and His111 Cu-binding sites, with residues in the vicinity of helix-1 (specifically 144-147), while the Cu-loaded OR region may also interact with helix-2, involving residues 174-185 [105]. The latter was confirmed by an elegant NMR and site-directed spin labeling EPR study that provided detailed structural information on how Cu binding at the low occupancy multiple His site (Component 3) in the OR region promotes electrostatic interactions with a highly conserved negatively charged region at helices 2 and 3 of the globular protein [103]. The C-term region of the protein engaged in the interaction with Cu-loaded OR overlaps with the region where neurotoxic PrP antibodies bind, and it also involves acidic residues associated with disease-related mutations, such as E200K. These observations underscore the important role of Cu^{2+} loading into the low-occupancy multiple-His coordination mode in promoting electrostatic interactions between the Cu-bound N-terminal region and the helical C-terminal domain, a stabilizing interaction that is considered to be regulatory for prion conversion [103,105].

On the other hand, in the non-OR region, the two His coordination modes identified at low pH for the PrP(90-231) construct were also found to induce stabilizing interactions with the globular C-terminal domain, whereas Cu^{2+} binding solely to the His111 site induced local beta-sheet structure [91]. Consistently, copper binding to the non-OR sites in the amyloidogenic fragment 90-126 induces a β-sheet-like transition [106]. These observations underscore the important role that Cu binding to the non-OR region may play in amyloid aggregation and prion conversion.

Cu^{2+}-PrP^C interactions and their perturbation by disease-related mutations have been suggested to play a role for Hu/Mo PrP^C aggregation and prion disease progression [107]. Specifically, the GSS-linked Q211P PM [60] (Q212P in Hu numbering) in the $HuPrP^C$ GD can influence the Cu^{2+} binding coordination at H96 and H111 [108], implicating a role of abnormal Cu^{2+} binding in the pathology of PMs in $HuPrP^C$. As discussed above, the multiple His Cu-binding modes induce stabilizing interactions of the N-term_$HuPrP^C$ with the globular C-terminal domain, whereas Cu^{2+} binding solely to the His111 site induces local beta-sheet structure [91,103,105]. Thus, any perturbation of the local conformation around the Cu-binding sites may have an impact on the stability of the protein. Consistently, structural analysis by molecular simulations of the N-term_$HuPrP^C$ indicates that some disease-linked mutations may affect the local conformation and intramolecular interactions around the Cu^{2+} binding sites, including His111 [31,32], providing a molecular basis to understand their impact on disease progression. Although further studies are needed to understand how Cu binding impacts the folding and conformation of the flexible N-term_$HuPrP^C$, it is clear that the different Cu^{2+} coordination modes formed at the His anchoring sites can favor distinct metal-induced conformations, while disease-related mutations may also impact the conformation of the N-term region, its Cu-binding properties, and its interactions with the C-terminal globular region of $HuPrP^C$.

Copper binding may also be affected by a specific post-translational modification, namely the proteolytic processing at specific sites of the N-term region [109]. This includes the following: (i) the β-cleavage of the OR region, leading to the N2 (23-89) and C2 (90-231) fragments [109]. This is induced by reactive oxygen species (ROS) produced in the presence of Cu^{2+} ions [110,111] (It can be also catalyzed by calpain and ADAM8—a member of the A Disintegrin And Metalloproteinase (ADAM) family of enzymes [112,113]). While the N2 fragment may be released, maintaining the Cu^{2+} binding properties of the OR sites as described above, the C2 fragment may remain anchored to the membrane, conserving the His96 and His111 sites, but with a free N-term group at residue 90 [109]. The free NH_2 moiety is expected to change significantly the Cu^{2+} coordination features of these non-OR sites. (ii) The α-cleavage occurs at several sites in the region encompassing residues 109-120, and it is a common feature in a wide range of cell lines [114]. The process, performed by members of the ADAM enzymes [112], increases in the presence of Cu^{2+} ions [115]. This metal also may modulate the relative amount of α-cleavage at each site, possibly by inducing local conformational changes that impact how the protein docks into the protease active site [112,113]. The most described α-cleavage site is located between Lys110 and His111 in the human sequence, producing two fragments N1 (23-110) and C1 (111-231) [109]. The released N1 fragment may still contain the OR region and the His96 site; however, the His111 site may be significantly disturbed, as the cleavage occurs between residues that participate in Cu^{2+} coordination, leaving a His111 with a free NH_2 terminal group at the membrane bound C1 fragment. A recent spectroscopic study determined the impact of α-cleavage processing on Cu^{2+} binding to His111, using a model peptide for the C1 fragment [116]. Indeed, in this fragment His111 and the free NH_2 terminal group act as the main anchoring sites for Cu^{2+}, resulting in coordination modes that are highly dependent on proton and copper concentrations, and are quite different from those characterized for the intact His111 site in the full protein (Figure 3). The Cu-binding affinity features and redox activity of this perturbed His111 site remain to be investigated. It is interesting to note that, while Cu ions can modulate the relative amount of α-cleavage at different sites of the 109-120 region of $HuPrP^C$ [112], the resulting membrane-bound C1 fragments and their Cu-binding properties could in turn determine the metal-induced conformation of the N-term region and its ability to interact with important receptors, such as the HuNMDAR and HuAMPAR [113,117].

4. Conclusions

Recent advances in computational biophysics [31,32,52] have led, for the first time, to a description of the conformational ensemble on the full-length N-term $MoPrP^C$, a fully disordered domain of 125 amino acids, with high similarity to the human domain. This has made it possible to probe the impact of disease-related mutations on the structural properties of this flexible region of the protein.

N-term_$HuPrP^C$ binds copper ions in vivo [24]. It yields a diverse range of Cu coordination modes, each with distinct redox properties and binding affinity features. The Cu-binding properties of the N-term region provide $HuPrP^C$ with the ability to respond to the wide range of Cu concentrations that the protein is exposed to at the synapse, adopting different metal-induced conformations, which in turn may have distinct functional implications. On the other hand, Cu^{2+}-PrP^C interactions and their perturbation by disease-related mutations may play a role in protein aggregation and prion disease progression.

While the interplay between metal ion binding and conformational flexibility in the entire N-term remains to be understood, it is well established that copper displays site-specific effects on its folding, either by promoting stabilizing interactions or inducing conversion to beta-sheet folds. Conversely, molecular simulations suggest that some disease-related mutations may affect the local conformation around the Cu anchoring sites, thus affecting the Cu-binding properties of the N-term_$HuPrP^C$ and the stability of the protein.

Combined computational and experimental studies on the structural impact of Cu^{2+} binding and disease-related mutations at the N-term_$HuPrP^C$, such as those on copper(II)-alpha-synuclein—an intrinsically disordered protein undergoing fibril formation in Parkinson's disease [118]—could advance dramatically our understanding of the functional role of Cu^{2+}-PrP^C interactions in health and disease.

Author Contributions: Conceptualization, all the authors; Writing-Review & Editing, all the authors.

Acknowledgments: C.S.L. acknowledges SECITI CDMX, Mexico for the postdoctoral fellowship.

References

1. Prusiner, S.B. Molecular biology and pathogenesis of prion diseases. *Trends Biochem. Sci.* **1996**, *21*, 482–487. [CrossRef]

2. Prusiner, S.B. Prions. *Proc. Natl. Acad. Sci. USA* **1998**, *95*, 13363–13383. [CrossRef] [PubMed]

3. Prusiner, S.B. Novel proteinaceous infectious particles cause scrapie. *Science* **1982**, *216*, 136–144. [CrossRef] [PubMed]

4. Gendoo, D.M.A.; Harrison, P.M. Discordant and chameleon sequences: Their distribution and implications for amyloidogenicity. *Protein Sci.* **2011**, *20*, 567–579. [CrossRef] [PubMed]

5. Kuznetsov, I.B.; Rackovsky, S. Comparative computational analysis of prion proteins reveals two fragments with unusual structural properties and a pattern of increase in hydrophobicity associated with disease-promoting mutations. *Protein Sci.* **2004**, *13*, 3230–3244. [CrossRef] [PubMed]

6. Dima, R.; Thirumalai, D. Exploring the propensities of helices in PrP^c to form beta sheet using NMR structures and sequence alignments. *Biophys. J.* **2002**, *83*, 1268–1280. [CrossRef]

7. Dima, R.I.; Thirumalai, D. Probing the instabilities in the dynamics of helical fragments from mouse PrP^C. *Proc. Natl. Acad. Sci. USA* **2004**, *101*, 15335–15340. [CrossRef] [PubMed]

8. Adrover, M.; Pauwels, K.; Prigent, S.; de Chiara, C.; Xu, Z.; Chapuis, C.; Pastore, A.; Rezaei, H. Prion Fibrillization Is Mediated by a Native Structural Element That Comprises Helices H2 and H3. *J. Biol. Chem.* **2010**, *285*, 21004–21012. [CrossRef]

9. Priola, S.A.; Meade-White, K.; Lawson, V.A.; Chesebro, B. Flexible N-terminal region of prion protein influences conformation of protease-resistant prion protein isoforms associated with cross-species scrapie infection in vivo and in vitro. *J. Biol. Chem.* **2004**, *279*, 13689–13695. [CrossRef]

10. Rudd, P.M.; Merry, A.H.; Wormald, M.R.; Dwek, R.A. Glycosylation and prion protein. *Curr. Opin. Struct. Biol.* **2002**, *12*, 578–586. [CrossRef]

11. Abid, K.; Soto, C. The intriguing prion disorders. *Cell. Mol. Life Sci.* **2006**, *63*, 2342–2351. [CrossRef] [PubMed]

12. Swietnicki, W.; Petersen, R.; Gambetti, P.; Surewicz, W.K. pH-dependent stability and conformation of the recombinant human prion protein PrP(90–231). *J. Biol. Chem.* **1997**, *272*, 27517–27520. [CrossRef] [PubMed]

13. Jackson, G.S.; Hosszu, L.L.; Power, A.; Hill, A.F.; Kenney, J.; Saibil, H.; Craven, C.J.; Waltho, J.P.; Clarke, A.R.; Collinge, J. Reversible conversion of monomeric human prion protein between native and fibrilogenic conformations. *Science* **1999**, *283*, 1935–1937. [CrossRef] [PubMed]

14. Choi, C.J.; Kanthasamy, A.; Anantharam, V.; Kanthasamy, A.G. Interaction of metals with prion protein: Possible role of divalent cations in the pathogenesis of prion diseases. *Neurotoxicology* **2006**, *27*, 777–787. [CrossRef] [PubMed]

15. Lehmann, S. Metal ions and prion diseases. *Curr. Opin. Chem. Biol.* **2002**, *6*, 187–192. [CrossRef]

16. Jin, T.; Gu, Y.; Zanusso, G.; Sy, M.; Kumar, A.; Cohen, M.; Gambetti, P.; Singh, N. The chaperone protein BiP binds to a mutant prion protein and mediates its degradation by the proteasome. *J. Biol. Chem.* **2000**, *275*, 38699–38704. [CrossRef]

17. Hachiya, N.S.; Imagawa, M.; Kaneko, K. The possible role of protein X, a putative auxiliary factor in pathological prion replication, in regulating a physiological endoproteolytic cleavage of cellular prion protein. *Med. Hypotheses* **2007**, *68*, 670–673. [CrossRef]

18. Dossena, S.; Imeri, L.; Mangieri, M.; Garofoli, A.; Ferrari, L.; Senatore, A.; Restelli, E.; Baiducci, C.; Fiordaliso, F.; Salio, M.; et al. Mutant Prion Protein Expression Causes Motor and Memory Deficits and Abnormal Sleep Patterns in a Transgenic Mouse Model. *Neuron* **2008**, *60*, 598–609. [CrossRef]

19. Antonyuk, S.V.; Trevitt, C.R.; Strange, R.W.; Jackson, G.S.; Sangar, D.; Batchelor, M.; Cooper, S.; Fraser, C.; Jones, S.; Georgiou, T.; et al. Crystal structure of human prion protein bound to a therapeutic antibody. *Proc. Natl. Acad. Sci. USA* **2009**, *106*, 2554–2558. [CrossRef]

20. Friedman-Levi, Y.; Meiner, Z.; Canello, T.; Frid, K.; Kovacs, G.G.; Budka, H.; Avrahami, D.; Gabizon, R. Fatal Prion Disease in a Mouse Model of Genetic E200K Creutzfeldt-Jakob Disease. *PLoS Pathog.* **2011**, *7*. [CrossRef]

21. Kovacs, G.G.; Trabattoni, G.; Hainfellner, J.A.; Ironside, J.W.; Knight, R.S.G.; Budka, H. Mutations of the prion protein gene phenotypic spectrum. *J. Neurol.* **2002**, *249*, 1567–1582. [CrossRef] [PubMed]

22. Campana, V.; Sarnataro, D.; Zurzolo, C. The highways and byways of prion protein trafficking. *Trends Cell Biol.* **2005**, *15*, 102–111. [CrossRef] [PubMed]

23. Linden, R.; Martins, V.R.; Prado, M.A.M.; Cammarota, M.N.; Izquierdo, I.N.; Brentani, R.R. Physiology of the Prion Protein. *Physiol. Rev.* **2008**, *88*, 673–728. [CrossRef] [PubMed]

24. Brown, D.; Qin, K.; Herms, J.W.; Madlung, A.; Manson, J.C.; Strome, R.; Fraser, P.E.; Kruck, T.; Von Bohlen, A.; Schulz-Schaeffer, W.; et al. The cellular prion protein binds copper in vivo. *Nature* **1997**, *390*, 684–687. [CrossRef] [PubMed]

25. Zahn, R.; Liu, A.Z.; Luhrs, T.; Riek, R.; von Schroetter, C.; Garcia, F.L.; Billeter, M.; Calzolai, L.; Wider, G.; Wuthrich, K. NMR solution structure of the human prion protein. *Proc. Natl. Acad. Sci. USA* **2000**, *97*, 145–150. [CrossRef] [PubMed]

26. Baskakov, I.V.; Legname, G.; Prusiner, S.B. Folding of prion protein to its native α-helical conformation is under kinetic control. *J. Biol. Chem.* **2001**, *276*, 19687–19690. [CrossRef] [PubMed]

27. Surewicz, W.K.; Apostol, M.I. Prion Protein and Its Conformational Conversion: A Structural Perspective. In *Prion Proteins*; Springer: Berlin/Heidelberg, Germany, 2011; Volume 305, pp. 135–167.

28. Govaerts, C.; Wille, H.; Prusiner, S.B.; Cohen, F.E. Evidence for assembly of prions with left-handed beta-helices into trimers. *Proc. Natl. Acad. Sci. USA* **2004**, *101*, 8342–8347. [CrossRef] [PubMed]

29. Cobb, N.J.; Sonnichsen, F.D.; McHaourab, H.; Surewicz, W.K. Molecular architecture of human prion protein amyloid: A parallel, in-register beta-structure. *Proc. Natl. Acad. Sci. USA* **2007**, *104*, 18946–18951. [CrossRef]

30. DeMarco, M.L.; Silveira, J.; Caughey, B.; Daggett, V. Structural properties of prion protein protofibrils and fibrils: An experimental assessment of atomic models. *Biochemistry* **2006**, *45*, 15573–15582. [CrossRef]

31. Rossetti, G.; Carloni, P. Structural Modeling of Human Prion Protein's Point Mutations. *Prog. Mol. Biol. Transl. Sci.* **2017**, *150*, 105–122. [CrossRef]

32. Rossetti, G.; Bongarzone, S.; Carloni, P. Computational studies on the prion protein. *Curr. Top. Med. Chem.* **2013**, *13*, 2419–2431. [CrossRef] [PubMed]

33. Diaz-Espinoza, R.; Soto, C. High-resolution structure of infectious prion protein: The final frontier. *Nat. Struct. Mol. Biol.* **2012**, *19*, 370–377. [CrossRef]

34. Aguzzi, A.; Calella, A.M. Prions: Protein aggregation and infectious diseases. *Physiol. Rev.* **2009**, *89*, 1105–1152. [CrossRef] [PubMed]

35. Aguzzi, A.; O'Connor, T. Protein aggregation diseases: Pathogenicity and therapeutic perspectives. *Nat. Rev. Drug Discov.* **2010**, *9*, 237–248. [CrossRef] [PubMed]

36. Nazabal, A.; Hornemann, S.; Aguzzi, A.; Zenobi, R. Hydrogen/deuterium exchange mass spectrometry identifies two highly protected regions in recombinant full-length prion protein amyloid fibrils. *J. Mass Spectrom.* **2009**, *44*, 965–977. [CrossRef]

37. Sim, V.; Caughey, B. Ultrastructures and strain comparison of under-glycosylated scrapie prion fibrils. *Neurobiol. Aging* **2009**, *30*, 2031–2042. [CrossRef] [PubMed]

38. Stoehr, J.; Weinmann, N.; Wille, H.; Kaimann, T.; Nagel-Steger, L.; Birkmann, E.; Panza, G.; Prusiner, S.B.; Eigen, M.; Riesner, D. Mechanisms of prion protein assembly into amyloid. *Proc. Natl. Acad. Sci. USA* **2008**, *105*, 2409–2414. [CrossRef]

39. Hegde, R.S.; Mastrianni, J.A.; Scott, M.R.; DeFea, K.A.; Tremblay, P.; Torchia, M.; DeArmond, S.J.; Prusiner, S.B.; Lingappa, V.R. A transmembrane form of the prion protein in neurodegenerative disease. *Science* **1998**, *279*, 827–834. [CrossRef]

40. Li, A.M.; Christensen, H.M.; Stewart, L.R.; Roth, K.A.; Chiesa, R.; Harris, D.A. Neonatal lethality in transgenic mice expressing prion protein with a deletion of residues 105–125. *EMBO J.* **2007**, *26*, 548–558. [CrossRef]

41. Ott, C.M.; Akhavan, A.; Lingappa, V.R. Specific features of the prion protein transmembrane domain regulate nascent chain orientation. *J. Biol. Chem.* **2007**, *282*, 11163–11171. [CrossRef]

42. Chakrabarti, O.; Ashok, A.; Hegde, R.S. Prion protein biosynthesis and its emerging role in neurodegeneration. *Trends Biochem. Sci.* **2009**, *34*, 287–295. [CrossRef] [PubMed]

43. Beland, M.; Roucou, X. The prion protein unstructured N-terminal region is a broad-spectrum molecular sensor with diverse and contrasting potential functions. *J. Neurochem.* **2012**, *120*, 853–868. [CrossRef] [PubMed]

44. Silva, J.L.; Vieira, T.C.R.G.; Gomes, M.P.B.; Rangel, L.P.; Scapin, S.M.N.; Cordeiro, Y. Experimental approaches to the interaction of the prion protein with nucleic acids and glycosaminoglycans: Modulators of the pathogenic conversion. *Methods* **2011**, *53*, 306–317. [CrossRef] [PubMed]

45. Hajj, G.N.M.; Lopes, M.H.; Mercadante, A.F.; Veiga, S.S.; da Silveira, R.B.; Santos, T.G.; Ribeiro, K.C.B.; Juliano, M.A.; Jacchieri, S.G.; Zanata, S.M.; et al. Cellular prion protein interaction with vitronectin supports axonal growth and is compensated by integrins. *J. Cell Sci.* **2007**, *120*, 1915–1926. [CrossRef]

46. Zanata, S.M.; Lopes, M.H.; Mercadante, A.F.; Hajj, G.N.M.; Chiarini, L.B.; Nomizo, R.; Freitas, A.R.O.; Cabral, A.L.B.; Lee, K.S.; Juliano, M.A.; et al. Stress-inducible protein 1 is a cell surface ligand for cellular prion that triggers neuroprotection. *EMBO J.* **2002**, *21*, 3307–3316. [CrossRef]

47. Lauren, J.; Gimbel, D.A.; Nygaard, H.B.; Gilbert, J.W.; Strittmatter, S.M. Cellular prion protein mediates impairment of synaptic plasticity by amyloid-beta oligomers. *Nature* **2009**, *457*, 1128–1184. [CrossRef]

48. Nicoll, A.J.; Panico, S.; Freir, D.B.; Wright, D.; Terry, C.; Risse, E.; Herron, C.E.; O'Malley, T.; Wadsworth, J.D.; Farrow, M.A.; et al. Amyloid-beta nanotubes are associated with prion protein-dependent synaptotoxicity. *Nat. Commun.* **2013**, *4*, 2416. [CrossRef]

49. Chen, S.G.; Yadav, S.P.; Surewicz, W.K. Interaction between Human Prion Protein and Amyloid-beta (A beta) Oligomers Role of N-Terminal Residues. *J. Biol. Chem.* **2010**, *285*, 26377–26383. [CrossRef]

50. Parkyn, C.J.; Vermeulen, E.G.; Mootoosamy, R.C.; Sunyach, C.; Jacobsen, C.; Oxvig, C.; Moestrup, S.; Liu, Q.; Bu, G.; Jen, A.; et al. LRP1 controls biosynthetic and endocytic trafficking of neuronal prion protein. *J. Cell Sci.* **2008**, *121*, 773–783. [CrossRef]

51. Schmitt-Ulms, G.; Legname, G.; Baldwin, M.A.; Ball, H.L.; Bradon, N.; Bosque, P.J.; Crossin, K.L.; Edelman, G.M.; DeArmond, S.J.; Cohen, F.E.; et al. Binding of neural cell adhesion molecules (N-CAMs) to the cellular prion protein. *J. Mol. Biol.* **2001**, *314*, 1209–1225. [CrossRef]

52. Cong, X.; Casiraghi, N.; Rossetti, G.; Mohanty, S.; Giachin, G.; Legname, G.; Carloni, P. Role of Prion Disease-Linked Mutations in the Intrinsically Disordered N-Terminal Domain of the Prion Protein. *J. Chem. Theory Comput.* **2013**, *9*, 5158–5167. [CrossRef] [PubMed]

53. Calzolai, L.; Zahn, R. Influence of pH on NMR structure and stability of the human prion protein globular domain. *J. Boil. Chem.* **2003**, *278*, 35592–35596. [CrossRef] [PubMed]

54. Degioia, L.; Selvaggini, C.; Ghibaudi, E.; Diomede, L.; Bugiani, O.; Forloni, G.; Tagliavini, F.; Salmona, M. Conformational Polymorphism of the Amyloidogenic and Neurotoxic Peptide Homologous to Residues-106–126 of the Prion Protein. *J. Biol. Chem.* **1994**, *269*, 7859–7862.

55. Miura, T.; Yoda, M.; Takaku, N.; Hirose, T.; Takeuchi, H. Clustered negative charges on the lipid membrane surface induce beta-sheet formation of prion protein fragment 106–126. *Biochemistry* **2007**, *46*, 11589–11597. [CrossRef] [PubMed]

56. Satheeshkumar, K.S.; Jayakumar, R. Conformational polymorphism of the amyloidogenic peptide homologous to residues 113–127 of the prion protein. *Biophys. J.* **2003**, *85*, 473–483. [CrossRef]

57. Zahn, R. The octapeptide repeats in mammalian prion protein constitute a pH-dependent folding and aggregation site. *J. Mol. Biol.* **2003**, *334*, 477–488. [CrossRef] [PubMed]

58. Van der Kamp, M.W.; Daggett, V. The consequences of pathogenic mutations to the human prion protein. *Protein Eng. Des. Sel.* **2009**, *22*, 461–468. [CrossRef]

59. Rossetti, G.; Cong, X.J.; Caliandro, R.; Legname, G.; Carloni, P. Common Structural Traits across Pathogenic Mutants of the Human Prion Protein and Their Implications for Familial Prion Diseases. *J. Mol. Biol.* **2011**, *411*, 700–712. [CrossRef]

60. Ilc, G.; Giachin, G.; Jaremko, M.; Jaremko, L.; Benetti, F.; Plavec, J.; Zhukov, I.; Legname, G. NMR structure of the human prion protein with the pathological Q212P mutation reveals unique structural features. *PLoS ONE* **2010**, *5*, e11715. [CrossRef]

61. Van der Kamp, M.W.; Daggett, V. Pathogenic Mutations in the Hydrophobic Core of the Human Prion Protein Can Promote Structural Instability and Misfolding. *J. Mol. Biol.* **2010**, *404*, 732–748. [CrossRef]

62. Apetri, A.C.; Surewicz, K.; Surewicz, W.K. The effect of disease-associated mutations on the folding pathway of human prion protein. *J. Biol. Chem.* **2004**, *279*, 18008–18014. [CrossRef] [PubMed]

63. Liemann, S.; Glockshuber, R. Influence of amino acid substitutions related to inherited human prion diseases on the thermodynamic stability of the cellular prion protein. *Biochemistry* **1999**, *38*, 3258–3267. [CrossRef] [PubMed]

64. Swietnicki, W.; Petersen, R.B.; Gambetti, P.; Surewicz, W.K. Familial mutations and the thermodynamic stability of the recombinant human prion protein. *J. Biol. Chem.* **1998**, *273*, 31048–31052. [CrossRef] [PubMed]

65. Evans, E.G.B.; Millhauser, G.L. Copper- and Zinc-Promoted Interdomain Structure in the Prion Protein: A Mechanism for Autoinhibition of the Neurotoxic N-Terminus. In *Prion Protein*; Elsevier: Amsterdam, The Netherlands, 2017; Volume 150, pp. 35–56.

66. Hegde, R.S.; Kang, S.W. The concept of translocational regulation. *J. Cell Biol.* **2008**, *182*, 225–232. [CrossRef] [PubMed]

67. Hornemann, S.; von Schroetter, C.; Damberger, F.F.; Wuthrich, K. Prion Protein-Detergent Micelle Interactions Studied by NMR in Solution. *J. Biol. Chem.* **2009**, *284*, 22713–22721. [CrossRef] [PubMed]

68. Kardos, J.; Kovacs, I.; Hajos, F.; Kalman, M.; Simonyi, M. Nerve endings from rat brain tissue release copper upon depolarization. A possible role in regulating neuronal excitability. *Neurosci. Lett.* **1989**, *103*, 139–144. [CrossRef]

69. Walter, E.D.; Chattopadhyay, M.; Millhauser, G.L. The affinity of copper binding to the prion protein octarepeat domain: Evidence for negative cooperativity. *Biochemistry* **2006**, *45*, 13083–13092. [CrossRef]

70. Lee, K.S.; Magalhaes, A.C.; Zanata, S.M.; Brentani, R.R.; Martins, V.R.; Prado, M.A.M. Internalization of mammalian fluorescent cellular prion protein and N-terminal deletion mutants in living cells. *J. Neurochem.* **2001**, *79*, 79–87. [CrossRef]

71. Pauly, P.C.; Harris, D.A. Copper Stimulates Endocytosis of the Prion Protein. *J. Biol. Chem.* **1998**, *273*, 33107–33110. [CrossRef]

72. Ren, K.; Gao, C.; Zhang, J.; Wang, K.; Xu, Y.; Wang, S.-B.; Wang, H.; Tian, C.; Shi, Q.; Dong, X.-P. Flotillin-1 Mediates PrP^C Endocytosis in the Cultured Cells During Cu^{2+} Stimulation Through Molecular Interaction. *Mol. Neurobiol.* **2013**, *48*, 631–646. [CrossRef]

73. Sumudhu, W.; Perera, S.; Hooper, N.M. Ablation of the metal ion-induced endocytosis of the prion protein by disease-associated mutation of the octarepeat region. *Curr. Biol.* **2001**, *11*, 519–523. [CrossRef]

74. Huang, S.; Chen, L.; Bladen, C.; Stys, P.K.; Zamponi, G.W. Differential modulation of NMDA and AMPA receptors by cellular prion protein and copper ions. *Mol. Brain* **2018**, *11*. [CrossRef] [PubMed]

75. You, H.; Tsutsui, S.; Hameed, S.; Kannanayakal, T.J.; Chen, L.; Xia, P.; Engbers, J.D.T.; Lipton, S.A.; Stys, P.K.; Zamponi, G.W. Aβ neurotoxicity depends on interactions between copper ions, prion protein, and *N*-methyl-D-aspartate receptors. *Proc. Natl. Acad. Sci. USA* **2012**, *109*, 1737–1742. [CrossRef] [PubMed]

76. Stys, P.K.; You, H.; Zamponi, G.W. Copper-dependent regulation of NMDA receptors by cellular prion protein: Implications for neurodegenerative disorders. *J. Physiol.* **2012**, *590*, 1357–1368. [CrossRef]

77. Gasperini, L.; Meneghetti, E.; Pastore, B.; Benetti, F.; Legname, G. Prion Protein and Copper Cooperatively Protect Neurons by Modulating NMDA Receptor Through S-nitrosylation. *Antioxid. Redox Signal.* **2015**, *22*, 772–784. [CrossRef]

78. Aronoff-Spencer, E.; Burns, C.S.; Avdievich, N.I.; Gerfen, G.J.; Peisach, J.; Antholine, W.E.; Ball, H.L.; Cohen, F.E.; Prusiner, S.B.; Millhauser, G.L. Identification of the Cu^{2+} binding sites in the N-terminal domain of the prion protein by EPR and CD spectroscopy. *Biochemistry* **2000**, *39*, 13760–13771. [CrossRef]

79. Millhauser, G.L. Copper Binding in the Prion Protein. *Acc. Chem. Res.* **2004**, *37*, 79–85. [CrossRef]

80. Millhauser, G.L. Copper and the Prion Protein: Methods, Structures, Function, and Disease. *Annu. Rev. Phys. Chem.* **2007**, *58*, 299–320. [CrossRef]

81. Quintanar, L.; Rivillas-Acevedo, L.; Grande-Aztatzi, R.; Gómez-Castro, C.Z.; Arcos-López, T.; Vela, A. Copper coordination to the prion protein: Insights from theoretical studies. *Coord. Chem. Rev.* **2013**, *257*, 429–444. [CrossRef]

82. Wells, M.A.; Jelinska, C.; Hosszu, L.L.; Craven, C.J.; Clarke, A.R.; Collinge, J.; Waltho, J.P.; Jackson, G.S. Multiple forms of copper(II) co-ordination occur throughout the disordered N-terminal region of the prion protein at pH 7.4. *Biochem. J.* **2006**, *400*, 501–510. [CrossRef]

83. Wopfner, F.; Weidenhöfer, G.; Schneider, R.; von Brunn, A.; Gilch, S.; Schwarz, T.F.; Werner, T.; Schätzl, H.M. Analysis of 27 mammalian and 9 avian PrPs reveals high conservation of flexible regions of the prion protein. *J. Mol. Biol.* **1999**, *289*, 1163–1178. [CrossRef] [PubMed]

84. Chattopadhyay, M.; Walter, E.D.; Newell, D.J.; Jackson, P.J.; Aronoff-Spencer, E.; Peisach, J.; Gerfen, G.J.; Bennett, B.; Antholine, W.E.; Millhauser, G.L. The octarepeat domain of the prion protein binds Cu(II) with three distinct coordination modes at pH 7.4. *J. Am. Chem. Soc.* **2005**, *127*, 12647–12656. [CrossRef] [PubMed]

85. Burns, C.S.; Aronoff-Spencer, E.; Legname, G.; Prusiner, S.B.; Antholine, W.E.; Gerfen, G.J.; Peisach, J.; Millhauser, G.L. Copper coordination in the full-length, recombinant prion protein. *Biochemistry* **2003**, *42*, 6794–6803. [CrossRef] [PubMed]

86. Burns, C.S.; Aronoff-Spencer, E.; Dunham, C.M.; Lario, P.; Avdievich, N.I.; Antholine, W.E.; Olmstead, M.M.; Vrielink, A.; Gerfen, G.J.; Peisach, J.; et al. Molecular features of the copper binding sites in the octarepeat domain of the prion protein. *Biochemistry* **2002**, *41*, 3991–4001. [CrossRef] [PubMed]

87. Liu, L.; Jiang, D.; McDonald, A.; Hao, Y.; Millhauser, G.L.; Zhou, F. Copper redox cycling in the prion protein depends critically on binding mode. *J. Am. Chem. Soc.* **2011**, *133*, 12229–12237. [CrossRef] [PubMed]

88. Zhou, F.; Millhauser, G.L. The rich electrochemistry and redox reaacions of the copper sites in the cellular prion protein. *Coord. Chem. Rev.* **2012**, *256*, 2285–2296. [CrossRef] [PubMed]

89. Jackson, G.S.; Murray, I.; Hosszu, L.L.; Gibbs, N.; Waltho, J.P.; Clarke, A.R.; Collinge, J. Location and properties of metal-binding sites on the human prion protein. *Proc. Natl. Acad. Sci. USA* **2001**, *98*, 8531–8535. [CrossRef]

90. Walter, E.D.; Stevens, D.J.; Spevacek, A.R.; Visconte, M.P.; Dei Rossi, A.; Millhauser, G.L. Copper binding extrinsic to the octarepeat region in the prion protein. *Curr. Protein Pept. Sci.* **2009**, *10*, 529–535. [CrossRef]

91. Giachin, G.; Mai, P.T.; Tran, H.N.; Salzano, G.; Benetti, F.; Migliorati, V.; Arcovito, A.; Della Longa, S.; Mancini, G.; D'Angelo, P.; et al. The non-octarepeat copper binding site of the prion protein is a key regulator of prion conversion. *Sci. Rep.* **2015**, *5*, 15253. [CrossRef]

92. Klewpatinond, M.; Davies, P.; Bowen, S.; Brown, D.R.; Viles, J.H. Deconvoluting the Cu(2+) binding modes of full-length prion protein. *J. Biol. Chem.* **2008**, *283*, 1870–1881. [CrossRef]

93. Grande-Aztatzi, R.; Rivillas-Acevedo, L.; Quintanar, L.; Vela, A. Structural models for Cu(II) bound to the fragment 92-96 of the human prion protein. *J. Phys. Chem. B* **2013**, *117*, 789–799. [CrossRef] [PubMed]

94. Hureau, C.; Charlet, L.; Dorlet, P.; Gonnet, F.; Spadini, L.; Anxolabéhère-Mallart, E.; Girerd, J.J. A spectroscopic and voltammetric study of the pH-dependent Cu(II) coordination to the peptide GGGTH: Relevance to the fifth Cu(II) site in the prion protein. *J. Boil. Inorg. Chem.* **2006**, *11*, 735–744. [CrossRef] [PubMed]

95. Hureau, C.; Mathé, C.; Faller, P.; Mattioli, T.A.; Dorlet, P. Folding of the prion peptide GGGTHSQW around the copper(II) ion: Identifying the oxygen donor ligand at neutral pH and probing the proximity of the tryptophan residue to the copper ion. *J. Biol. Inorg. Chem.* **2008**, *13*, 1055–1064. [CrossRef] [PubMed]

96. Rivillas-Acevedo, L.; Grande-Aztatzi, R.; Lomelí, I.; García, J.E.; Barrios, E.; Teloxa, S.; Vela, A.; Quintanar, L. Spectroscopic and electronic structure studies of copper(II) binding to His111 in the human prion protein fragment 106–115: Evaluating the role of protons and methionine residues. *Inorg. Chem.* **2011**, *50*, 1956–1972. [CrossRef] [PubMed]

97. Arcos-López, T.; Qayyum, M.; Rivillas-Acevedo, L.; Miotto, M.C.; Grande-Aztatzi, R.; Fernández, C.O.; Hedman, B.; Hodgson, K.O.; Vela, A.; Solomon, E.I.; et al. Spectroscopic and Theoretical Study of Cu(I) Binding to His111 in the Human Prion Protein Fragment 106–115. *Inorg. Chem.* **2016**, *55*, 2909–2922. [CrossRef] [PubMed]

98. Berti, F.; Gaggelli, E.; Guerrini, R.; Janicka, A.; Kozlowski, H.; Legowska, A.; Miecznikowska, H.; Migliorini, C.; Pogni, R.; Remelli, M.; et al. Structural and dynamic characterization of copper(II) binding of the human prion protein outside the octarepeat region. *Chem. Eur. J.* **2007**, *13*, 1991–2001. [CrossRef] [PubMed]

99. DiNatale, G.; Ösz, K.; Nagy, Z.; Sanna, D.; Micera, G.; Pappalardo, G.; Sóvágó, I.; Rizzarell, E. Interaction of Copper(II) with the Prion Peptide Fragment HuPrP(76–114) Encompassing Four Histidyl Residues within and outside the Octarepeat Domain. *Inorg. Chem.* **2009**, *48*, 4239–4250. [CrossRef] [PubMed]

100. Jones, C.E.; Klewpatinond, M.; Abdelraheim, S.R.; Brown, D.R.; Viles, J.H. Probing copper^{2+} binding to the prion protein using diamagnetic nickel^{2+} and ^{1}H NMR: The unstructured N terminus facilitates the coordination of six copper^{2+} ions at physiological concentrations. *J. Mol. Biol.* **2005**, *346*, 1393–1407. [CrossRef] [PubMed]

101. Nadal, R.C.; Davies, P.; Brown, D.R.; Viles, J.H. Evaluation of copper^{2+} affinities for the prion protein. *Biochemistry* **2009**, *48*, 8929–8931. [CrossRef] [PubMed]

102. Sánchez-López, C.; Rivillas-Acevedo, L.; Cruz-Vásquez, O.; Quintanar, L. Methionine 109 plays a key role in Cu(II) binding to His111 in the 92–115 fragment of the human prion protein. *Inorg. Chim. Acta* **2018**, *481*, 87–97. [CrossRef]

103. Evans, E.G.B.; Pushie, M.J.; Markham, K.A.; Lee, H.-W.; Millhauser, G.L. Interaction between Prion Protein's Copper-Bound Octarepeat Domain and a Charged C-Terminal Pocket Suggests a Mechanism for N-Terminal Regulation. *Structure* **2016**, *24*, 1057–1067. [CrossRef] [PubMed]

104. Spevacek, A.R.; Evans, E.G.B.; Miller, J.L.; Meyer, H.C.; Pelton, J.G.; Millhauser, G.L. Zinc drives a tertiary fold in the prion protein with familial disease mutation sites at the interface. *Structure* **2013**, *21*, 236–246. [CrossRef] [PubMed]

105. Thakur, A.K.; Srivastava, A.K.; Srinivas, V.; Chary, K.V.; Rao, C.M. Copper alters aggregation behavior of prion protein and induces novel interactions between its N- and C-terminal regions. *J. Biol. Chem.* **2011**, *286*, 38533–38545. [CrossRef] [PubMed]

106. Younan, N.D.; Klewpatinond, M.; Davies, P.; Ruban, A.V.; Brown, D.R.; Viles, J.H. Copper(II)-induced secondary structure changes and reduced folding stability of the prion protein. *J. Mol. Biol.* **2011**, *410*, 369–382. [CrossRef] [PubMed]

107. Leal, S.S.; Botelho, H.M.; Gomes, C.M. Metal ions as modulators of protein conformation and misfolding in neurodegeneration. *Coord. Chem. Rev.* **2012**, *256*, 2253–2270. [CrossRef]

108. D'angelo, P.; Della Longa, S.; Arcovito, A.; Mancini, G.; Zitolo, A.; Chillemi, G.; Giachin, G.; Legname, G.; Benetti, F. Effects of the Pathological Q212P Mutation on Human Prion Protein Non-Octarepeat Copper-Binding Site. *Biochemistry* **2012**, *51*, 6068–6079. [CrossRef]

109. Altmeppen, H.C.; Puig, B.; Dohler, F.; Thurm, D.K.; Falker, C.; Krasemann, S.; Glatzel, M. Proteolytic processing of the prion protein in health and disease. *Am. J. Neurodegener. Dis.* **2012**, *1*, 15–31.

110. McMahon, H.E.; Mangé, A.; Nishida, N.; Créminon, C.; Casanova, D.; Lehmann, S. Cleavage of the amino terminus of the prion protein by reactive oxygen species. *J. Biol. Chem.* **2001**, *276*, 2286–2291. [CrossRef]

111. Watt, N.T.; Taylor, D.R.; Gillott, A.; Thomas, D.A.; Perera, W.S.S.; Hooper, N.M. Reactive Oxygen Species-mediated β-Cleavage of the Prion Protein in the Cellular Response to Oxidative Stress. *J. Biol. Chem.* **2005**, *280*, 35914–35921. [CrossRef]

112. McDonald, A.J.; Dibble, J.P.; Evans, E.G.; Millhauser, G.L. A new paradigm for enzymatic control of alpha-cleavage and beta-cleavage of the prion protein. *J. Biol. Chem.* **2014**, *289*, 803–813. [CrossRef]

113. McDonald, A.J.; Millhauser, G.L. PrP overdrive. Does inhibition of alpha-cleavage contribute to PrP^C toxicity and prion disease? *Prion* **2014**, *8*, 183–191. [CrossRef]

114. Oliveira-Martins, J.B.; Yusa, S.; Calella, A.M.; Bridel, C.; Baumann, F.; Dametto, P.; Aguzzi, A. Unexpected tolerance of alpha-cleavage of the prion protein to sequence variations. *PLoS ONE* **2010**, *5*, e9107. [CrossRef] [PubMed]

115. Haigh, C.L.; Lewis, V.A.; Vella, L.J.; Masters, C.L.; Hill, A.F.; Lawson, V.A.; Collins, S.J. PrP^C-related signal transduction is influenced by copper, membrane integrity and the alpha cleavage site. *Cell Res.* **2009**, *19*, 1062–1078. [CrossRef] [PubMed]

116. Sánchez-López, C.; Fernández, C.O.; Quintanar, L. Neuroprotective alpha-cleavage of the human prion protein significantly impacts Cu(II) coordination at its His111 site. *Dalton Trans.* **2018**, *47*, 9274–9282. [CrossRef] [PubMed]

117. Black, S.A.G.; Stys, P.K.; Zamponi, G.W.; Tsutsui, S. Cellular prion protein and NMDA receptor modulation: Protecting against excitotoxicity. *Front. Cell Dev. Biol.* **2014**, *2*. [CrossRef] [PubMed]

118. Villar-Piqué, A.; Lopes da Fonseca, T.; Sant'Anna, R.; Szegö, E.M.; Fonseca-Ornelas, L.; Pinho, R.; Carija, A.; Gerhardt, E.; Masaracchia, C.; Abad Gonzalez, E.; et al. Environmental and genetic factors support the dissociation between α-synuclein aggregation and toxicity. *Proc. Natl. Acad. Sci. USA* **2016**, *113*, 6506–65115. [CrossRef] [PubMed]

Permissions

The contributors of this book come from diverse backgrounds, making this book a truly international effort. This book will bring forth new frontiers with its revolutionizing research information and detailed analysis of the nascent developments around the world.

We would like to thank all the contributing authors for lending their expertise to make the book truly unique. They have played a crucial role in the development of this book. Without their invaluable contributions this book wouldn't have been possible. They have made vital efforts to compile up to date information on the varied aspects of this subject to make this book a valuable addition to the collection of many professionals and students.

This book was conceptualized with the vision of imparting up-to-date information and advanced data in this field. To ensure the same, a matchless editorial board was set up. Every individual on the board went through rigorous rounds of assessment to prove their worth. After which they invested a large part of their time researching and compiling the most relevant data for our readers.

The editorial board has been involved in producing this book since its inception. They have spent rigorous hours researching and exploring the diverse topics which have resulted in the successful publishing of this book. They have passed on their knowledge of decades through this book. To expedite this challenging task, the publisher supported the team at every step. A small team of assistant editors was also appointed to further simplify the editing procedure and attain best results for the readers.

Apart from the editorial board, the designing team has also invested a significant amount of their time in understanding the subject and creating the most relevant covers. They scrutinized every image to scout for the most suitable representation of the subject and create an appropriate cover for the book.

The publishing team has been an ardent support to the editorial, designing and production team. Their endless efforts to recruit the best for this project, has resulted in the accomplishment of this book. They are a veteran in the field of academics and their pool of knowledge is as vast as their experience in printing. Their expertise and guidance has proved useful at every step. Their uncompromising quality standards have made this book an exceptional effort. Their encouragement from time to time has been an inspiration for everyone.

The publisher and the editorial board hope that this book will prove to be a valuable piece of knowledge for researchers, students, practitioners and scholars across the globe.

List of Contributors

Beata Szabo, Tamas Horvath, Eva Schad, Nikoletta Murvai, Agnes Tantos and Lajos Kalmar
Institute of Enzymology, Center of Natural Sciences, Hungarian Academy of Sciences, 1117 Budapest, Hungary

Lucía Beatriz Chemes
Instituto de Investigaciones Biotecnológicas IIB-INTECH, Consejo Nacional de Investigaciones Científicas y Técnicas (CONICET), Universidad Nacional de San Martín, Buenos Aires 1650, Argentina

Kyou-Hoon Han
Genome Editing Research Center, Division of Biomedical Science, Korea Research Institute of Bioscience and Biotechnology (KRIBB), Daejeon 34113, Korea
Department of Nano and Bioinformatics, University of Science and Technology (UST), Daejeon 34113, Korea

Peter Tompa
Institute of Enzymology, Center of Natural Sciences, Hungarian Academy of Sciences, 1117 Budapest, Hungary
VIB Center for Structural Biology, Vrije Univresiteit Brussel, 1050 Belgium, Brussel

Erik H. A. Rikkerink
The New Zealand Institute for Plant & Food Research Ltd., 120 Mt. Albert Rd., Auckland 1025, New Zealand

Roberta Corti
School of Medicine and Surgery, Nanomedicine Center NANOMIB, University of Milan-Bicocca, 20900 Monza, Italy
Department of Materials Science, University of Milan-Bicocca, 20125 Milan, Italy

Claudia A. Marrano, Domenico Salerno, Francesco Mantegazza and Valeria Cassina
School of Medicine and Surgery, Nanomedicine Center NANOMIB, University of Milan-Bicocca, 20900 Monza, Italy

Stefania Brocca, Antonino Natalello, Carlo Santambrogio and Rita Grandori
Department of Biotechnology and Biosciences, University of Milan-Bicocca, 20126 Milan, Italy

Giuseppe Legname
Scuola Internazionale Superiore di Studi Avanzati, SISSA, 34136 Trieste, Italy

Izzy Owen and Frank Shewmaker
Department of Biochemistry, Uniformed Services University of the Health Sciences, Bethesda, MD 20814, USA

Sara Signorelli, Salvatore Cannistraro and Anna Rita Bizzarri
Biophysics & Nanoscience Centre, DEB, Università della Tuscia, 01100 Viterbo, Italy

Muhamad Fahmi, Gen Yasui and Kaito Seki
Advanced Life Sciences Program, Graduate School of Life Sciences, Ritsumeikan University, Kusatsu, Shiga 525-8577, Japan

Syouichi Katayama, Takako Kaneko-Kawano and Tetsuya Inazu
Department of Pharmacy, College of Pharmaceutical Sciences, Ritsumeikan University, Kusatsu, Shiga 525-8577, Japan

Yukihiko Kubota
Department of Bioinformatics, College of Life Sciences, Ritsumeikan University, Kusatsu, Shiga 525-8577, Japan

Masahiro Ito
Advanced Life Sciences Program, Graduate School of Life Sciences, Ritsumeikan University, Kusatsu, Shiga 525-8577, Japan
Department of Bioinformatics, College of Life Sciences, Ritsumeikan University, Kusatsu, Shiga 525-8577, Japan

Anne H. S. Martinelli
Department of Molecular Biology and Biotechnology & Department of Biophysics, Biosciences Institute-IB, (UFRGS), Porto Alegre CEP 91501-970, RS, Brazil

Fernanda C. Lopes and Elisa B. O. John
Center for Biotechnology, Universidade Federal do Rio Grande do Sul (UFRGS), Porto Alegre CEP 91501-970, RS, Brazil
Graduate Program in Cell and Molecular Biology, Universidade Federal do Rio Grande do Sul (UFRGS), Porto Alegre CEP 91501-970, RS, Brazil

Célia R. Carlini
Graduate Program in Cell and Molecular Biology, Universidade Federal do Rio Grande do Sul (UFRGS), Porto Alegre CEP 91501-970, RS, Brazil
Graduate Program in Medicine and Health Sciences, Pontifícia Universidade Católica do Rio Grande do Sul (PUCRS), Porto Alegre CEP 91410-000, RS, Brazil Brain Institute-InsCer, Laboratory of Neurotoxins, Pontifícia Universidade Católica do Rio Grande do Sul (PUCRS), Porto Alegre CEP 90610-000, RS, Brazil

Rodrigo Ligabue-Braun
Department of Pharmaceutical Sciences, Universidade Federal de Ciências da Saúde de Porto Alegre (UFCSPA), Porto Alegre CEP 90050-170, RS, Brazil

Andrey Machulin
Skryabin Institute of Biochemistry and Physiology of Microorganisms, Russian Academy of Sciences, Federal Research Center "Pushchino Scientific Center for Biological Research of the Russian Academy of Sciences, 142290 Pushchino, Russia

Evgenia Deryusheva
Institute for Biological Instrumentation, Federal Research Center "Pushchino Scientific Center for Biological Research of the Russian Academy of Sciences, 142290 Pushchino, Russia

Mikhail Lobanov and Oxana Galzitskaya
Institute of Protein Research, Russian Academy of Sciences, 142290 Pushchino, Russia

Maud Chan-Yao-Chong
BioCIS, University Paris-Sud, CNRS UMR 8076, University Paris-Saclay, 92290 Châtenay-Malabry, France
Institute for Integrative Biology of the Cell (I2BC), CEA, CNRS, University Paris-Sud, University Paris-Saclay, 91190 Gif-sur-Yvette, France

Dominique Durand
Institute for Integrative Biology of the Cell (I2BC), CEA, CNRS, University Paris-Sud, University Paris-Saclay, 91190 Gif-sur-Yvette, France

Tâp Ha-Duong
BioCIS, University Paris-Sud, CNRS UMR 8076, University Paris-Saclay, 92290 Châtenay-Malabry, France

Anikó Mentes and Csaba Magyar
Institute of Enzymology, Research Centre for Natural Sciences, Hungarian Academy of Sciences, Magyar Tudósok krt. 2., H-1117 Budapest, Hungary

Aneta Tarczewska and Beata Greb-Markiewicz
Department of Biochemistry, Faculty of Chemistry, Wroclaw University of Science and Technology, Wybrzeże Wyspiańskiego 27, 50-370 Wroclaw, Poland

André F. Faustino, Ana S. Martins, Nina Karguth, Vanessa Artilheiro, Francisco J. Enguita, Nuno C. Santos and Ivo C. Martins
Instituto de Medicina Molecular, Faculdade de Medicina, Universidade de Lisboa, Av. Prof. Egas Moniz, 1649-028 Lisbon, Portugal

Joana C. Ricardo
Centro de Química-Física Molecular, Instituto Superior Técnico, Universidade de Lisboa, 1049-001 Lisbon, Portugal

Bálint Mészáros
MTA-ELTE Momentum Bioinformatics Research Group, Department of Biochemistry, Eötvös Loránd University, Pázmány Péter stny 1/c, H-1117 Budapest, Hungary European Molecular Biology Laboratory, Structural and Computational Biology Unit, Meyerhofstraße 1, 69117 Heidelberg, Germany
Protein Structure Research Group, Institute of Enzymology, RCNS, HAS, Magyar Tudósok krt 2, H-1117 Budapest, Hungary

László Dobson
Membrane Protein Bioinformatics Research Group, Institute of Enzymology, RCNS, HAS, Magyar Tudósok krt 2, H-1117 Budapest, Hungary
Faculty of Information Technology and Bionics, Pázmány Péter Catholic University, Práter u. 50A, H-1083 Budapest, Hungary

Erzsébet Fichó and István Simon
Protein Structure Research Group, Institute of Enzymology, RCNS, HAS, Magyar Tudósok krt 2, H-1117 Budapest, Hungary

Carolina Sánchez-López and Liliana Quintanar
Department of Chemistry, Center for Research and Advanced Studies (Cinvestav), 07360 Mexico City, Mexico

Giulia Rossetti
Institute of Neuroscience and Medicine (INM-9) and Institute for Advanced Simulation (IAS-5), Forschungszentrum Jülich, Wilhelm-Johnen-Strasse, 52425 Jülich, Germany
Jülich Supercomputing Center (JSC), Forschungszentrum Jülich, 52428 Jülich, Germany Department of Oncology, Hematology and Stem Cell Transplantation, Faculty of Medicine, RWTH Aachen University, Pauwelsstraße 30, 52074 Aachen, Germany

Paolo Carloni
Institute of Neuroscience and Medicine (INM-9) and Institute for Advanced Simulation (IAS-5), Forschungszentrum Jülich, Wilhelm-Johnen-Strasse, 52425 Jülich, Germany

Department of Physics and Department of Neurobiology, RWTH Aachen University, 52078 Aachen, Germany
Institute for Neuroscience and Medicine (INM)-11, Forschungszentrum Jülich, 52428 Jülich, Germany

Index

Printed in the USA
CPSIA information can be obtained
at www.ICGtesting.com
JSHW061909131123
51979JS00006B/58

9 781646 475407